DYNAMICS IN MODELS OF COARSENING, COAGULATION, CONDENSATION AND QUANTIZATION

LECTURE NOTES SERIES
Institute for Mathematical Sciences, National University of Singapore

Series Editors: Louis H. Y. Chen and Ka Hin Leung
Institute for Mathematical Sciences
National University of Singapore

ISSN: 1793-0758

Published

Lecture Notes Series, Institute for Mathematical Sciences, National University of Singapore

Vol. 9

DYNAMICS IN MODELS OF COARSENING, COAGULATION, CONDENSATION AND QUANTIZATION

Editors

Weizhu Bao
National University of Singapore, Singapore

Jian-Guo Liu
University of Maryland, USA

World Scientific

NEW JERSEY • LONDON • SINGAPORE • BEIJING • SHANGHAI • HONG KONG • TAIPEI • CHENNAI

Published by

World Scientific Publishing Co. Pte. Ltd.

5 Toh Tuck Link, Singapore 596224

USA office: 27 Warren Street, Suite 401-402, Hackensack, NJ 07601

UK office: 57 Shelton Street, Covent Garden, London WC2H 9HE

British Library Cataloguing-in-Publication Data
A catalogue record for this book is available from the British Library.

Cover Images:
The images are sequences of STM images illustrating two-dimensional (single-atom high) islands of Ag on an Ag(100) surface at 300 K, undergoing coarsening via Smoluchowski ripening — reduction in island numbers through diffusion of islands and coalescence.

Adapted with permission from J. Phys. Chem. B 104 (2000) 1663, Figures 5 and 8 c & d.
Copyright © 2000 American Chemical Society.

Lecture Notes Series, Institute for Mathematical Sciences, NUS — Vol. 9
DYNAMICS IN MODELS OF COARSENING, COAGULATION, CONDENSATION AND QUANTIZATION

ISBN-13 978-981-270-850-2
ISBN-10 981-270-850-2

Printed by Mainland Press Pte Ltd

CONTENTS

FOREWORD

The Institute for Mathematical Sciences at the National University of Singapore was established on 1 July 2000. Its mission is to foster mathematical research, particularly multidisciplinary research that links mathematics to other disciplines, to nurture the growth of mathematical expertise among research scientists, to train talent for research in the mathematical sciences, and to provide a platform for interaction and collaboration between local and foreign mathematical scientists, in support of national development.

The Institute organizes thematic programs which last from one month to six months. The theme or themes of a program will generally be of a multidisciplinary nature, chosen from areas at the forefront of current research in the mathematical sciences and their applications, and in accordance with the scientific interests and technological needs in Singapore.

Generally, for each program there will be tutorial lectures on background material followed by workshops at the research level. Notes on these lectures are usually made available to the participants for their immediate benefit during the program. The main objective of the Institute's Lecture Notes Series is to bring these lectures to a wider audience. Occasionally, the Series may also include the proceedings of workshops and expository lectures organized by the Institute.

The World Scientific Publishing Company has kindly agreed to publish the Lecture Notes Series. This Volume, "Dynamics in Models of Coarsening, Coagulation, Condensation and Quantization", is the ninth of this Series. We hope that through the regular publication of these lecture notes the Institute will achieve, in part, its objective of promoting research in the mathematical sciences and their applications.

April 2007

Louis H. Y. Chen
Ka Hin Leung
Series Editors

PREFACE

A two-month program on Nanoscale Material Interfaces: Experiment, Theory and Simulation was held at the Institute for Mathematical Sciences (IMS) at the National University of Singapore (NUS) from November 2004 to January 2005. The goals of this program were to: (i) review the recent development in the research on materials surfaces and interfaces, from experiment to theory to simulation; (ii) identify critical scientific issues in the understanding of the fundamental principles and basic mechanisms of interfacial dynamics in different kinds of materials systems, particularly those that are characterized by fluctuation, multiscale, and non-equilibrium; and (iii) accelerate the interaction of applied mathematics and computational science with physics and material science, and promote the highly interdisciplinary research on new material interface problems with emerging applications.

As part of the program, tutorials were conducted by leading experts in the fields. These tutorials covered dynamics in models of coarsening, coagulation, condensation and quantization as well as complex fluids and were meant for graduate students and researchers who would like to prepare themselves for original research in the fields. The current volume collects four expanded lecture notes with each self-contained tutorials. In the following, we give a brief introduction to these tutorials here:

Part I. Lectures on dynamics in models of coarsening and coagulation by Robert Pego: It starts with a hierarchy of different domain coarsening models in one space dimension, explores different models to domain coarsening in two and three dimensions, derives rigorous power-law bounds on coarsening rates by using the Kohn-Otto method, and studies the Smolushowski's coagulation equations and dynamics on the scaling attractors.

Part II. Quantized vortices in superfluids – a mathematical and computational study by Qiang Du: It provides a concise description of the physical background, several relevant mathematical models, and the numerical methods developed for the study of the motion and interaction of quantized vortices in the celebrated Ginzburg-Landau models of superconductivity and the mean field Gross-Pitaevskii equations of superfluidity.

Part III. The nonlinear Schrödinger equation and applications in Bose-Einstein condensation and plasma physics by Weizhu Bao: It begins with the derivation of the nonlinear Schrödinger equation (NLS) from wave propagation and Bose-Einstein condensation (BEC); reviews variational formulation, plane and solitary wave solutions, existence/blowup results, WKB expansion and quantum hydrodynamics, Wigner transform and semiclassical limit of NLS; presents different numerical methods for computing ground states and dynamics of Gross-Pitaevskii equation with applications in BEC. Derivation of the Zakharov system with application in plasma physics is provided and different numerical methods are proposed for efficient computation.

Part IV. Introduction to constitutive modeling of macromolecular fluids by Qi Wang: It gives a crash course on the basics needed to model the complex fluids in fundamental thermodynamics, statistic mechanics, polymer physics and continuum mechanics, surveys the existing models for various polymeric liquids and explores a systematic approach for flexible polymers and liquid crystal polymers within the framework of the kinetic theory.

Besides us, the other members of the Organizing Committee are Gan-Moog Chow (Material Sciences Department, NUS), Weinan E (Princeton University), Yuanping Feng (Physics Departmnet, NUS), Bo Li (University of California at San Diego and Co-Chairman), Ping Lin (Mathematics Department, NUS and Co-Chairman), Chun Lu (Institute of High Performance Computing, NUS), Xingbin Pan (East China Normal University), Chang Shu (Mechanical Engineering Department, NUS), Eitan Tadmor (University of Maryland), Xuesen Wang (Physics Department, NUS), Kaiyang Zeng (Institute of Materials Research Engineering, NUS), Yongwei Zhang (Material Sciences Department, NUS). We are very much grateful to their invaluable services. Thanks also to all the participants of this program for their support and stimulating interactions during the two months!

We would like to take this opportunity to thank Professor Louis Chen, Director of IMS, for his leadership in creating an exciting environment for mathematical research in IMS and for his guidance throughout our program. The expertise and dedication of all IMS staff contributed essentially to the success of this program. Last but not least, we would like to acknowledge IMS for providing financial support to the program.

April 2007

Weizhu Bao
National University of Singapore

Jian-Guo Liu
University of Maryland

Lectures on Dynamics in Models of Coarsening and Coagulation

Robert L. Pego

Department of Mathematical Sciences
Carnegie Mellon University
Pittsburgh, PA 15213, USA
E-mail: rpego@cmu.edu

Contents

1. Introduction

1.1. These lectures are intended to serve as a rough introduction to a variety of mathematical problems aimed at understanding dynamic behavior in certain complex nonlinear physical systems, systems described by an evolution equation

$$\partial_t u = A(u), \tag{1.1}$$

where $u(t)$ is the state of the system at time t, with $u(t) \in X$, a high-dimensional state space. In general, one may think of problems to do with statistical mechanics (dim $X \sim 10^{24}$), fluid turbulence, polycrystalline grain structure of typical metals, etc. There are many ways to approach the question of dynamics in such systems—experimental, computational, statistical, etc. We focus on the role of mathematical analysis, whose appropriate uses include: identifying general principles for dynamic behavior; detailed study of prototypical examples and critical cases; developing and justifying procedures to reduce complex systems to simpler ones (of lower dimension, perhaps); and identifying and studying significant structure (Hamiltonian dynamics, gradient flow, thermodynamic compatibility).

From the physical point of view, our particular focus is on models of kinetic behavior for systems whose spatial structure develops a pattern of domains or clusters that coarsen as time increases, in ways that seem to be *statistically self-similar*. This kind of behavior is seen, for example, in foams (bubble bath), grain structure in alloys, and many agglomeration and clustering processes. Although many of these systems are subject to the second law of thermodynamics—entropy increases (or at constant temperature, free energy decreases)—they do not reach equilibrium on the time scales of interest, and so it is an interesting problem to understand the regularities observed. If there is one universal theme of these lectures, it is that ultimately the universe may be doomed to heat death, but the path it takes along the way could be interesting nevertheless.

From the mathematical point of view, the emphasis is on studying dynamical phenomena peculiar to infinite-dimensional, spatially extended systems, such as weak convergence and self-similarity (dynamic scaling behavior). In infinite dimensions, the choice of an appropriate topology becomes nontrivial, and sometimes depends upon both mathematical and physical considerations.

Note. These lecture notes grew from tutorial lectures given in January 2005 in conjunction with the program on Nanoscale Material Interfaces held at the Institute for Mathematical Sciences, University of Singapore, and were delivered in a course of lectures given at Humboldt University in Berlin in summer 2005. The notes were prepared with the assistance of Apostolos Damialis, Simone Hock, and Dirk Peschka. The lectures were developed for an audience including graduate and advanced undergraduate students; therefore a minimum of background in PDE theory is assumed, and the derivation of models and discussion of PDE results is largely done in a heuristic way. But a number of new rigorous results for models of the dynamics of domain size distributions are presented. Gradient structure turns out to play a surprisingly interesting and significant role throughout.

1.2. First models of domain formation and an open problem.

(i) A scalar ODE $\partial_t u = -f(u)$ with $f : \mathbb{R} \to \mathbb{R}$ C^1 always gives a gradient flow: $\partial_t u = -W'(u)$ where $W(u) = \int f(u)\,du$. Every solution $t \mapsto u(t)$ is monotonic, and every bounded solution converges as $t \to \infty$.

(ii) We introduce spatial variation, considering solutions of

$$\partial_t u(x, t) = -f(u(x, t)), \quad x \in \Omega = [0, 1], \ t \geq 0. \tag{1.2}$$

What happens as $t \to \infty$ is simple enough to describe: For any bounded solution, $u_\infty(x) = \lim_{t\to\infty} u(x, t)$ exists for every x, with $f(u_\infty(x)) = 0$ for all x. If f has multiple stable zeros (for example, $f(u) = u^3 - u$), the limiting state u_∞ is typically non-constant—domains will form in Ω as time proceeds, corresponding to different limiting values of $u_\infty(x)$.

Even for this simple equation, it is problematic to choose an appropriate space and topology for studying these infinite-dimensional dynamics. For state space we could try $X = C(\Omega)$ with norm $\|u\|_X = \sup_x |u(x)|$. This is fine for proving local solvability of the initial value problem, but is a poor choice for studying long-time behavior. The pointwise limit u_∞ may be discontinuous, thus not in X. One can achieve $u_\infty \in X$ by taking $X = B(\Omega)$, the space of bounded functions on Ω with the sup norm. But then it is false in general that $\|u(\cdot, t) - u_\infty\|_X \to 0$ as t gets large. One has $\int_\Omega |u(\cdot, t) - u_\infty|^p \to 0$ for any finite $p \geq 1$, which suggests taking $X = L^p(\Omega)$. But this is not wonderful either—the Nemytskii operator $f : L^p(\Omega) \to L^p(\Omega)$ is not usually C^1, and local solvability for all initial data may require restrictive hypotheses on f (globally Lipschitz, say).

(iii) A small change in the problem leads us quickly to the realm of the unknown. We alter the previous model by introducing a scalar "mean-field" coupling parameter $\theta(t)$, determined by imposing the global constraint that total "mass" $\int_\Omega u(x, t)\,dx$ should remain conserved in time. Thus we consider

$$\partial_t u(x,t) = -f(u(x,t)) + \theta(t), \quad x \in \Omega = [0,1], \ t \geq 0.$$

Mass conservation formally implies $\theta(t) = \int_\Omega f(u(x,t))\,dx$. Assuming $f : \mathbb{R} \to \mathbb{R}$ is C^1, the map $u \mapsto \int_\Omega f(u)$ is C^1 on the state space $X = B(\Omega)$, and Picard iteration gives us local-in-time solvability for initial-value problems

$$\partial_t u = -f(u) + \int_\Omega f(u), \quad u(t_0) = u_0 \in X.$$

One can rather easily prove global existence forward in time for bounded solutions under the mild hypothesis that

$$\liminf_{z \to -\infty} f(z) < f(s) < \limsup_{z \to +\infty} f(z) \quad \text{for every } s \in \mathbb{R}.$$

This leads us directly to the following innocent-sounding

> **Open question:** Under only these mild hypotheses on f, must every bounded solution converge (pointwise a.e.) as $t \to \infty$?

Various partial results are known from studying related PDEs from viscoelasticity [4] and anomalous diffusion [43]. But an answer to the general question remains elusive. Tantalizing is the fact that "free energy" is decreasing:

$$\frac{d}{dt} \int_\Omega W(u) = \int_\Omega W'(u)(\partial_t u) = \int_\Omega \left(f(u) - \int_\Omega f(u) \right) \partial_t u = - \int_\Omega (\partial_t u)^2,$$

thus

$$\int_\Omega W(u(t)) + \int_{t_0}^t \int_\Omega (\partial_t u)^2 = \int_\Omega W(u(t_0)).$$

Then it follows that as t increases, $\int_\Omega W(u(t))$ decreases and has a limit, and one can show $\int_\Omega (\partial_t u)^2$ is Lipschitz in t and tends to zero as $t \to \infty$. This rules out many types of recurrent dynamics such as asymptotically periodic behavior, but ever-slower drift remains a mathematical possibility. The main difficulty appears to be that it is not known whether $\lim_{t \to \infty} \int_\Omega f(u(x,t))\,dx$ must exist.

In [47] it is proved that *if the initial data u_0 has finite range*, then indeed $\lim_{t \to \infty} u(x,t)$ exists for all $x \in \Omega$. In this case, the problem reduces to one for a finite-dimensional ODE system for the unknown $\vec{u}(t) = (u_j(t)) \in \mathbb{R}^N$, where

$$u_j(t) = u(x,t) \quad \text{for } x \in A_j \subset \Omega, \quad \cup_{j=1}^N A_j = \Omega.$$

The proof exploits this finite-dimensionality in an essential way. The main ideas go as follows (see [47] for details): The ω-limit set of the solution trajectory,

$$\omega(\vec{u}) = \cap_{t_0 > 0} \overline{\{\vec{u}(t)|t \geq t_0\}} = \{\vec{v} \in \mathbb{R}^N : \exists t_n \to \infty, \ \vec{u}(t_n) \to \vec{v}\},$$

is a connected, compact set. By Sard's theorem, f has an open, dense set of regular values. If $\omega(\vec{u})$ is not a single point, then one can show that it contains a hyperbolic curve of equilibria. Using a theorem of Hale and Massat [25] related to center manifold theory, it follows that $\omega(\vec{u})$ must contain a non-equilibrium point. But for this system, $\omega(\vec{u})$ contains only equilibria, giving a contradiction.

2. A hierarchy of domain coarsening models in one space dimension

2.1. *Domain walls in the Allen-Cahn equation*

In a variety of physical processes, domains that form in multi-stable systems slowly change in time, with the overall pattern becoming coarser. Important examples in materials science include the growth of single-crystal grains in polycrystalline materials, phase separation in alloys, and anti-phase boundary motion in antiferromagentic materials. One of the simplest mathematical models of this behavior arises as a modification of (1.2) above. Namely we consider the Allen-Cahn equation (or scalar Ginzburg-Landau equation)

$$\partial_t u(x,t) = -f(u) + \varepsilon^2 \partial_x^2 u, \qquad 0 < x < 1, \ t > 0. \tag{2.1}$$

The term $\varepsilon^2 \partial_x^2 u$ can arise from diffusion, or through continuum modeling of nearest-neighbor coupling effects in atomic lattices, for example [3]. We have in mind that ε could be quite small, representing a ratio of microscopic to macroscopic length scales, for example. We fix attention on the bi-stable nonlinearity

$$f(u) = W'(u), \qquad \text{where} \quad W(u) = \frac{1}{8}(u-1)^2(u+1)^2, \tag{2.2}$$

and for convenience impose the Neumann boundary conditions

$$\partial_x u = 0 \quad \text{at } x = 0 \text{ and } 1, \ t > 0. \tag{2.3}$$

Our aim in this section is to describe the process of domain wall formation that occurs for this system, and how coarsening by domain wall motion and annihilation can be described via a hierarchy of models at varying levels of description, leading to a *universal kind of statistical behavior* on very large scales.

From the PDE-theoretic point of view it is convenient to take the state space X as the Sobolev space $H^1([0,1])$ of functions on $[0,1]$ with square-integrable first derivatives, but for these notes, we ignore PDE technicalities. For the most part, one may suppose solutions are smooth and take $X = C^\infty([0,1])$.

For smooth solutions of the Allen-Cahn equation, the "free energy"

$$\mathcal{F}(u) = \int_0^1 \left(W(u(x)) + \frac{\varepsilon^2}{2}(\partial_x u)^2 \right) dx \qquad (2.4)$$

decreases in time:

$$\frac{d}{dt}\mathcal{F}(u(t)) = \int_0^1 W(u)\partial_t u + \varepsilon^2 \partial_x u \, \partial_{tx} u = -\int_0^1 (\partial_t u)^2.$$

A fact of significance to us is that the Allen-Cahn equation has the structure of gradient flow for this free energy, with respect to the L^2 inner-product on $[0,1]$ given by $\langle u, v \rangle_{L^2} = \int_0^1 u(x)v(x)\,dx$. Namely, if $u, v \in X$ and u satisfies the boundary conditions, then

$$d\mathcal{F}(u)v = \frac{d}{d\tau}\mathcal{F}(u + \tau v)|_{\tau=0} = \int_0^1 (W'(u)v + \varepsilon^2 \partial_x u \, \partial_x v)\,dx = \langle \nabla_u \mathcal{F}, v \rangle_{L^2},$$

where the formal gradient $\nabla_u \mathcal{F}(u) = W'(u) - \varepsilon^2 \partial_x^2 u$. So formally,

$$\partial_t u = -\nabla_u \mathcal{F}(u). \qquad (2.5)$$

Theory developed in the 1980's established some satisfying facts regarding the behavior of solutions in the long-time limit $t \to \infty$: For every solution, $u_\infty = \lim_{t\to\infty} u(t)$ exists and satisfies the equation of equilibrium:

$$W'(u) - \varepsilon^2 \partial_x^2 u = 0. \qquad (2.6)$$

There are exactly two stable equilibrium states, and no more: $u \equiv 1$ and $u \equiv -1$. Unstable equilibria are restrictions of periodic solutions of the equation of equilibrium, which has first integral $\varepsilon^2 v^2/2 - W(u) = W_0$, $v = \partial_x u$.

Domain wall formation. These results contrast sharply with one's expectations based on the model with $\varepsilon = 0$ in (ii) above, and with numerical computations performed with small $\varepsilon > 0$. One expects and computes that u approaches 1 where $u > 0$ initially, and u approaches -1 where $u < 0$ initially. "Domain walls" or transition layers form between these domains, at positions corresponding roughly to zeros in the initial data (so not necessarily periodically arranged).

Structure of a domain wall. The structure of a domain wall, located near position $h \in (0,1)$, has a characteristic width of order ε, and can be described crudely in terms of the stretched variable $y = (x - h)/\varepsilon$ as a solution $u = \pm\Theta(y)$ of the rescaled equilibrium equation

$$W'(\Theta) - \partial_y^2\Theta = 0, \quad -\infty < y < \infty, \quad \Theta(y) \to \pm 1 \text{ as } y \to \pm\infty.$$

Explicity one finds $\Theta(y) = \tanh(y/2)$. Note that as $x \to \infty$,

$$\tanh(x/2\varepsilon) = 1 - 2\exp(-x/\varepsilon) + O(\exp(-2x/\varepsilon)).$$

Thus the domain structure that one expects to develop consists of arbitrarily placed domain walls of characteristic width ε, separating domains in which u is exponentially close to the stable states ± 1. (For recent rigorous results on the various stages of domain wall formation and evolution, see [13].)

2.2. *Domain wall dynamics by restricted gradient flow*

Numerics indicates that the kind of domain wall pattern that develops as just described essentially stops[a] evolving, contrary to what theory says should happen as $t \to \infty$. What actually happens is that the domain walls typically move, albeit *extremely slowly*. A delicate formal analysis of the domain wall dynamics was made by J. Neu (unpublished notes), and this prompted subsequent development of rigorous geometric theory by Fusco and Hale [21] and in [11, 9].

The result, which we shall derive here formally using a restricted gradient flow approach, is that given N domain walls initially located at given positions $h_1 < h_2 < \ldots < h_N$ in $(0, 1)$, the positions will evolve in time according to equations well-approximated by exponentially small nearest-neighbor interactions:

$$\partial_t h_j = 12\varepsilon \left(\exp\left(-\frac{h_{j+1} - h_j}{\varepsilon} \right) - \exp\left(-\frac{h_j - h_{j-1}}{\varepsilon} \right) \right). \tag{2.7}$$

Here the wall positions $h_0 = -h_1$ and $h_{N+1} - 1 = 1 - h_N$ are obtained by reflection through the boundaries.

Geometric description. The Fusco-Hale idea for a geometric description of these slow dynamics is to describe solutions containing N domain walls in terms of an N-dimensional manifold of "metastable" states in X. We will parametrize a state $u^h \in X$ on this manifold by a vector

---

[a]This depends upon the fact that the relative minima of W are equal. Different relative minima produce domain walls that move at speeds of order ε times the difference in minima.

$h = (h_1, \ldots, h_N) \in \Delta_N$ of arbitrarily-placed well-separated domain wall positions. Here we set

$$\Delta_N = \left\{ h \in \mathbb{R}^N \mid 0 < h_1 < \ldots < h_N < 1, \ (h_{j+1} - h_j)/\varepsilon > K \right\},$$

where K is a large constant. Then

$$\mathcal{M}_N = \left\{ u^h \mid h \in \Delta_N \right\} \subset X$$

is an N-dimensional manifold which should approximate the metastable states. Slowly evolving solutions will have a representation

$$u(x, t) = u^{h(t)}(x) + v(x, t) \tag{2.8}$$

where the error $v(x, t)$ is ideally zero or approaching zero if $\mathcal{M}$ is invariant.

Restricted gradient flow. Our formal approach here will be to construct $\mathcal{M}_N$ as "approximately invariant" and compute equations of motion for the domain wall positions by restricting the gradient flow from (2.5) to the manifold $\mathcal{M}_N$.

Geometrically this can be interpreted in a couple of equivalent ways. One way is to simply project the right-hand side of the Allen-Cahn PDE (2.1) onto the tangent space $T\mathcal{M}_N$, using the L^2 inner product on $[0, 1]$: $\langle u, v \rangle_{L^2} = \int_0^1 u(x)v(x)\, dx$. Thus we require

$$\langle \partial_t(u^h), v \rangle_{L^2} = \langle -\nabla_u \mathcal{F}(u^h), v \rangle_{L^2} \quad \text{for all } v \in T\mathcal{M}_N. \tag{2.9}$$

Since $\partial_t(u^h) = \sum_{j=1}^N (\partial u^h/\partial h_j)\partial_t h_j$, we get N equations for N unknowns $\partial_t h_j$.

An equivalent, more geometric, interpretation is useful to describe, directly in terms of a gradient flow

$$\partial_t h = -\nabla_h \widetilde{\mathcal{F}}(h), \qquad \text{where} \ \ \widetilde{\mathcal{F}}(h) := \mathcal{F}(u^h). \tag{2.10}$$

This is a gradient flow on $\Delta_N \subset \mathbb{R}^N$, and the gradient is computed using the appropriate Riemannian metric pulled back from the tangent space $T\mathcal{M}_N$. Namely, given two tangent vectors $a, \tilde{a} \in \mathbb{R}^N = T\Delta_N$, we let

$$v = \sum_j a_j \frac{\partial u^h}{\partial h_j}, \qquad \tilde{v} = \sum_j \tilde{a}_j \frac{\partial u^h}{\partial h_j}, \tag{2.11}$$

and define a metric $g_h : T\Delta_N \times T\Delta_N \to \mathbb{R}$ by

$$g_h(a, \tilde{a}) = \langle v, \tilde{v} \rangle_{L^2} = \sum_{i,k} g_{ik}\, a_i \tilde{a}_k, \qquad g_{ik} = \left\langle \frac{\partial u^h}{\partial h_i}, \frac{\partial u^h}{\partial h_k} \right\rangle_{L^2}.$$

The gradient $\nabla_h \widetilde{\mathcal{F}}(h)$ is determined from

$$g_h(\nabla_h \widetilde{\mathcal{F}}(h), a) = \mathrm{d}\widetilde{\mathcal{F}}(h)(a) = \frac{\mathrm{d}}{\mathrm{d}\tau} \mathcal{F}(u^{h+\tau a})|_{\tau=0} = \sum_{k=1}^{N} \left\langle \nabla_u \mathcal{F}(u^h), \frac{\partial u^h}{\partial h_k} \right\rangle_{L^2} a_k.$$

The equations (2.9) and (2.10) are easily seen equivalent using the correspondence (2.11):

$$g_h(\partial_t h, a) = g_h(-\nabla_h \widetilde{\mathcal{F}}(h), a) = \langle -\nabla_u \mathcal{F}(u^h), v \rangle_{L^2} = \langle \partial_t(u^h), v \rangle_{L^2}.$$

We can adequately approximate the metric coefficients g_{ik} formally using the domain-wall approximation that for x near h_j,

$$u^h(x) \approx \pm\Theta\left(\frac{x - h_j}{\varepsilon}\right), \qquad \frac{\partial u^h}{\partial h_j} \approx -\partial_x u^h \approx \mp\frac{1}{\varepsilon}\,\Theta'\left(\frac{x - h_j}{\varepsilon}\right).$$

Then

$$g_{ij} = \int_0^1 \frac{\partial u^h}{\partial h_i} \frac{\partial u^h}{\partial h_j}\,\mathrm{d}x \approx 0 \qquad (i \neq j), \tag{2.12}$$

$$g_{jj} = \int_0^1 \left(\frac{\partial u^h}{\partial h_j}\right)^2 \mathrm{d}x \approx \frac{1}{\varepsilon}\int_{-\infty}^{+\infty} \Theta'(y)^2\,\mathrm{d}y \tag{2.13}$$

Since $\frac{1}{2}\Theta_y^2 = W(\Theta) + 0$ one concludes

$$\int_{-\infty}^{+\infty} \Theta_y^2\,\mathrm{d}y = \int_{-1}^{1} \sqrt{W(\theta)}\,d\theta = 2\int_0^1 \frac{1 - \theta^2}{2}\,d\theta = \frac{2}{3}. \tag{2.14}$$

Thus the restricted metric is approximated by a diagonal matrix:

$$g_{ij} \approx \frac{2}{3\varepsilon}\,\delta_{ij}.$$

Approximating manifold of metastable states. The essential properties of the slowly evolving solutions are that $W'(u) - \varepsilon^2 \partial_x^2 u \approx 0$ away from $x = h_j$ and $u(h_j, x) \approx 0$ for $x \approx h_j$. Thus we define $\mathcal{M}_N$ as follows: Given $h = (h_1, \ldots, h_N) \in \Delta_N$ require for $j = 0, \ldots, N$ that

$$W'(u^h) - \varepsilon^2 \partial_x^2 u^h = 0 \text{ for } x \in I_j := (h_j, h_{j+1}), \tag{2.15}$$

$$u^h(x) = 0 \text{ for } x = h_j \text{ and } h_{j+1}, \tag{2.16}$$

$$(-1)^j u^h(x) > 0 \text{ in } I_j. \tag{2.17}$$

That is, in each interval I_j we have a piece of a periodic equilibrium with period $2l_j$, where $l_j = (h_{j+1} - h_j)/\varepsilon$ measures domain length: Requiring

$\mathcal{U}(y, l)$ to satisfy

$$W'(\mathcal{U}) - \partial_y^2 \mathcal{U}(y, l) = 0, \tag{2.18}$$

$$\mathcal{U}(0, l) = \mathcal{U}(l, l) = 0, \tag{2.19}$$

$$\mathcal{U}(y, l) < 0 \quad \text{for } 0 < y < l, \tag{2.20}$$

we have

$$u^h(x) = (-1)^j \, \mathcal{U}\left(\frac{x - h_j}{\varepsilon}, \frac{h_{j+1} - h_j}{\varepsilon}\right).$$

Thus described, the states u^h are not smooth. They are continuous but $\partial_x u^h$ is generally discontinuous at h_j. For this reason, the states u^h in [11, 9] were smoothed, trading one nuisance for another. (Results obtained without smoothing have been reported by Reznikoff (in preparation).)

Now, the restricted free energy turns out to be a sum of terms that depend only on the domain lengths: With

$$V(l) = \int_0^l \frac{1}{2}(\partial_y \mathcal{U})^2 + W(\mathcal{U}(y, l)) \, dy, \tag{2.21}$$

we can write

$$\widetilde{\mathcal{F}}(h) = \mathcal{F}(u^h) = \sum_j \int_{I_j} \left(\frac{\varepsilon^2}{2}(\partial_x u^h)^2 + W(u^h)\right) dy = \sum_j \varepsilon \, V\left(\frac{h_{j+1} - h_j}{\varepsilon}\right).$$

We compute the gradient $\nabla_h \widetilde{\mathcal{F}}(h)$ as follows: Given $a \in \mathbb{R}^N$,

$$\frac{d}{d\tau}\widetilde{\mathcal{F}}(h + \tau a)\big|_{\tau=0} = \sum_k V'\left(\frac{h_{k+1} - h_k}{\varepsilon}\right)(a_{k+1} - a_k)$$

$$= \sum_k \left(V'\left(\frac{h_k - h_{k-1}}{\varepsilon}\right) - V'\left(\frac{h_{k+1} - h_k}{\varepsilon}\right)\right) a_k$$

$$= \sum_k \frac{\partial \widetilde{\mathcal{F}}}{\partial h_k} a_k = \sum_{j,k} g_{jk} \nabla_h \widetilde{\mathcal{F}}(h)_j \, a_k.$$

Using the diagonal approximation of the metric from above gives the approximate equations of motion

$$\partial_t h_j = -\nabla_h \widetilde{\mathcal{F}}(h)_j \approx \frac{3\varepsilon}{2}\left(V'\left(\frac{h_{k+1} - h_k}{\varepsilon}\right) - V'\left(\frac{h_k - h_{k-1}}{\varepsilon}\right)\right) \tag{2.22}$$

The interpretation is that each domain wall moves as if attracted by each of its two nearest neighbors, with "force" determined by the potential V.

Approximation of the force. With $V(l)$ given by (2.21), one obtains

$$V'(l) = \left(\frac{1}{2}(\partial_y \mathcal{U})^2 + W(\mathcal{U}) \right)_{y=l} + \int_0^l \left((\partial_y \mathcal{U})(\partial_{yl}^2 \mathcal{U}) + W'(\mathcal{U}) \partial_l \mathcal{U} \right) \, dy$$

$$= \left(\frac{1}{2}(\partial_y \mathcal{U})^2 + W(\mathcal{U}) \right)_{y=l} + (\partial_y \mathcal{U})(\partial_l \mathcal{U})\Big|_0^l$$

after an integration by parts using (2.18). Recall that $U(l,l) = 0$, hence we have $(\partial_y \mathcal{U} + \partial_l \mathcal{U})(l,l) = 0$. Also $\partial_l \mathcal{U}(0,l) = 0$. Thus

$$V'(l) = \overbrace{\left(-\frac{1}{2}(\partial_y \mathcal{U})^2 + W(\mathcal{U}) \right)}^{\text{const. on } [0,l]}\Bigg|_{y=l} = W\left(\mathcal{U}\left(\frac{l}{2}, l \right) \right).$$

To approximate this, we use that for large l, we have $\mathcal{U}(y + l/2, l) \approx -1$ when $|y| \ll l/2$ (meaning $0 \ll y + l/2 \ll l$), whence

$$\mathcal{U}\left(y + \frac{l}{2}, l \right) \approx -1 + \widetilde{\mathcal{U}}(y) \quad \text{where} \quad \partial_y^2 \widetilde{\mathcal{U}} - W''(-1)\widetilde{\mathcal{U}} = 0.$$

Since $W''(-1) = 1$ and $\partial_y \mathcal{U}(l/2, l) = 0$, we have $\widetilde{\mathcal{U}}(y) = \alpha(e^y + e^{-y})$. On the other hand, near the right endpoint of $[0, l]$, $\mathcal{U}$ is approximated by the domain wall structure:

$$\mathcal{U}(\tilde{y} + l, l) \approx \Theta(\tilde{y}) = \tanh(\tilde{y}/2) \approx -1 + 2\,e^{\tilde{y}} \quad \text{for } l/2 \ll \tilde{y} + l \ll l.$$

To identify $\alpha = \alpha(l)$, we match these approximations in the regime $0 \ll y = \tilde{y} + l/2 \ll l/2$. This means we require $\alpha e^{\tilde{y}+l/2} = 2e^{\tilde{y}}$, whence $\alpha = 2\,e^{-l/2}$ and consequently for large l we get the approximation

$$V'(l) = W(\mathcal{U}(l/2, l)) \approx W(-1) + W'(-1)\,\tilde{\mathcal{U}} + \frac{1}{2}W''(-1)\,\tilde{\mathcal{U}}^2$$

$$= \frac{(2\alpha)^2}{2} = 8\,e^{-l}.$$

Together with (2.22) this leads to the equations of motion in (2.7).

2.3. *Punctuated equilibrium and 1D bubble bath*

Lifetime of metastable states. Let us imagine a system in which $1/\varepsilon$, the ratio of macroscopic domain size to microscopic domain wall thickness, is quite large (say $> 10^5$), and suppose that at some point the system settles into a metastable state with a great number of domain walls. Then due to the exponential dependence of terms in (2.7) on domain size, one can expect the dynamics to be dominated by the smallest domains. Say $l = \min(h_j - h_{j-1})$, then approximately

$$\partial_t l = -24\varepsilon\,\mathrm{e}^{-l/\varepsilon}, \quad \text{or} \quad \partial_t \mathrm{e}^{l/\varepsilon} = -24.$$

According to this equation, the domain size $l(t)$ shrinks to zero in a finite time T determined from $\exp(l_0/\varepsilon) = 1 - 24\,T$. The following table gives some indication of how the lifetime T of the metastable state depends strongly on the minimum initial domain size, l_0/ε (measured in units of domain wall thickness):

l_0/ε	5	10	20	50	100
T	4	918	2×10^7	2×10^{20}	10^{42}

In geometric terms, we have described the nature of N-wall metastable states using an approximately invariant manifold $\mathcal{M}_N$. (Of course there are two of these, using $-u^h$.) Interestingly, however, there is indeed an invariant manifold close to $\mathcal{M}_N$, which nearby solutions approach at a uniform rate as long as domain walls remain well separated [9]. That metastable states should correspond to such an invariant manifold was conjectured by Fusco and Hale [21], who also suggested that this manifold is part of the global *unstable manifold* of the unstable N-wall equilibrium state with equal domain sizes $h_{j+1} - h_j$. Indeed it was established in [9] that this unstable manifold is given in terms of (2.8) (with smoothed u^h) as a graph $h \mapsto v$ globally over Δ_N with exponentially small Lipschitz constant.

Punctuated equilibrium. The analysis so far indicates that the story of how gradient systems relax to equilibrium is not as simple as looking at stable steady states (here only $u \equiv 1$ and $u \equiv -1$) and finding the local rate of approach to these states (which here is $O(1)$). Instead, dynamics in the simple PDE (2.1) can be expected to exhibit a cascading behavior reminiscent of the "punctuated equilibrium" description of species evolution advocated by Stephen Jay Gould. (Some rigorous results along these lines were established in [18].) We might expect a typical solution trajectory to behave as follows. Domain walls develop and the system approaches an N-dimensional metastable manifold of states with N domain walls positioned arbitrarily as determined by initial data. It flows very slowly along this invariant manifold until two domain walls come close together (or one approaches the boundary). The two walls (or the one and its reflection) rapidly annihilate each other on an order 1 time scale, and the solution then settles into approaching an $N - 2$ (or $N - 1$) dimensional metastable manifold. The slow motion grows dramatically slower as fewer walls remain, with greater distance between them. Note that the solution may be "near equilibrium" in the sense of being nearly stationary, but need never be near an equilibrium *state*. In the limit $t \to \infty$, one can expect to approach one of the stable states $u \equiv \pm 1$—but it may not be practical to wait that long!

1D bubble bath. So far we have reduced the study of long-time behavior in the Allen-Cahn PDE to the motion of domain walls according to the ODEs in (2.7), and have further simplified by noticing that the largest term in these ODEs should strongly dominate and produce collapse of the smallest domain. This suggests an even simpler model that we can use to investigate the statistical behavior of the coarsening process.

Starting by partitioning the interval $[0, 1]$, placing domain walls randomly according to some scheme. Then coarsen this domain pattern according to the following recipe:

(1) The smallest domain joins its two neighbors.
(2) Nobody else moves.
(3) Repeat.

This process is a kind of 1D model of bubble bath—the smallest bubble pops first, and the foam becomes coarser in time. This 1D model of coarsening is easy to simulate by computer with many thousands of domains. (See [10]. Many results for related models also exist in the physical literature; see [17] for a review.) Results show a remarkable thing: After scaling by mean size, the distribution of domain sizes develops toward a *universal self-similar form*. This raises the interesting question:

Why?

2.4. *Mean-field model of domain growth — The Gallay-Mielke transform*

To try and understand this phenomenon of universal self-similarity, we formulate a model of this domain coalescence process that aims to describe how the domain size distribution evolves in time. The main idea is to develop a rate equation for the domain size distribution function, based upon the mean-field assumption that the sizes of domains undergoing coalescence are accurately characterized by the overall size distribution of all domains. For the model that results, a remarkable solution procedure was recently developed by Gallay and Mielke [22], which we will use in this section to prove a theorem regarding universal approach to self-similar form.

Derivation of the model. We let x denote domain size, and at first take x discrete: $x = j\Delta x$, $j \in \mathbb{N}$. We introduce a function to describe the domain size distribution by

$$f(x,t)\Delta x = \text{expected number of domains of size } x$$
$$\text{(normalized by initial total)}$$

Let $\mathcal{L}(t)$ denote the size of the smallest domain remaining at time t; thus $f(x,t) = 0$ for $x < \mathcal{L}(t)$. And let $N(t) = \sum_x f(x,t)\Delta x$ be the total number at time t.

In the time interval $(t, t + \Delta t)$, the total number of coalescence events (involving the smallest remaining domain combining with its two neighbors) is expected to be

$$f(\mathcal{L}(t), t)\Delta\mathcal{L} = f(\mathcal{L}(t), t)\frac{\Delta\mathcal{L}}{\Delta t}\,\Delta t.$$

The change in number of size-x domains will equal the total number of coalescence events times the sum over subevents of the relative probability of the subevent times the change in number of size-x domains in the subevent. Three types of subevents affect size-x domains:

(1) Sizes $(x, \mathcal{L}, y)$ combine to form $x + \mathcal{L} + y$.
(2) Sizes $(y, \mathcal{L}, x)$ combine to form $y + \mathcal{L} + x$.
(3) Sizes $(y, \mathcal{L}, x - y - \mathcal{L})$ combine to form x.

Under the mean field assumption, these events respectively have the relative probabilities

$$\frac{f(x)\Delta x}{N}\frac{f(y)\Delta x}{N}, \quad \frac{f(y)\Delta x}{N}\frac{f(x)\Delta x}{N}, \quad \frac{f(y)\Delta x}{N}\frac{f(x - y - \mathcal{L})\Delta x}{N}.$$

From these ideas, one finds the change in number of size-x domains is

$$\Delta(f(x,y)\Delta x) =$$
$$f(\mathcal{L},t)\Delta\mathcal{L}\sum_y \left(\frac{f(y,t)\Delta x}{N}\frac{f(x - y - \mathcal{L}, t)\Delta x}{N} - 2\frac{f(x)\Delta x}{N}\frac{f(y)\Delta x}{N}\right)$$

Dividing by $\Delta x \Delta t$ and passing formally to the continuum limit, one obtains

$$\partial_t f(x,t) = \frac{f(\mathcal{L},t)\dot{\mathcal{L}}}{N^2}\int_0^\infty \Big(f(y,t)f(x - y - \mathcal{L}, t) - 2\,f(x,t)f(y,t)\Big)\,dy \quad (2.23)$$

This is the model rate equation we seek [36, 10].

Proceeding formally, we derive a useful moment identity and reformulate the rate equation. In the discrete case, we can write a general moment identity as follows:

$$\Delta\left(\sum_x a(x)f(x,t)\Delta x\right) = -a(\mathcal{L})f(\mathcal{L},t)\Delta\mathcal{L}$$

$$+ \frac{f(\mathcal{L},t)\Delta\mathcal{L}}{N^2}\sum\left(a(\tilde{x}+y+\mathcal{L})f(y)f(\tilde{x}) - 2\,a(x)f(x)f(y)\right)\Delta x\,\Delta y$$

$$= \frac{f(\mathcal{L},t)\Delta\mathcal{L}}{N^2}\sum\left(a(x+y+\mathcal{L}) - a(x) - a(y) - a(\mathcal{L})\right)f(x)f(y)\Delta x\Delta y.$$

In the continuum limit this yields

$$\partial_t \int_{\mathcal{L}}^{\infty} a(x)f(x,t)\,\mathrm{d}x \tag{2.24}$$

$$= \frac{f(\mathcal{L},t)\dot{\mathcal{L}}}{N(t)^2}\int_0^{\infty}\int_0^{\infty}\left(a(x+y+\mathcal{L}) - a(x) - a(y) - a(\mathcal{L})\right)f(x,t)f(y,t)\,\mathrm{d}x\,\mathrm{d}y$$

Considering $a(x) = x$ yields

$$\partial_t \int_0^{\infty} xf(x,t)\,\mathrm{d}x = 0,$$

thus total size is preserved (if finite). Taking $a(x) = 1$ next we get

$$\partial_t N(t) = -2\,f(\mathcal{L},t)\dot{\mathcal{L}}.$$

Thus total number decreases, and average domain size $\bar{x} = \int xf/\int f$ increases. Also $a(x) = x^2$ yields a growth law for second moment:

$$\partial_t \int_0^{\infty} x^2 f(x,t)\mathrm{d}x = 2f(\mathcal{L},t)\dot{\mathcal{L}}(\bar{x}^2 + 2\bar{x}\mathcal{L}).$$

Our model is invariant under reparametrization in time: If one changes variables via $t = T(\tilde{t})$, $\tilde{f}(x,\tilde{t}) = f(x,t)$, $\tilde{\mathcal{L}}(\tilde{t}) = \mathcal{L}(t)$, then the equation retains its form since

$$\partial_{\tilde{t}}\tilde{f} = \dot{T}\partial_t f, \quad \partial_{\tilde{t}}\tilde{\mathcal{L}} = \dot{T}\partial_t \mathcal{L}.$$

The model has no intrinsic time scale since the process is simply driven by the rate of collapse of smallest domains.

Following Gallay and Mielke, it is convenient to parametrize time by the size of smallest domain, and take $\mathcal{L}(t) = t$. Also it is convenient to rewrite the model in terms of the probability density for domain size:

$$\rho(x,t) = \frac{f(x,t)}{N(t)}.$$

Since for $\mathcal{L} = t$ we have

$$\partial_t \left(\frac{f(x,t)}{N(t)} \right) = \frac{\partial_t f}{N} - \frac{f \partial_t N}{N^2} = \frac{\partial_t f(x,t)}{N} + \frac{f(x,t)}{N} \cdot \frac{2f(t,t)}{N},$$

the model takes the form

$$\partial_t \rho(x,y) = \rho(t,t) \int_0^{x-t} \rho(y,t)\rho(x-y-t,t)\,\mathrm{d}y \qquad \text{for } x > t, \qquad (2.25)$$

with $\rho(x,t) = 0$ for $x < t$. Note that due to the latter condition, the integrand vanishes unless $t < y < x - 2t$, requiring $x > 3t$.

The Gallay-Mielke global linearizing transform. An amazing solution procedure for this model was found by Gallay and Mielke in [22], and used to establish several results regarding convergence to self-similar form at various rates depending upon the tail of the initial data. In these notes our aim is to give a simple proof of universal weak convergence to self-similar form for all classical solutions with finite total number and size.

Consider the initial value problem for the model (2.25), with initial data given at a time when smallest domain size $t = 1$, say. Leaving aside til later the question of solvability of this initial value problem, let us describe the solution procedure of Gallay and Mielke. For brevity we use the (probabilists') notation $\rho_t(x) = \rho(x,t)$ to denote the solution at time t, and denote its Fourier transform by

$$\mathcal{F}\rho_t = \hat{\rho}(\xi,t) = \int_{\mathbb{R}} e^{-i\xi x} \rho(x,t)\,\mathrm{d}x.$$

With $\Phi(z) = \frac{1}{2}\ln\frac{1+z}{1-z} = \tanh^{-1} z$, introduce the change of variables

$$v_t(x) = \mathcal{F}^{-1} \circ \Phi \circ \mathcal{F}\rho_t, \qquad \text{so} \quad \hat{v}_t(\xi) = \frac{1}{2}\ln\left(\frac{1+\hat{\rho}_t}{1-\hat{\rho}_t}\right).$$

In terms of v_t the solution is given by the simple formula (!)

$$v_t(x) = H(x-t)v_1(x) = \begin{cases} v_1(x), & x \geq t, \\ 0, & x \leq t, \end{cases}$$

where H is the Heaviside function. Thus, to find the solution to the nonlinear model (2.25) at time t, the procedure is:

(i) Transform the initial data $\rho_1(x) = \rho(x,1)$: Let $v_1(x) = \mathcal{F}^{-1} \circ \Phi \circ \mathcal{F}\rho_1$.
(ii) Set to zero for $x \leq t$: Let $v_t(x) = H(x-t)v_1(x)$.
(iii) Invert the transformation: $\rho_t = \mathcal{F}^{-1} \circ \Phi^{-1} \circ \mathcal{F}v_t = \mathcal{F}^{-1}(\tanh \mathcal{F}v_t)$.

Let us now formally derive this method. I prefer to work with the Laplace transform to simplify rigorous analysis later. Denote the Laplace transform of ρ_t by

$$R_t(q) = \mathcal{L}\rho_t(q) = \int_0^\infty \mathrm{e}^{-qx}\rho_t(x)\,\mathrm{d}x.$$

The integral term on the right-hand side of (2.25) is a shifted convolution, and its Laplace transform is given by

$$\begin{aligned}
\mathcal{L}(\rho * \rho(\cdot - t)) &= \int_0^\infty \int_0^{x-t} \mathrm{e}^{-q(x-t-y+t+y)}\rho_t(y)\rho_t(x-t-y)\,\mathrm{d}y\,\mathrm{d}x \\
&= \int_0^\infty \int_0^\infty \mathrm{e}^{-q\tilde{x}}\mathrm{e}^{-q\tilde{y}}\mathrm{e}^{-qt}\rho_t(\tilde{y})\rho_t(\tilde{x})\,d\tilde{y}\,d\tilde{x} \\
&= \mathrm{e}^{-qt}R_t(q)^2.
\end{aligned}$$

Since $\partial_t R_t = \partial_t \int_t^\infty \mathrm{e}^{-qt}\rho(x,t)\,\mathrm{d}x = -\mathrm{e}^{-qt}\rho(t,t) + \int_0^\infty \mathrm{e}^{-qx}\partial_t\rho\,\mathrm{d}x$, taking the Laplace transform of (2.25) yields

$$\partial_t R_t = \alpha(t)\mathrm{e}^{-qt}(-1 + R_t^2), \qquad \text{where } \alpha(t) = \rho(t,t). \tag{2.26}$$

This yields

$$\partial_t \Phi(R_t) = \frac{\partial_t R_t}{1 - R_t^2} = -\alpha(t)\mathrm{e}^{-qt},$$

whence upon integration,

$$\Phi(R_t) - \Phi(R_1) = -\int_1^t \alpha(s)\mathrm{e}^{-qs}\,\mathrm{d}s. \tag{2.27}$$

Note that for $q > 0$,

$$R_t(q) = \int_t^\infty \mathrm{e}^{-qx}\rho_t(x)\,\mathrm{d}x \le \mathrm{e}^{-qt}\int_t^\infty \rho_t(x)\,\mathrm{d}x = \mathrm{e}^{-qt}\cdot 1 \to 0 \text{ as } t \to \infty.$$

Hence taking $t \to \infty$ in (2.27) above yields

$$\Phi(R_1) = \int_1^\infty \alpha(s)\mathrm{e}^{-qs}\,\mathrm{d}s = \mathcal{L}\alpha. \tag{2.28}$$

This formula determines $\alpha = \rho(\cdot,\cdot)$ in terms of the initial data, according to

$$\alpha = \mathcal{L}^{-1}\circ\Phi\circ\mathcal{L}\circ\rho_1.$$

Plugging back into (2.27) yields

$$\Phi(R_t) = \int_t^\infty \mathrm{e}^{-qs}\alpha(s)\,\mathrm{d}s = \mathcal{L}(H(\cdot - t)\alpha) = \mathcal{L}(\alpha_t), \tag{2.29}$$

where we set $\alpha_t(x) = H(x-t)\alpha(x)$ for all x. Since $\tanh(\Phi(z)) = z$, it follows

$$R_t = \tanh(\mathcal{L}\alpha_t), \tag{2.30}$$

that is, $\rho_t = \mathcal{L}^{-1} \circ \tanh \circ \mathcal{L}\alpha_t$. This finishes the derivation. (In fact, $\alpha_t = v_t$.)

Initial value problem. Here we sketch a proof of existence for classical solutions of the initial value problem for the model (2.25), and a rigorous justification of the solution formula (2.30). We fix $\tau = 1$ and suppose that $\rho_\tau : [\tau, \infty) \to \mathbb{R}$ is given as a continuous function with $\int_\tau^\infty \rho_\tau(x)\,dx = 1$.

Note that since the solution is to satisfy $\rho_t(x) = 0$ for $x < t$, the convolution term on the right-hand side of (2.25) will depend only upon values of $\rho_t(y)$ for $\tau < y < x - 2t \le x - 2\tau$. In particular, the right-hand side vanishes for $x < 3\tau$.

This means we can construct the solution for $\tau < t < 3\tau$ by an inductive procedure as follows: For $\tau < t \le x < 3\tau$ we have $\rho_t(x) = \rho_\tau(x)$ and in particular $\rho_t(t) = \rho_\tau(t)$. For $\tau < t \le 3\tau$, successively on strips $x \in [k\tau, (k+2)\tau]$, for $k = 3, 5, \ldots$, by simple integration in time we can now compute

$$\rho_t(x) = \rho_\tau(x) + \int_\tau^t \rho_s(s) \int_0^{x-s} \rho_s(y)\rho_s(x - y - s)\,dy\,ds,$$

where the right-hand side is always known from a previous step. This determines $\rho_t(x)$ for $\tau \le t \le 3\tau$ and all x.

To determine the solution globally for all $t > 1$, the idea is to replace 3τ by τ and repeat. But in order to justify this we need to verify that ρ_t remains integrable and conserves total probability. In particular we need to justify (2.26). Let us introduce the distribution function

$$F_t(x) = \int_0^x \rho_t(y)\,dz. \tag{2.31}$$

This is the probability that a domain has size $\le x$ at time t. Integrating the convolution term in (2.25), we get

$$\int_0^x \int_0^\infty \rho_t(y)\rho_t(z - y - t)\,dy\,dz = \int_\mathbb{R} F_t(x - y - t)\,F_t(\,dy)$$

$$\le F_t(x) \int_0^x F_t(\,dy) = F_t(x)^2$$

since $x \mapsto F_t(x)$ is increasing. Thus $\partial_t F_t(x) \le \rho_t(t)(F_t(x)^2 - 1)$, and since $\rho_t(t) \ge 0$ and $F_t(x) \le 1$ initially, $F_t(x)$ is decreasing in t for fixed x. It follows $F_t(\infty) \le 1$, and so the Laplace-Stieltjes transform

$$R_t(q) = \int_0^\infty e^{-qx} F_t(\,dx) = \int_t^\infty e^{-qx} \rho_t(x)\,dx$$

is well defined and $\leq \mathrm{e}^{-qt}$. Since $\partial_t F_t(x)$ is continuous in t for all x, $R_t(q)$ is C^1 in t for all $q > 0$. This justifies (2.26) and the solution formula (2.30) in the previous subsection.

Background on Laplace transforms. Our rigorous study of dynamic scaling behavior will make use of some basic facts regarding Laplace transforms of measures on $[0, \infty)$. For this material we refer to Feller's excellent book [19]. In particular we recall the following main results from chapters VIII and XIII of [19]:

1. (Selection theorem) Every sequence of probability distributions has a subsequence that converges (weakly, i.e., in distribution) to some limit distribution (possibly defective).
2. (Continuity theorem) Weak convergence of measures is equivalent to pointwise convergence of the corresponding Laplace transforms.
3. (Tauberian theorem) Let U be a measure on $[0, \infty)$ with $U(0) = 0$ and suppose its Laplace-Stieltjes transform is

$$\omega(q) = \int_0^\infty \mathrm{e}^{-qx} U(\,\mathrm{d}x).$$

Let $p \in [0, \infty)$, and let L be a function *slowly varying at* ∞, meaning $\lim_{t \to \infty} L(tx)/L(t) = 1$ for all $x > 0$. Then the following are equivalent:

(i) $U(t) = U([0, t]) \sim t^p L(t)$ as $t \to \infty$.
(ii) $\omega(q) \sim q^{-p} L(1/q) \Gamma(1 + p)$ as $q \to 0$.

For later reference we also now recall a fundamental lemma on scaling limits:

4. (Rigidity of scaling limits) Suppose $U : [0, \infty) \to \mathbb{R}$ is positive and increasing, and suppose the following limit exists:

$$\lim_{t \to \infty} \frac{U(tx)}{U(t)} = \psi(x) \leq \infty,$$

for all x in some set S dense in $[0, \infty)$. Then necessarily the limit is a power law: $\psi(x) = x^p$ for some $p \in [0, \infty]$, and furthermore, U is *regularly varying at* ∞ *with exponent* p, meaning $U(t) = t^p L(t)$ where L is slowly varying at ∞.

2.5. *Proof of universal self-similar behavior*

The goal here is to prove that every solution of the coarsening model (2.25) with initially finite expected size $\int_0^\infty x \rho_1(x)\,\mathrm{d}x$ will converge in distribution to a universal self-similar form. One can think of this as a dynamic analog of the *central limit theorem* in probability. See section 5.2 below for a further development of this analogy.

Rescaling. In studying dynamic scaling behavior for this model, it is natural to rescale to keep the smallest domain size fixed. Hence we introduce the rescaled probability distribution function

$$\eta_t(x) = F_t(tx) = \int_0^{tx} \rho_t(y)\,\mathrm{d}y. \tag{2.32}$$

We have $\eta_t(x) = 0$ for $x < 1$, and $\eta_t(\infty) = 1$. Its Laplace transform is

$$N_t(q) = \int_1^\infty \mathrm{e}^{-qx}\eta_t(\,\mathrm{d}x) = \int_t^\infty \mathrm{e}^{-qx/t} F_t(\,\mathrm{d}x) = \mathcal{L}F_t(q/t)$$

We introduce the notation

$$A(t) = \int_1^t \alpha(s)\,\mathrm{d}s = \int_1^t \rho_s(s)\,\mathrm{d}s, \qquad \text{so} \quad A(\,\mathrm{d}s) = \alpha(s)\,\mathrm{d}s.$$

By the solution formula (2.30) we have that

$$N_t(q) = \tanh\left(\int_1^\infty \mathrm{e}^{-qs} A(t\,\mathrm{d}s)\right). \tag{2.33}$$

Self-similar solutions. For self-similarily, $N_t \equiv N$ is independent of t. Since then we have $H(s - t)A(t\,\mathrm{d}s) = A(\,\mathrm{d}s)$ for all t, we must have

$$A(\,\mathrm{d}s) = \frac{\beta}{s}H(s - 1)\,\mathrm{d}s$$

for some constant $\beta > 0$. This means that the Laplace transform of the profile

$$N(q) = \tanh\left(\beta\int_1^\infty \frac{\mathrm{e}^{-qs}}{s}\,\mathrm{d}s\right) = \tanh(\beta\,\mathrm{Ei}(q)), \tag{2.34}$$

where $\mathrm{Ei}(q) = \int_q^\infty (\mathrm{e}^{-s}/s)\,\mathrm{d}s$ is the exponential integral function.

Only certain values of β are physically meaningful here. Note that:

(i) To have $\bar{y} = \int_1^\infty y\eta(\,\mathrm{d}y) < \infty$ we need

$$-\partial_q N(q) = \int_1^\infty \mathrm{e}^{-qy} y\,\eta(\,\mathrm{d}y) \longrightarrow \bar{y} < \infty \quad \text{as } q \to 0.$$

(ii) $\dfrac{\mathrm{d}}{\mathrm{d}x}\tanh x = \mathrm{sech}^2 x = \dfrac{4}{(\mathrm{e}^x + \mathrm{e}^{-x})^2} = \dfrac{4\mathrm{e}^{-2x}}{(1 + \mathrm{e}^{-2x})^2}.$

(iii) $\mathrm{Ei}(q) = \displaystyle\int_q^\infty \frac{\mathrm{e}^{-s}}{s}\,\mathrm{d}s = -\ln q + \gamma(q)$, where $\gamma(0) \approx 0.577216$ is Euler's constant.

(iv) $\exp(-\beta \operatorname{Ei}(q)) = q^\beta e^{\beta\gamma(q)}$, hence as $q \to 0$,

$$-\partial_q N(q) = \operatorname{sech}^2(\beta \operatorname{Ei}(q)) \frac{\beta e^{-q}}{q} = \frac{4\beta e^{-q}}{q} \frac{q^{2\beta} e^{2\beta\gamma}}{(1 + q^{2\beta} e^{2\beta\gamma})^2} \sim c q^{2\beta - 1}.$$

Thus, we find that it is *necessary* that

$$\beta > 0, \quad \text{for } N(q) > 0,$$
$$\beta \le \tfrac{1}{2}, \quad \text{for } -\partial_q N(q) = \int_0^\infty e^{-qy} y \eta(\mathrm{d}y) \quad \text{to decrease in } q,$$
$$\beta = \tfrac{1}{2}, \quad \text{for } -\partial_q N(q) \to \tilde{y} \in (0, \infty) \text{ as } q \to 0.$$

So there is a *unique possibility* for a self-similar solution with finite expected domain size, namely with $\beta = 1/2$ and

$$N(q) = \tanh\left(\frac{1}{2} \operatorname{Ei}(q)\right). \tag{2.35}$$

Main result. At this point we need to address the following questions:

- Does a positive self-similar solution really exist satisfying (2.35)?
- Is it stable, and does it attract every solution?

The answers are positive.

Theorem 2.1: *Suppose that the initial data for model (2.25) satisfy* $\int_0^\infty x \rho_1(\mathrm{d}x) < \infty$. *Then with* $\eta_t(x)$ *given by (2.32), we have* $\lim_{t \to \infty} \eta_t(x) = \eta_*(x)$ *for all* $x > 1$, *where* η_* *is a probability distribution function satisfying* $\mathcal{L}\eta_*(q) = \tanh(\tfrac{1}{2} \operatorname{Ei}(q))$.

Proof: 1. By the selection and continuity theorems for Laplace transforms, it suffices to show $N_t(q) \to \tanh(\tfrac{1}{2} \operatorname{Ei}(q))$ for all $q > 0$, or equivalently by (2.33),

$$\int_1^\infty e^{-qs} A(t \, \mathrm{d}s) \longrightarrow \int_1^\infty \frac{e^{-qs}}{2s} \, \mathrm{d}s \quad \text{as } t \to \infty \text{ for all } q > 0. \tag{2.36}$$

2. Let $\bar{x} = \int_1^\infty y \eta_1(\mathrm{d}y)$. Then as $q \to 0$, $-\partial_q N_1(q) = \int_1^\infty e^{-qx} x \rho_1(\mathrm{d}x) \to \bar{x}$, and thus $N_1(q) = 1 - q\bar{x}(1 + o(1))$. We have $\mathcal{L}A(q) = \int_1^\infty e^{-qs} A(\mathrm{d}s) = \Phi(N_1(q))$ by (2.28), hence

$$-\partial_q \mathcal{L}A(q) = \int_1^\infty e^{-qs} s A(\mathrm{d}s) = \frac{-\partial_q N_1(q)}{(1 + N_1(q))(1 - N_1(q))}$$

$$= \frac{\bar{x}}{(1+1)(q\bar{x})} \cdot (1 + o(1)) = \frac{1}{2q}(1 + o(1))$$

as $q \to 0$. By the Tauberian theorem, it follows that the distribution function for the measure $sA(\,ds)$ satisfies

$$\int_0^t sA(\,ds) = \frac{t}{2}(1 + \sigma(1)) \quad \text{as } t \to \infty.$$

3. Thus the distribution function for the measure $sA(t\,ds)$ satisfies

$$\mu_t(x) = \int_0^x sA(t\,ds) = \frac{1}{t}\int_0^{tx} sA(\,ds) \to \frac{x}{2} \quad \text{as } t \to \infty,$$

i.e., $\mu_t \to \frac{1}{2}$ in distribution. It remains to show that for all $q > 0$,

$$\int_1^\infty \frac{e^{-qs}}{s}\mu_t(\,ds) \to \int_1^\infty \frac{e^{-qs}}{s}\frac{ds}{2} \quad \text{as } t \to \infty.$$

We establish this in two steps: (i) By the weak convergence theorem for probability measures, given $\bar{x} > 0$, since the probability distribution function

$$\nu_t(x) = \min(\mu_t(x)/\mu_t(\bar{x}), 1) \to \min(x/\bar{x}, 1) \quad \text{as } t \to \infty,$$

we have $\int_0^\infty u(s)\mu_t(\,ds)/\mu_t(\bar{x}) \to \int_0^\infty u(s)\,ds/\bar{x}$ for all $u \in \mathcal{C}_c([0, \bar{x}))$ (continuous u with compact support in $[0, \bar{x})$, and hence

$$\int_0^\infty u(s)\mu_t(\,ds) \to \int_0^\infty u(s)\frac{ds}{2}.$$

This holds also for discontinuous u, provided that $\int_0^\infty u = \inf \int_0^\infty v_+ = \sup \int_0^\infty v_-$, where the inf and sup are taken over all $v_- \le u \le v_+$ with $v_+, v_- \in \mathcal{C}_c([0, \bar{x}))$. Hence for all $x > 1$, we can conclude that

$$\int_1^x \frac{e^{-qs}}{s}\mu_t(\,ds) \to \int_1^x \frac{e^{-qs}}{s}\frac{ds}{2}.$$

(ii) We compute that

$$\int_0^\infty e^{-qs}\mu_t(\,ds) = \int_0^\infty e^{-qs}sA(t\,ds) = \frac{1}{t}\int_0^\infty e^{-qs/t}sA(\,ds)$$

$$= \frac{1}{t}\frac{-\partial_q N_1(q/t)}{(1 + N_1)(1 - N_1(q/t))} \le \frac{1}{t}\frac{1}{q/t} = \frac{1}{q}$$

independent of t, since $-\partial_q N_1$ decreases. (By the mean value theorem, $1 - N_1(q/t) = -\partial_q N_1(c)(q/t)$ for some $c < q/t$.) Now

$$\int_x^\infty \frac{e^{-qs}}{s}\mu_t(\,ds) \le \frac{1}{x}\int_0^\infty e^{-qs}\mu_t(\,ds) \le \frac{1}{x}\cdot\frac{1}{q} < \varepsilon$$

for $x > 1/q\varepsilon$; this estimate controls the tail. It follows that

$$\int_1^\infty \frac{e^{-qs}}{s} \mu_t(\,ds) \to \int_1^\infty \frac{e^{-qs}}{s} \frac{ds}{2}$$

as $t \to \infty$. This finishes the proof. $\qquad\square$

3. Models of domain coarsening in two and three dimensions

Universal scaling behavior. One of the classic dynamic scaling phenomena observed in material systems is Ostwald ripening, a process that occurs during the condensation of a supersaturated vapor, for example (think of clouds or fog), or during phase separation in metallic alloys. Many nuclei of the new phase appear and grow until there is a rough equilibrium with a complex arrangement of particles. The system continues to evolve in the late stages of this phase transition, however, driven by fluxes generated by curvature variation. In certain mixtures of metals, Ostwald observed that the typical particle size grows like $t^{1/3}$. The total phase fraction is conserved; large particles grow while small particles shrink and disappear.

An important paradigm for understanding this power-law scaling behavior is the model of Lifshitz and Slyozov [31] and Wagner [50] for the evolution of the particle size distribution. In this section we aim to describe how the LSW model fits in a hierarchy of multidimensional domain coarsening models similar to the one-dimensional hierarchy of the previous section, and indicate how recent mathematical analysis has helped to clarify why the LSW model is an unsatisfactory explanation for power-law scaling behavior as observed in practice.

Diffuse and sharp interfaces. We start with diffuse interface models (though there is research relating these to even more microscopic stochastic Ising models). Domain walls are "diffuse interfaces" which become "sharp interfaces" in the limit that their characteristic width divided by a macroscopic scale is taken to zero. The free energy concentrates on domain walls and becomes proportional to the interface surface area. Formally, curvature is the gradient of surface area (as we will see), and gradient flow means that coarsening in multidimensional systems is driven in many cases by the curvature of interfaces.

Allen & Cahn [3] argued by physical considerations that weakly curved domain walls in the multidimensional PDE

$$\partial_t u = -f(u) + \Delta u \tag{3.1}$$

should move with normal velocity v proportional to mean curvature κ. This result was derived a few years later within the systematic formal approximation procedure of matched asymptotic expansions by Rubinstein, Sternberg & Keller [48]. The same method was used to produce many interesting singular limits in the phase field system (a model of solidification) by Caginalp [8].

The hierarchy that leads to the LSW model of Ostwald ripening starts with the Cahn-Hilliard equation (a generalized diffusion equation)

$$\partial_t u = \Delta(f(u) - \varepsilon^2 \Delta u). \tag{3.2}$$

The sharp-interface limit of the Cahn-Hilliard equation turns out to be a model of phase transition kinetics found by Mullins & Sekerka to generate shape instabilities [46]: Given an interface set Γ separating two phases of material, steady-state diffusion of material (subject to a Gibbs-Thomson boundary condition for chemical potential) produces a jump in flux that drives the interface to move. Thus, the normal velocity v of $\Gamma(t)$ is determined by solving a boundary value problem of the following form (setting physical constants to 1):

$$\Delta u = 0 \quad \text{in } \mathbb{R}^3 \setminus \Gamma, \tag{3.3}$$

$$u = \kappa \quad \text{on } \Gamma, \tag{3.4}$$

$$v = [\partial_\nu u]_-^+ \quad \text{on } \Gamma. \tag{3.5}$$

The LSW model arises from this model in a dilute regime in which particles are widely separated and the potential field u is approximated by a sum of "monopoles" $\sum a_j/|x - x_j|$ plus a constant mean field. Below we will describe how the monopole model arises naturally by restriction from the gradient structure of the Mullins-Sekerka model. The LSW model then inherits a gradient structure that turns out to be useful for some things—see section 4.2 for an example.

3.1. *Diffuse and sharp-interface models of nanoscale island coarsening*

Rather than discuss the simplest situation in detail, it is interesting to consider a recent treatment of a problem of current interest related to nanoscale material interfaces. Understanding the dynamics of nanoscale structures is important for much emerging technology (such as the production of gem diamonds(!) by chemical vapor deposition [16]).

We study the motion of "steps" on surfaces of pure crystals in the vicinity of atomically flat (so-called *vicinal* surfaces). On the atomic level, material surfaces can be modeled discretely in terms of atoms occupying a lattice, or they can be modeled on the large scale at the continuum level by smooth surfaces when the surface is atomically rough. In an intermediate range described nicely in a review article of Jeong and Williams [26], crystalline materials can have atomically flat surfaces with partially filled layers of atoms on top. The edges of these layers ("steps") are atomically rough and can be described by smooth curves that bound a raised "terrace" on the surface. Motion of these steps occurs due to thermal agitation of atoms along step edges and attachment and detachment of single atoms (adatoms) that diffuse on the terraces.

A classic continuum model of this step motion, that considers steps as smooth curves forming sharp interfaces between terraces, is the Burton-Cabrera-Franck (BCF) model [7]; see the treatment by Bales & Zangwill [5]. We will first describe the BCF model and then discuss a diffuse-interface approximation developed by Otto et al. [45] that yields the BCF model in the sharp-interface limit.

BCF model. The BCF model is based on the step-terrace description of the surface together with a number of assumptions: On the terraces, adatoms: (i) are deposited at rate F per site; (ii) desorb from the terrace with lifetime τ; and (iii) diffuse, hopping to neighboring sites with rate D. Let $\rho(x,t)$ denote the adatom density (expected number of atoms at site x) on a terrace (a region in $\mathbb{R}^2$ bounded by a union of smooth curves). The processes can be accounted for by a discrete model. In a short time interval Δt, the adatom density at site x on the terrace changes according to

$$\rho(x, t+\Delta t) \sim \rho(x,t) + F\Delta t - \rho(x,t)\Delta t/\tau + D\Delta t \sum_{x' \in N(x)} (\rho(x',t) - \rho(x,t)),$$

where the sum goes over a set $N(x)$ of neighbors of site x. If the lattice spacing is a, passing to a continuum model yields the PDE

$$\partial_t \rho(x,t) = Da^2 \Delta \rho + F - \rho/\tau \quad \text{in } \mathbb{R}^2 \setminus \Gamma \tag{3.6}$$

where $\Delta = \partial_{x_1}^2 + \partial_{x_2}^2$ is the Laplacian, and Γ is a set of curves comprising the steps.

Near steps, BCF suppose that adatoms attach to or detach from terrace edges at different rates from the upper and lower terraces. (This models an effect known as the Ehrlich-Schwoebel barrier effect— adatoms on the upper terrace experience a higher-than-usual potential barrier to get over the edge of the step.) Attachment produces a net normal velocity v for the step in the direction of the lower terrace (denoted as the plus side here), and is driven by the difference between the terrace step density $\rho(x,t)$ and an equilibrium constant ρ_* corrected by a term proportional to step curvature $\kappa(x,t)$:

$$\frac{v}{a} = k^+(\rho^+ - \rho_*(1+\xi\kappa)) + k^-(\rho^- - \rho_*(1+\xi\kappa)) \quad \text{on } \Gamma. \tag{3.7}$$

Here $\rho^+(x,t)$ and $\rho^-(x,t)$ are limits of $\rho(y,t)$ as y approaches position x on the step from the lower and upper terraces respectively. The quantities k^+, k_-, ρ_* and ξ are constants.

The continuum model is completed by an equation that states that the flux of adatoms diffusing to the step edges balances the rate at which they attach to the step: Letting ∂_ν denote the derivative along the unit normal ν that points from the upper to the lower terrace,

$$Da\, \partial_\nu\rho^+ = k^+(\rho^+ - \rho_*(1+\xi\kappa)), \tag{3.8}$$

$$-Da\, \partial_\nu\rho^- = k^-(\rho^- - \rho_*(1+\xi\kappa)). \tag{3.9}$$

Nondimensionalization and quasistatic limit. We nondimensionalize time and space by $t = T\tilde{t}$, $x = L\tilde{x}$. We take the time scale of interest to be $T = 1/F$, the mean rate that layers are deposited. We presume the lifetime $\tau \gg T$ is long by comparison.

The length scale of interest L is taken to balance the effects that curvature and deposition have upon density variations—we require

$$\frac{\rho_*\xi}{L} \sim \frac{FL^2}{Da^2}, \quad \text{so} \quad L = \left(\frac{Da^2\rho_*\xi}{F}\right)^{1/3},$$

and scale excess density according to

$$w = \rho - \rho_* = \frac{\rho_*\xi}{L}\tilde{w}.$$

Then in terms of new variables (and dropping the tildes), the BCF model equations take the form

$$\Delta w + 1 = 0 \tag{3.10}$$

on terraces, and

$$v = \partial_\nu w^+ - \partial_\nu w^-, \tag{3.11}$$

$$\zeta_+ \partial_\nu w^+ = w^+ - \kappa, \tag{3.12}$$

$$-\zeta_- \partial_\nu w^- = w^- - \kappa \tag{3.13}$$

at steps.

The time derivative $\partial_t w$ in (3.10) is neglected by supposing the diffusion time L^2/Da^2 is small compared to T. When $\zeta_\pm = 0$, the conditions in (3.12)–(3.13) reduce to a Gibbs-Thomson boundary condition at steps. In this case, in the absence of deposition flux (replacing (3.10) by $\Delta w = 0$), the interface dynamics reduces to the Mullins-Sekerka model.

Diffuse-interface approximation. We now describe the diffuse-interface approximation to this BCF model that was constructed by Otto, Penzler, Rätz, Rump, and Voigt [45]. The height of terraces in the sharp-interface model is an integer times the thickness of an atomic layer (taken as 1 here). In the diffuse approximation, the material surface height $z(x,t)$ is modeled as a smooth function whose "free energy"

$$\mathcal{F}(z) = \int_{\mathbb{R}^2} \frac{\varepsilon}{2}|\nabla z|^2 + \frac{1}{\varepsilon}G(z)\,\mathrm{d}x$$

is bounded. Here $G : \mathbb{R} \to \mathbb{R}$ is a smooth function that we take to be like $\sin^2(\pi z)$: periodic with period 1, zero at integers, and positive otherwise. We take it normalized so $\int_0^1 \sqrt{2G(z)}\,\mathrm{d}z = 1$. If $\mathcal{F}(z)$ is small, it should mean z is close to an integer in large regions corresponding to terraces, separated by narrow transition zones between.

The equation governing the evolution of $z(x,t)$ will take the form of a modified Cahn-Hilliard equation. This takes the form of a diffusion equation

$$\partial_t z + \nabla \cdot \boldsymbol{j} = 1 \tag{3.14}$$

to hold everywhere in space, with the "flux" $\boldsymbol{j}$ given in terms of a "chemical potential" w according to

$$(1 + \varepsilon^{-1}\zeta_2(z))\boldsymbol{j} = -\nabla w, \tag{3.15}$$

$$w = \varepsilon\zeta_1(z)\partial_t z - \varepsilon\Delta z + \varepsilon^{-1}G'(z). \tag{3.16}$$

The drag coefficient $\zeta_1(z)$ and mobility coefficient $\zeta_2(z)$ are non-negative and 1-periodic. We take $\zeta_2(z)$ to vanish on integers and be positive otherwise (like G); this is to force the flux $\boldsymbol{j}$ to be small in the transition layers as $\varepsilon \to 0$.

Equations (3.14)–(3.16) comprise the OPRRV model. For a closed system (replace 1 by 0 in (3.14)) with no-flux boundary conditions $\nu \cdot \boldsymbol{j} = 0$, the "free energy" $\mathcal{F}(z)$ decreases:

$$\frac{\mathrm{d}}{\mathrm{d}t}\mathcal{F}(z) = \frac{\mathrm{d}}{\mathrm{d}t}\int \left(\frac{\varepsilon}{2}|\nabla z|^2 + \frac{1}{\varepsilon}G(z)\right)\mathrm{d}x = \int \left(-\varepsilon\Delta z + \varepsilon^{-1}G'(z)\right)\partial_t z$$

$$= \int (w - \varepsilon\zeta_1\,\partial_t z)\partial_t z = \int \nabla w \cdot \boldsymbol{j} - \varepsilon\zeta_1(\partial_t z)^2 \leq 0.$$

Method of matched asymptotic expansions. We describe how the evolution of z by motion of transition layers yields the BCF interface dynamics formally in the limit $\varepsilon \to 0$, using the method of matched asymptotic expansions. Mathematically, the method involves constructing an *approximate solution* to the OPRRV equations in two overlapping zones: One constructs (i) an inner expasion to be valid distances $\mathcal{O}(\varepsilon)$ from the evolving steps $\Gamma(t)$; and (ii) an outer expansion to be valid distances $\mathcal{O}(1)$ from the steps. The two expansions are linked by a matching procedure at intermediate distances from the steps. Maximizing the order of accuracy of the approximation will require step evolution to be given by the sharp-interface BCF model. Further insight on the formal matching procedure can be found in [20, 8].

We are mainly interested in interface dynamics, so we will neglect the deposition flux, replacing 1 by 0 on the right hand sides of (3.10) and (3.14).

Consider the outer expansion first: We seek $z(x,t)$ in the form

$$z(x,t) = z_0(x,t) + \varepsilon z_1(x,t) + \mathcal{O}(\varepsilon^2),$$

where z_0, z_1 are independent of ε, with similar expansions for w and $\boldsymbol{j}$. Plugging these into the equations, we require the $\mathcal{O}(\varepsilon^{-1})$ terms balance, yielding

$$G'(z_0) = 0 \quad \text{and} \quad \zeta_2(z_0)\boldsymbol{j}_0 = 0.$$

By our hypotheses, the first equation forces $z_0(x,t)$ to take integer values, constant in components complementary to the steps $\Gamma(t)$. Then $\zeta_2(z_0) = 0$ and $\boldsymbol{j}_0$ is not restricted.

Balancing terms of order $\mathcal{O}(1)$ yields three equations:

$$\partial_t z_0 + \nabla \cdot \boldsymbol{j}_0 = 0,$$
$$(1 + \zeta_2'(z_0)z_1)\boldsymbol{j}_0 + \zeta_2(z_0)\boldsymbol{j}_1 = -\nabla w_0,$$
$$w_0 = G''(z_0)\, z_1.$$

Since $0 = \partial_t z_0 = \zeta_2(z_0) = \zeta_2'(z_0)$, this simplifies to $\boldsymbol{j}_0 = -\nabla w_0$ and

$$-\Delta w_0 = 0, \tag{3.17}$$

with $w_0 = G''(0)z_1$. This is all we need from the outer expansion.

Next consider the inner expansion, distances $\mathcal{O}(\varepsilon)$ from the steps $\Gamma(t)$, which we take to be a union of smooth curves independent of ε. We suppose without loss that the step models a single atomic layer, with $z_0^+ = 0$, $z_0^- = 1$. We will need to "stretch" the coordinates normal to the steps. In the neighborhood of some point on Γ, introduce the signed distance

$$\Phi(x,t) := \pm\mathrm{dist}(x,\Gamma(t))$$

from x to $\Gamma(t)$, where we take the sign as $+$ on the lower terrace and $-$ on the upper. Then $\nu(x,t) = \nabla\Phi(x,t)$ is a unit vector normal to Γ if $x \in \Gamma$. The normal velocity of Γ at x is

$$v(x,t) = -\partial_t\Phi.$$

We let $r = \Phi/\varepsilon$, and for x near Γ change variables via $x = y + \varepsilon\,r\nu(y,t)$ for $y \in \Gamma(t)$. In principle, this yields a local change of variables $(x,t) \mapsto (y,r,t)$. It is a convenient rule, however, to regard quantities $Q(y,r,t)$ as having the form $Q(x,r,t)$, in which x is not restricted to lie on Γ but Q must be constant as x varies along ν in the first argument, i.e., $\nu \cdot \nabla_x Q = 0$.

We seek our inner expansion in the form

$$\begin{aligned}
z &= Z_0(x,r,t) + \varepsilon\,Z_1(x,r,t) + \mathcal{O}(\varepsilon^2), \\
\boldsymbol{j} &= J_0(x,r,t)\,\nu(x,t) + \mathcal{O}(\varepsilon), \\
w &= W_0(x,r,t) + \mathcal{O}(\varepsilon), \qquad r = \Phi(x)/\varepsilon.
\end{aligned}$$

Here, the indicated functions of (x,r,t) are to be independent of ε and need to be determined for all $r \in (-\infty,\infty)$. Evaluating derivatives by the chain rule, we have

$$\begin{aligned}
\partial_t z &= -\varepsilon^{-1} v\,\partial_r Z_0 + \mathcal{O}(1), \\
\nabla \cdot \boldsymbol{j} &= \varepsilon^{-1}\nabla\Phi \cdot \nu\,\partial_r J_0 + \mathcal{O}(1) = \varepsilon^{-1}\partial_r J_0 + \mathcal{O}(1), \\
\nabla w &= \varepsilon^{-1}\nabla\Phi\,\partial_r W_0 + \mathcal{O}(1), \\
\Delta z &= \varepsilon^{-2}\partial_r^2 Z_0 + \varepsilon^{-1}\Delta\Phi\,\partial_r Z_0 + \mathcal{O}(1).
\end{aligned}$$

A fact that must be left to the reader to check is that for $x \in \Gamma(t)$,

$$\Delta\Phi(x,t) = \kappa(x,t)$$

is the curvature of $\Gamma(t)$.

We now use these expressions in the OPRRV model equations (3.14)–(3.16) and match terms of order $\mathcal{O}(\varepsilon^{-1})$ to find:

$$-v\partial_r Z_0 + \partial_r J_0 = 0, \tag{3.18}$$

$$-\partial_r^2 Z_0 + G'(Z_0) = 0, \tag{3.19}$$

$$\zeta_2(Z_0)J_0 = -\partial_r W_0. \tag{3.20}$$

From these equations it follows that

$$-v\,Z_0 + J_0 = \lambda(x,t) \tag{3.21}$$

independent of $r \in (-\infty, \infty)$. We match to the outer expansion in the regime where $r = \Phi/\varepsilon$ is large but Φ is small. This leads us to require

$$Z_0(r) \to 1 \quad \text{as } r \to -\infty, \quad 0 \quad \text{as } r \to \infty, \tag{3.22}$$

and we satisfy (3.19) by taking $Z_0 = Z_0(r)$ to be a *domain wall* independent of x, t and centered so that $Z_0(0) = 1/2$, say. From (3.19) we find Z_0 satisfies

$$\partial_r Z_0 = -\sqrt{2G(Z_0)}, \quad Z_0(0) = 1/2. \tag{3.23}$$

Taking $r \to +\infty$ then $-\infty$ in (3.21) now leads to the matching conditions

$$\lambda = J_0(x, +\infty, t) = \nu \cdot \boldsymbol{j}_0^+ = -\partial_\nu w_0^+,$$
$$v = J_0(x, -\infty, t) - \lambda = \partial_\nu w_0^+ - \partial_\nu w_0^-.$$

At this point, equations (3.17) and (3.24) correspond respectively to (3.10) and (3.11), and it remains to recover the equations (3.12)–(3.13) that govern the attachment kinetics. From terms of order $\mathcal{O}(1)$ we need only observe that

$$W_0 = \zeta_1(Z_0)(-v\,\partial_r Z_0) - \kappa\,\partial_r Z_0 - \partial_r^2 Z_1 + G''(Z_0)Z_1. \tag{3.24}$$

Since $(-\partial_r^2 + G''(Z_0))\partial_r Z_0 = 0$, we can say

$$\int_{-\infty}^{\infty} (\partial_r Z_0)(-\partial_r^2 + G''(Z_0))Z_1 \; \mathrm{d}r = 0.$$

Thus, a *necessary condition* for the solvability of (3.24) for Z_1 is that

$$\int_{-\infty}^{\infty} (\partial_r Z_0)(W_0 + \kappa\,\partial_r Z_0 + v\zeta_1\,\partial_r Z_0)\,\mathrm{d}r = 0.$$

Now, using (3.20) and (3.21) we compute

$$\kappa \int_{-\infty}^{\infty} (\partial_r Z_0)^2 + v \int_{-\infty}^{\infty} \zeta_1(\partial_r Z_0)^2 = -\int_{-\infty}^{\infty} W_0(\partial_r Z_0)\,\mathrm{d}r$$
$$= -[W_0 Z_0]_{-\infty}^{\infty} - \int_{-\infty}^{\infty} (\zeta_2 J_0)Z_0 = w_0^- - v\int_{-\infty}^{\infty} \zeta_2 Z_0^2 - \lambda\int_{-\infty}^{\infty} \zeta_2 Z_0,$$

and

$$w_0^+ - w_0^- = \int_{-\infty}^{\infty} \partial_r W_0\,\mathrm{d}r = -\int_{-\infty}^{\infty} \zeta_2 J_0 = -v\int_{-\infty}^{\infty} \zeta_2 Z_0 - \lambda\int_{-\infty}^{\infty} \zeta_2.$$

By (3.24) and (3.24), and since G is normalized so that

$$\int_0^1 \sqrt{2G(z)}\,\mathrm{d}z = \int_{-\infty}^{\infty} (\partial_r Z_0)^2\,\mathrm{d}r = 1,$$

this yields the BCF equations

$$\zeta_+ \, \partial_\nu w_0^+ = w_0^+ - \kappa, \tag{3.25}$$

$$-\zeta_- \, \partial_\nu w_0^- = w_0^- - \kappa, \tag{3.26}$$

provided the following hold:

$$\zeta_- = \int_{-\infty}^{\infty} \left(\zeta_1(Z_0)(\partial_r Z_0)^2 + \zeta_2(Z_0)Z_0^2 \right) \mathrm{d}r = \int_{-\infty}^{\infty} \zeta_2(Z_0)Z_0 \, \mathrm{d}r, \tag{3.27}$$

$$\zeta_+ = \int_{-\infty}^{\infty} \zeta_2(Z_0)(1 - Z_0) \, \mathrm{d}r. \tag{3.28}$$

By changing variables using (3.23), these constraints can be written in the form

$$\zeta_- = \int_0^1 \left(\zeta_1(z)\sqrt{2G(z)} + \frac{\zeta_2(z)z^2}{\sqrt{2G(z)}} \right) \mathrm{d}z = \int_0^1 \frac{\zeta_2(z)z}{\sqrt{2G(z)}} \, \mathrm{d}z, \tag{3.29}$$

$$\zeta_+ = \int_0^1 \frac{\zeta_2(z)(1 - z)}{\sqrt{2G(z)}} \, \mathrm{d}z. \tag{3.30}$$

This completes the formal derivation. We point out that deriving the Mullins-Sekerka sharp-interface model (3.3)–(3.5) from the Cahn-Hilliard equation (3.2) is just a special case of the above, taking $\zeta_1 = \zeta_2 = 0$. A rigorous justification of the sharp-interface limit in this case was given by Alikakos, Bates, and Chen [1]. For work relating the Cahn-Hilliard equation to the LSW model, see the recent paper [2] and the references therein.

3.2. *Gradient structure for Mullins-Sekerka flow*

For use below, here we explain (essentially following Niethammer & Otto [39]) how the Mullins-Sekerka model can be described formally as gradient flow for surface area with respect to a certain metric structure on a "manifold" of smooth surfaces in $\mathbb{R}^3$. (We remark that for the closely related problem of Hele-Shaw flow between parallel plates, this type of gradient flow structure, as described by Otto [44], was exploited by Glasner [24] to derive a corresponding diffuse-interface model in a very interesting manner.)

First, recall the general structure of gradient flow for an energy functional $\mathcal{E} : \mathcal{M} \to \mathbb{R}$ where $\mathcal{M}$ is a Riemannian manifold with metric g: A solution trajectory is a curve $t \mapsto z(t) \in \mathcal{M}$ with tangent vector $\partial_t z \in T_{z(t)}\mathcal{M}$ that satisfies

$$g_{z(t)}(\partial_t z, v) = -\mathrm{d}\mathcal{E}(z(t))v \quad \text{for all } v \in T_{z(t)}\mathcal{M}.$$

Now, we consider $\mathcal{M}$ to be the "manifold" of smooth bounded surfaces Γ that are boundaries of bounded domains in $\mathbb{R}^3$. Formally, elements of the tangent space $T_\Gamma \mathcal{M}$ correspond to "normal velocity fields" $v : \Gamma \to \mathbb{R}$. We define a "metric" on $T_\Gamma \mathcal{M}$ as follows: To each such $v \in T_\Gamma \mathcal{M}$ we associate a harmonic "potential" $u = \mathcal{T}v : \mathbb{R}^3 \to \mathbb{R}$ by solving[b] the PDE boundary value problem

$$\Delta u = 0 \text{ in } \mathbb{R}^3 \setminus \Gamma,$$
$$\nu \cdot \nabla u^+ - \nu \cdot \nabla u^- = -v \text{ on } \Gamma,$$
$$u(x) \to 0 \text{ as } |x| \to \infty.$$

(Here ν is the unit outward normal to the domain enclosed by Γ, and u^+, u^- are respective limits on Γ along ν from the outside and inside respectively.) Given $v_1, v_2 \in T_\Gamma \mathcal{M}$, let $u_1 = \mathcal{T}v_1$, $u_2 = \mathcal{T}v_2$ and put

$$g_\Gamma(v_1, v_2) = \int_{\mathbb{R}^3} \nabla u_1 \nabla u_2 = \int_\Gamma u_1 \, [n\nabla u_2]_+^- = \int_\Gamma u_1 v_2 = \int_\Gamma v_1 u_2 \quad (3.31)$$

We let $\mathcal{E}(\Gamma)$ be the surface area of Γ. It is known that if $\Gamma(t)$ is smoothly evolving with normal velocity $v(t)$, then

$$\frac{\mathrm{d}}{\mathrm{d}t} \mathcal{E}(\Gamma(t)) = \int_\Gamma \kappa v =: \mathrm{d}\mathcal{E}(\Gamma)v,$$

where κ is the sum of principal curvatures of Γ (positive for spheres), and the volume of the domain $\Omega(t)$ enclosed by Γ evolves by

$$\frac{\mathrm{d}}{\mathrm{d}t} \, \mathrm{vol}(\Omega(t)) = \int_\Gamma v.$$

Mullins-Sekerka flow is gradient flow for surface area with *enclosed volume conserved*. Let $\mathcal{M}_0$ be a submanifold of $\mathcal{M}$ corresponding to surfaces with *fixed* enclosed volume. Velocity fields $v \in T_\Gamma \mathcal{M}_0$ should satisfy $\int_\Gamma v = 0$. Gradient flow requires that $\Gamma(t)$ evolves so that its normal velocity v satisfies

$$g_\Gamma(v, \tilde{v}) = -\mathrm{d}\mathcal{E}(\Gamma)\tilde{v}, \quad \text{or} \quad \int u\tilde{v} = -\int_\Gamma \kappa\tilde{v}, \quad \text{for all } \tilde{v} \in T_\Gamma \mathcal{M}_0.$$

---

[b]How to do this technically: Let $\mathcal{H} = \{u \in L^6(\mathbb{R}^3) | \nabla u \in L^2(\mathbb{R}^3)^3\}$ be a Hilbert space with $\|u\|_\mathcal{H} = (\int |\nabla u|^2)^{1/2}$. $\mathcal{H}$ is complete due to a critical Sobolev embedding theorem. Find u so

$$\langle u, v \rangle_\mathcal{H} = \int_{\mathbb{R}^3} \nabla u \cdot \nabla w = \int_\Gamma vw \quad \text{for all } w \in \mathcal{H}.$$

Since this holds for all smooth $\tilde{v}$ with $\int_\Gamma \tilde{v} = 0$, we can infer that $u = -\kappa + \theta(t)$ on $\Gamma(t)$, where $\theta(t)$ is constant in space. This yields the Mullins-Sekerka motion law, since we have

$$\Delta(-u + \theta) = 0 \quad \text{in } \mathbb{R} \setminus \Gamma(t),$$
$$-u + \theta = \kappa \quad \text{on } \Gamma,$$
$$v = [\nu \cdot \nabla(-u + \theta)]_-^+ \quad \text{on } \Gamma.$$

Note that $\theta = \lim_{|x| \to \infty}(-u + \theta)$ is the "mean field," the limit at infinity of the harmonic function $-u + \theta$ determined by the curvature of $\Gamma(t)$.

3.3. *Monopole models by restricted gradient flow of surface energy*

The morphology of domains coarsening according to Mullins-Sekerka flow can be complex. Singularities may occur frequently through the shrinking of small blobs to zero size, or the "pinch-off" of necks in dumb-bell shaped regions, for example. When the minority phase occupies a small fraction of a sample region, however, frequently the morphology seen is that of a dilute suspension of approximately spherical domains. (Presumably one sees spheres due to some sort of local minimization of surface area with constrained enclosed volume).

This leads us to consider a geometrically simplified model in which the evolving surface Γ is constrained to consist of a collection of spheres. As we show below following S. Dai's Ph.D. thesis [14], Mullins-Sekerka gradient flow constrained geometrically to spheres exactly yields the classic *monopole model*, in which the harmonic potential (diffusion field) $u = \mathcal{T}v$ is a superposition of monopole fields $1/|x - x_i|$. The monopole model is important due to the fact that it is amenable to large-scale simulation; computations involving hundreds of thousands of spheres have been performed (taking some shortcuts).

We restrict attention to a submanifold $\mathcal{M}_N$ of surfaces Γ consisting of a collection of spheres Γ_i bounding a fixed number $N > 0$ of non-overlapping balls B_i where $|x - x_i| < R_i$, $i = 1, \ldots, N$. We consider the centers $x_i \in \mathbb{R}^3$ fixed, and fix total volume, so $\sum R_i^3 = Q$ is constant. The manifold $\mathcal{M}_N$ is $N - 1$ dimensional and each tangent vector $v \in T_\Gamma \mathcal{M}_N$ corresponds to a normal velocity field with $v = v_i = \dot{R}_i$ constant on the sphere Γ_i where $|x - x_i| = R_i$.

We will find $u = \mathcal{T}v$ explicitly as a superposition of monopoles:

$$u(x) = \begin{cases} \displaystyle\sum_{j=1}^{N} \frac{a_j}{|x - x_j|}, & \text{if } |x - x_j| \geq R_j \text{ for all } j, \\[2em] \displaystyle\sum_{j \neq i} \frac{a_j}{|x - x_j|} + \frac{a_i}{R_i}, & \text{if } |x - x_i| < R_i. \end{cases}$$

Evidently $\Delta u = 0$ in $\mathbb{R}^3 \setminus \Gamma$ and $u(x) \to 0$ as $|x| \to \infty$. The jump condition on Γ_i reads $[\nu \cdot \nabla u]_-^+ = -a_i/R_i^2 = -v_i$, hence we must have

$$a_i = R_i^2 v_i,$$

and consequently,

$$\sum a_i = 0.$$

To describe how the $v_i = \dot{R}_i$ are determined, we must show how the monopole amplitudes a_i can be determined from the sphere radii R_i by solving a linear system of equations. Given a tangent vector $\tilde{v} \in T_\Gamma \mathcal{M}_N$, we compute

$$g_\Gamma(v, \tilde{v}) = \int_\Gamma u\tilde{v} = \sum_i \int_{\Gamma_i} \tilde{v}_i \sum_j \frac{a_j}{|x - x_j|}$$

$$= \sum_i 4\pi R_i^2 \tilde{v}_i \left(\frac{a_i}{R_i} + \sum_{j \neq i} \frac{a_i}{|x_i - x_j|} \right),$$

since by the mean value property of harmonic functions ($\Delta(1/|x|) = 0$),

$$\int_{\Gamma_i} \frac{1}{|x - x_j|} = \frac{4\pi R_i^2}{|x_i - x_j|}.$$

For the surface area $\mathcal{E} = \sum 4\pi R_i^2$ we can write

$$d\mathcal{E}(\Gamma)\tilde{v} = \sum 8\pi R_i \tilde{v}_i = \sum 4\pi R_i^2 \left(\frac{2}{R_i} \tilde{v}_i \right) = \int_\Gamma \kappa \tilde{v}.$$

Hence, gradient flow ($g_\Gamma(v, \tilde{v}) = -d\mathcal{E}(\Gamma)\tilde{v}$ for all $\tilde{v} \in T_\Gamma \mathcal{M}_N$) means that

$$\sum_i 4\pi R_i^2 \tilde{v}_i \left(\frac{2}{R_i} + \frac{a_i}{R_i} + \sum_{j \neq i} \frac{a_i}{|x_i - x_j|} \right) = 0 \quad \text{whenever} \quad \sum_i R_i^2 \tilde{v}_i = 0.$$

Hence the term in parentheses must be independent of i, and we denote it by $\tilde{\theta}(t)$. We now have $N + 1$ equations to determine $N + 1$ unknowns

$v_1, \ldots, v_N$ and $\tilde{\theta}$, namely:

$$R_i^3 v_i + \sum_{j \neq i} \frac{R_i^2 R_j^2 v_j}{|x_i - x_j|} = R_i^2 \left(\tilde{\theta} - \frac{2}{R_i} \right), \qquad \sum_i R_i^2 v_i = 0. \qquad (3.32)$$

In matrix-vector form this reads:

$$A\vec{v} = \tilde{\theta}\vec{s} - \vec{b}, \qquad \vec{s} \cdot \vec{v} = 0,$$

with $s_i = R_i^2$, $b_i = R_i^2(\tilde{\theta} - 2/R_i)$. The matrix A is symmetric and positive definite since

$$\vec{v} \cdot A\vec{v} = \frac{g_\Gamma(v, v)}{4\pi} > 0.$$

The solution can be expressed in the form

$$\vec{v} = \tilde{\theta} A^{-1}\vec{s} - A^{-1}\vec{b}, \qquad \tilde{\theta} = \frac{\vec{s} \cdot A^{-1}\vec{b}}{\vec{s} \cdot A^{-1}\vec{s}}.$$

This completes the derivation of the monopole model:

$$\partial_t R_i = v_i(\vec{R}).$$

This equation applies up to a time when one or more particle radii vanish $(R_i \to 0)$, after which the system is continued with fewer particles, or until particles collide.

An advantage of this derivation of the monopole model by restricted gradient flow is that it shows that *the velocities $\dot{R}_i$ in the monopole model are well-defined as long as the balls B_i are non-overlapping*. Moreover, the derivation shows how the velocities can be determined in terms of a positive definite matrix; this apparently has not been recognized before and could be useful in numerical computations.

3.4. *Lifshitz-Slyozov-Wagner mean-field model*

The first quantitative explanation of the $t^{1/3}$ power-law growth of typical domain size observed during Ostwald ripening was provided by work of Lifshitz and Slyozov [31] and Wagner [50], based upon arguments involving self-similar behavior for a Liouville equation governing the particle size distribution in a regime where all the interactions of particles are subsumed in one mean-field coupling term. In this subsection we describe this LSW model and indicate how recent rigorous analysis has clarified the fact that the self-similar nature of experimentally observed size distributions is not completely explained by the mean-field model, and must depend upon factors not taken into account by it.

In order to derive the LSW model, we look at at the dilute limit in which the typical particle radius R_i is very small compared to the interparticle distances $|x_i - x_j|$. Replacing x_i by $\bar{x}x_i$ and letting $\bar{x} \to \infty$, the off-diagonal terms in the system vanish $(A \to \mathrm{diag}(R_i^3))$, and we are left with the system

$$R_i^3 v_i = R_i^2 \left(\tilde{\theta} - \frac{2}{R_i} \right),$$

that is,

$$\partial_t R_i = \frac{1}{R_i} \left(\tilde{\theta} - \frac{2}{R_i} \right).$$

Since $\sum R_i^2 \partial_t R_i = 0$, the mean field $\tilde{\theta}$ is determined by

$$\tilde{\theta} = \frac{\sum 2}{\sum R_i} = \frac{2}{R_{\mathrm{av}}},$$

where $R_{\mathrm{av}} = \sum R_i / N$ is the average particle radius.

For later reference, we note that the metric g_Γ degenerates to diagonal form and the surface area decays as follows: Using that $\sum 4\pi R_i^2 v_i = 0$,

$$\frac{\mathrm{d}E}{\mathrm{d}t} = \sum 8\pi R_i v_i = \sum 4\pi R_i^2 \left(\frac{2}{R_i} - \tilde{\theta} \right) v_i = -4\pi \sum R_i^3 v_i^2 = -g_\Gamma(v, v).$$

Coarsening proceeds according to the LSW model in the following way. Particles larger than average grow, and ones smaller than average shrink: If $R_i > R_{\mathrm{av}}$ then $\partial_t R_i > 0$; if $R_i < R_{\mathrm{av}}$ then $\partial_t R_i < 0$. Very small particles will vanish in a finite time: If the smallest particle is the ith and R_i is small, then $\partial_t R_i^3 \sim -6$, and $R_i^3 \sim 6(T_i - t_i)$ where $R_i(T_i) = 0$. Beyond this time the system continues with fewer particles.

Theory of Lifshitz & Slyozov and Wagner. The arguments of LSW to explain $t^{1/3}$ growth of typical particle size involve writing a Liouville (or Fokker-Planck) equation for the size distribution function. We set

$$\varphi(R, t) = \frac{\text{\# of particles of radius } \geq R}{\text{\# at } t = 0}.$$

Since particle radii satisfy

$$\partial_t R = V(R, t) := \frac{1}{R} \left(\tilde{\theta} - \frac{2}{R} \right), \tag{3.33}$$

and this ODE preserves the order of particle sizes, conservation of particles implies φ satisfies a PDE:

$$\frac{\mathrm{d}}{\mathrm{d}t} \varphi(R(t), t) = \partial_t \varphi + V(R, t) \partial_R \varphi = 0. \tag{3.34}$$

In terms of a *number density* $n(R, t)$ for the size distribution, we have

$$\varphi(R, t) = \int_R^\infty n(s, t) \, ds,$$

and the governing equation is written as a conservation law,

$$\partial_t n + \partial_R(V(R, t)n) = 0. \tag{3.35}$$

This is the LSW mean-field model. (For rigorous derivations of such PDE models from monopole models, see [37, 38, 39].) Conservation of total volume means that $Q = \int_0^\infty R^3 n(R, t) \, dR$ is constant and leads to

$$\tilde\theta(t) = \frac{2 \int n \, dR}{\int Rn \, dR} = \frac{2}{R_{\mathrm{av}}(t)}.$$

Scaling and self-similarity. The particle growth law (3.33) has the scaling invariance $R = \lambda R'$, $t = \lambda^3 t'$. If the solution of (3.35) achieves scale invariant form, so that

$$n(R, t) = \lambda^p n(\lambda R, \lambda^3 t),$$

then we must have $p = 4$, since

$$Q = \int R^3 n(R, t) \, dR = \int R^3 \lambda^p n(\lambda R, \lambda^3 t) \, dR = \lambda^{p-3-1} \int R'^3 n(R', t') \, dR'.$$

Thus, a scale-invariant solution should take the self-similar form

$$n(R, t) = t^{-4/3} F(R/t^{1/3}).$$

Now one has the questions: Can we expect to see this? And what should F be?

The equation actually admits a one-parameter family of self-similar solutions. Lifshitz & Slyozov argued as follows to explain that F should have a particular explicit form that is smooth with bounded support. Change variables, normalizing radius by its average, and introduce

$$\rho = \frac{R}{R_{\mathrm{av}}}, \quad \tau = \log R_{\mathrm{av}}^3.$$

Then $\partial_t R^3 = 6(\rho - 1) = (\partial_t R_{\mathrm{av}}^3)(\rho^3 + \partial_\tau \rho^3)$, and we get

$$\partial_\tau \rho^3 = \gamma(\rho - 1) - \rho^3 \tag{3.36}$$

with $\gamma = \gamma(\tau) = 6/(\partial_t R_{\mathrm{av}}^3)$. Expecting the system to settle into a self-similar regime, we expect $\gamma(t) \to \gamma_\infty$. This means that for large t,

$$R_{\mathrm{av}}^3 \sim ct, \quad \text{with } c = 6/\gamma_\infty.$$

One argues that $\gamma_\infty = \frac{27}{4}$ (meaning $c = 8/9$, which differs from the classic $4/9$ only due to the factor 2 in (3.33)) based on the dynamics of (3.36): If γ_∞ is smaller, then the right-hand side is everywhere negative and ρ becomes zero in finite time for all particles, contradicting total volume conservation. If γ_∞ is larger, then (3.36) admits two equilibria, a stable ρ_2 above an unstable ρ_1. Self-similar solutions having bounded support exist in this case, but LSW argue they are unstable. Presuming some fraction of particle sizes lie above the unstable ρ_1 means their ρ will approach ρ_2 as t becomes large and this leads to growth of total volume, again contradicting conservation.

Non-self-similar behavior. These arguments of LSW are mathematically nonrigorous, but are physically precise and plausible and sparked a great deal of activity to investigate the predictions. It is fair to say the LSW analysis served as a paradigm for a range of related problems in materials science and solid state physics. Certain difficulties dogged the theory, however, especially the facts that observed size distributions are always broader than the predicted one, and that the assumption that the system is sufficiently dilute is never satisfied in real systems.

Rather recently, several groups of investigators [23, 12, 40] came to understand that solutions of mean-field LSW models such as (3.35) *need not* exhibit universal self-similar behavior, meaning perhaps that the LSW model lacks some feature of experimental systems which leads to observed self-similarity. A basic explanation is based upon equation (3.34). Evolution under this equation simply *stretches* the graph of φ according to the solution map $R(0) \mapsto R(t)$ for the characteristic equation (3.33). This map is smooth (analytic, in fact), and thus for finite time it produces a smooth distortion of the initial distribution function.

Suppose the system initially has a maximal particle size R_*, so that initially $\varphi(R, 0) = 0$ on (R_*, ∞) (maximal). The detailed way in which $\varphi(R, 0)$ vanishes near the end of support will be qualitatively preserved for finite time. Niethammer & Pego [40] proved two facts for (3.34):

(1) If there exist $c > 0$ and $p > 0$ such that initially

$$\varphi(R_* - r, 0) \geq cr^p,$$

then the rescaled solution *cannot* approach LSW's self-similar solution as $t \to \infty$.

(2) A *necessary* condition for the solution to approach *some* self-similar solution as $t \to \infty$ is that the initial distribution is "almost power-law"

near the maximal particle size. Namely,

$$\varphi(R_* - r, 0) \sim r^p L(r)$$

as $r \to 0^+$ for some $p \geq 0$ and some function L *slowly varying* near zero. (Recall this means $\lim_{t \to 0} L(tx)/L(t) = 1$ for all $x > 0$.)

The condition in (2) says that $r \mapsto \varphi(R_* - r, 0)$ is *regularly varying* at 0 with exponent p. This is the condition that figures in the necessary and sufficient conditions of the Tauberian theorem [19] and the rigidity lemma for scaling limits. Entire books have been written about it due to its importance in analysis and probability theory [6, 49]. In section 5 below we will show that regular variation is key for obtaining necessary and sufficient conditions to classify domains of attraction for self-similar behavior in Smoluchowski's coagulation equations.

Comments on analysis. Without getting into technical details, it is interesting to note that physical and structural considerations have some important consequences for the rigorous mathematical analysis of the LSW model. In the form (3.35) the model takes the form of a PDE conservation law as one has in shock wave theory. However, it is unwise to look to shock-wave theory for an appropriate topology to study the well-posedness of the initial-value problem. Instead, an appropriate topology that is physically meaningful should make it "difficult" to create large particles from nothing. On the other hand, small changes in particle size should be "easy," even if the size distribution is highly peaked like a Dirac mass. Mathematically, one would ideally like to allow arbitrary size distributions, which means arbitrary probability measures after normalization by number.

Looking back at (3.34), it is rather more natural to regard particle size as the actively evolving dependent variable, and describe the size distribution by inverting the map $R \mapsto \varphi$, regarding R as a function of $\varphi \in [0, 1]$, the initial particle "rank." The equation of evolution is just (3.33). The sup-norm distance between particle rankings corresponds to a transport metric called the L^∞ *Wasserstein distance* between probability distributions. (This is the minimal maximum size change needed to change one distribution to the other.) Based on this topology (in terms of volume $v = R^3$), well-posedness of the LSW initial-value problem was proved in [41], with size distribution allowed to be an arbitrary probability distribution of compact support. Also, computations are best based on (3.33) rather than (3.35). For this model, it is much easier to attain high accuracy for long times following characteristics than with shock-capturing schemes. Over many years, a good number of investigators computing solutions numerically from (3.35) determined (erroneously, as it turned out) that solutions always approached the smooth LSW form.

The restriction of compact support was removed in [42] using a gradient flow structure that the LSW model inherits from the (discrete) monopole model (despite the fact that (3.35) is a first-order PDE). The dissipation identity associated with this structure provides a compactness property useful for establishing the existence of solutions for any initial size distribution that is a probability measure giving finite expected volume.

4. Rigorous power-law bounds on coarsening rates — The Kohn-Otto method

The scientific achievement of Lifshitz and Slyozov and Wagner in producing an explanation for the $t^{1/3}$ power-law behavior for typical domain size in coarsening by phase separation was considerable; they spawned a large related literature that continues to expand rapidly. The LSW mean-field theory, however, is based upon a quite restrictive set of physical assumptions. In particular, the minority phase must be extremely dilute, and the particles spherical. These assumptions naturally fail in practical situations in metallic alloys with comparable phase fractions and anisotropy. So one hopes for a more general explanation.

Power-law heuristics. A general but vague idea is that power-law behavior is due to some kind of statistical self-similarity based on a simple scaling invariance principle. For example, consider the Mullins-Sekerka interface motion law:

$$\Delta w = 0 \text{ in } \Omega(t), \quad w = \kappa \text{ on } \Gamma(t) = \partial\Omega, \quad V = \frac{\partial w^+}{\partial \nu} - \frac{\partial w^-}{\partial \nu} \text{ on } \Gamma(t).$$

Changing scale according to $x = L\tilde{x}$, $t = T\tilde{t}$, $w = A\tilde{w}$ yields

$$\frac{A}{L^2}\Delta w = 0, \qquad A\tilde{w} = \frac{1}{L}\tilde{\kappa}, \qquad \frac{L}{T}\tilde{V} = \frac{A}{L}\left(\frac{\partial \tilde{w}^+}{\partial \tilde{\nu}} - \frac{\partial \tilde{w}^-}{\partial \tilde{\nu}}\right).$$

This yields a solution of the original system if $A = 1/L$ and $L^3 = T$. Suppose one observes a system undergoing coarsening and plots a length scale l vs. time t, so $l = f(t)$. Changing scale via $\tilde{l} = l/L$, $\tilde{t} = t/T$, one plots $\tilde{l} = f(T\tilde{t})/T^{1/3} = \tilde{f}(\tilde{t})$. If the behavior of the system is *independent of scale* (and this concept is very ill-defined for a system with complex morphology), then we can expect that $f = \tilde{f}$, and hence, putting $\tilde{t} = 1$ we get $f(T) = T^{1/3}f(1)$. This produces the power-law behavior $l = ct^{1/3}$.

Method of Kohn and Otto. Recently, a new and rigorous method for explaining power-law behavior was created by Kohn and Otto [27]. The method promises to be rather robust, as it depends only upon a few gross features of the system being considered. It produces *time-averaged power-law bounds* on the decay of a normalized energy $E(t)$ of the system. These bounds are:

- universal—they apply to every solution.
- one-sided—slower coarsening is possible (the solution can get 'stuck' at unstable equilibria, for example), but faster coarsening is *impossible.*
- independent of system complexity (size; morphology of patterns).
- independent of statistical assumptions about the system.

In many of the problems treated so far by the method [28, 29, 30] the power-law bounds are expected to be *typical,* as suggested by the heuristic scaling arguments. (At this time, an exception is that the $t^{-1/3}$ bounds achieved for a certain mound formation model with anisotropic surface energy having square symmetry do not correspond to the $t^{-1/4}$ behavior seen in simulations [35, 51].)

With regard to the last bullet above, it is interesting to note that no sort of statistical self-similarity is presumed. In fact, an experimental indication of the significance of this appears recently in work of Voorhees and co-workers [32]. This group experimentally studied 3D coarsening of domain structures in metallic alloys. They observed $t^{-1/3}$ decay of surface energy over a long range of times where statistics show that morphology and curvature distributions are evolving in a non-self-similar manner. Thus it appears that power-law energy decay is not necessarily indicative of scale-invariant structural behavior.

4.1. *Basic inequalities*

In these notes we will explain the Kohn-Otto method and apply it to the LSW mean-field model and to the monopole model. The essence of the argument involves the consequence of two inequalities that relate a normalized energy $E(t)$ and a dual quantity $L(t)$ that loosely characterizes a "length scale" for the system. (It is worth mentioning, though, that an even simpler approach works for the mound coarsening problem treated by Li and Liu in [30].)

                    Robert L. Pego

Lemma 4.1: *Suppose that for all $t > 0$, the functions $E(t)$ and $L(t)$ satisfy*

$$EL \geq 1 \quad and \quad \dot{L}^2 \leq -\dot{E}(t), \tag{4.1}$$

and suppose E is strictly decreasing. Then, with $\alpha_1 = \frac{1}{6}$ and $\alpha_2 = 16$ we have

$$\int_0^T E(t)^2 \, dt \geq \alpha_1 \int_0^T (t^{-1/3})^2 \, dt \quad for \; all \; T \geq \alpha_2 L_0^3. \tag{4.2}$$

Proof: 1. Since E is strictly decreasing, we may say $L(t) = l(\varepsilon)$, where $\varepsilon = E(t)$. Then

$$\left(\frac{dL}{dt}\right)^2 = \left(\frac{dl}{d\varepsilon}\frac{dE}{dt}\right)^2 \leq -\frac{dE}{dt} \quad \text{implies} \quad \left(\frac{dl}{d\varepsilon}\right)^2 \left(-\frac{dE}{dt}\right) \geq 1.$$

Multiplying by $E(t)^2$ and integrating, we get

$$f(T) := \int_0^T E(t)^2 dt \geq \int_0^T E^2 \left(\frac{dl}{d\varepsilon}\right)^2 \left(-\frac{dE}{dt}\right) dt = \int_{E_T}^{E_0} \varepsilon^2 \left(\frac{dl}{d\varepsilon}\right)^2 d\varepsilon.$$

2. Next, we fix T and *minimize* over $l(\varepsilon)$: Write $l = \hat{l} + \tilde{l}$, where

$$\tilde{l} = 0 \quad \text{at } \varepsilon = E_T \text{ and } E_0, \qquad \hat{l} = l = L_0 \text{ at } E_0, \; L_T \text{ at } E_T.$$

Then

$$\int_{E_T}^{E_0} \varepsilon^2 (\partial_\varepsilon(\hat{l} + \tilde{l}))^2 \, d\varepsilon = \int_{E_T}^{E_0} \varepsilon^2 ((\partial_\varepsilon \hat{l})^2 + (\partial_\varepsilon \tilde{l})^2) - 2\tilde{l}\partial_\varepsilon(\varepsilon^2 \partial_\varepsilon \hat{l}) \, d\varepsilon.$$

Choose $\hat{l}$ such that $\partial_\varepsilon(\varepsilon^2 \partial_\varepsilon \hat{l}) = 0$, requiring $\varepsilon^2 \partial_\varepsilon \hat{l} = \hat{C}$ constant (< 0). Then

$$L_0 - L_T = \int_{E_T}^{E_0} \frac{dl}{d\varepsilon} \, d\varepsilon = \hat{C} \int_{E_T}^{E_0} \varepsilon^{-2} \, d\varepsilon = \hat{C}(E_T^{-1} - E_0^{-1}).$$

Now it follows

$$f(T) \geq \int_{E_T}^{E_0} \hat{C} \left(\frac{dl}{d\varepsilon}\right) d\varepsilon = \hat{C}(L_0 - L_T) = \frac{(L_T - L_0)^2}{(E_T^{-1} - E_0^{-1})},$$

consequently

$$f(T) \geq E_T(L_T - L_0)^2.$$

3. Next, observe that

$$f'(T) = E_T^2 \geq \frac{1}{L_T 2}, \quad f'(T)f(T)^2 \geq E_T^4(L_T - L_0)^4 \geq \left(1 - \frac{L_0}{L_T}\right)^4.$$

The value L_T is either large or small:

- If $L_T \geq 2L_0$, then $f' f^2(T) \geq (\frac{1}{2})^4 = \frac{1}{16}$.
- If $L_T \leq 2L_0$, then $f'(T) L_0^2 \geq \frac{1}{4}$.

Hence for all t,

$$\frac{\mathrm{d}}{\mathrm{d}t}\left(\frac{f^3}{3} + \frac{f L_0^2}{4} \right) \geq \frac{1}{16},$$

and thus

$$\frac{f(T)^3}{3} + \frac{f(T) L_0^2}{4} \geq \frac{T}{16}.$$

4. We finish as follows. Let $f = L_0 F$, then $L_0^3(\frac{1}{3}F^3 + \frac{1}{4}F) \geq \frac{1}{16}T$. Hence, if $T \geq 16 L_0^3$ it follows $F^2 \geq \frac{3}{2}$, since $F^2 < \frac{3}{2}$ leads to a contradiction:

$$\frac{F^3}{3} + \frac{F}{4} < F\left(\frac{1}{2} + \frac{1}{4} \right) < \sqrt{\frac{3}{2}\frac{3}{4}} < 1 \leq \frac{T}{16 L_0^2}.$$

Then if $T \geq 16 L_0^3$ we infer $\frac{1}{2}F^3 \geq \frac{1}{3}F^3 + \frac{1}{4}F$, therefore $\frac{1}{2}f^3 \geq \frac{1}{16}T$. Thus

$$\int_0^T E(t)^2 \mathrm{d}t \geq \frac{T^{1/3}}{2} = \frac{1}{6}\int_0^T (t^{-1/3})^2 \mathrm{d}t. \qquad \square$$

4.2. *Bounds on coarsening rates for the LSW mean-field model*

Kohn and Otto applied their method to two variants of the Cahn-Hilliard equation in [27]. The method was applied to LSW mean-field models in [15], but Barbara Niethammer has suggested the following argument which exploits the gradient structure of the equations and is very simple.

Suppose we have a collection of spheres of radius R_i coarsening according to the mean-field law

$$\partial_t R_i = v_i = \frac{1}{R_i}\left(\tilde{\theta} - \frac{2}{R_i} \right).$$

Normalize total volume volume so $\sum R_i^3 = 1$. We take

$$E = \sum_i 4\pi R_i^2, \qquad L = \sum_i \frac{R_i^4}{4\pi}$$

as energy (total surface area) and "length scale" respectively (normalized by volume). Then by Cauchy-Schwartz we obtain the "interpolation inequality"

$$1 = \left(\sum R_i^3 \right)^2 \leq \left(\sum R_i^2 \right)\left(\sum R_i^4 \right) = EL.$$

The idea to obtain a dissipation inequality is that $-\dot{E} = g_\Gamma(v,v)$, coming from the gradient structure of the model. We write

$$-\dot{E} = -\sum 8\pi R_i \dot{R}_i = \sum 4\pi R_i^2 \left(\tilde{\theta} - \frac{2}{R_i} \right) \dot{R}_i = 4\pi \sum R_i^3 \dot{R}_i^2.$$

Then

$$(\dot{L})^2 = \left(\sum \frac{R_i^3 \dot{R}_i}{\pi} \right)^2 \leq \left(\sum R_i^3 \right) \left(\sum R_i^3 \dot{R}_i^2 \right) \leq -\dot{E}.$$

Then we simply apply the Lemma to obtain time-averaged bounds on the coarsening rate, universally valid for any size distribution of spheres:

$$\int_0^T E(t)^2 \, \mathrm{d}t \geq \frac{1}{6} \int_0^T (t^{-1/3})^2 \, \mathrm{d}t \quad \text{for } T \geq 16 L_0^2.$$

The same calculation can be made for the Lifshitz–Slyozov conservation law:

$$\partial_t n(R,t) + \partial_R(Vn) = 0, \quad V(R,t) = \frac{1}{R} \left(\tilde{\theta} - \frac{2}{R} \right),$$

using the *characteristic map* $\hat{R}(r,t)$ satisfying

$$\partial_t \hat{R} = V(\hat{R},t), \quad \hat{R}(r,0) = r,$$

and the fact that $n(R,t)\,\mathrm{d}R = n_0(r)\,\mathrm{d}r$ with $R = \hat{R}(r,t)$. We omit the details.

4.3. *Bounds on coarsening rates for the monopole model*

As discussed above, the monopole model is important because of its status as a model of multi-dimensional coarsening for which extensive computer simulations are feasible and have been carried out. The spatial distribution of particle positions affects coarsening dynamics in ways that are not well understood; presumably correlations develop between neighboring particles in a time-dependent fashion. As we now show, however, it is not difficult to establish a basic power-law bound on the coarsening rate using the Kohn-Otto method.

We recall that $\dot{R}_i = v_i$, where

$$R_i^3 v_i + \sum_{i \neq j} \frac{R_i^2 R_j^2 v_j}{|x_i - x_j|} = R_i^2 \left(\tilde{\theta} - \frac{2}{R_i} \right), \qquad \sum R_i^2 v_i = 0.$$

Note that the field $u = \mathcal{T}v$ satisfies

$$\Delta u = 0 \text{ in } \mathbb{R} \setminus \Gamma, \quad [n \cdot \nabla u]_-^+ = v \text{ on } \Gamma, \quad u \to 0 \text{ as } |x| \to \infty, \qquad (4.3)$$

and that

$$\int_\Gamma (u + \kappa)\tilde{v} = 0 \quad \text{for all } \tilde{v} \text{ with } \int_\Gamma \tilde{v} = 0.$$

Again, the surface area (surface energy) $E = \sum_i 4\pi R_i^2 = \int_\Gamma 1$ satisfies

$$-\dot{E} = -\sum 8\pi R_i \dot{R}_i = \sum 4\pi R_i^2 \left(-\frac{2}{R_i} v_i\right)$$

$$= \int_\Gamma -\kappa v = \int_\Gamma uv = \int_{\mathbb{R}^3} |\nabla u|^2 = g_\Gamma(v, v).$$

We use a rather fancy "length scale" L determined by

$$L^2 = \int_{\mathbb{R}^3} |\nabla \Psi|^2, \quad \text{where } -\Delta \Psi = \mathbf{1}_{\cup B_i},$$

meaning $-\Delta \Psi = 1$ on the union of balls $\cup_i B_i$ and 0 on the complement. The function Ψ is a superposition $\Psi = \sum \Psi_i$, where by scaling,

$$\Psi_i(x) = \phi\left(\frac{|x - x_i|}{R_i}\right) R_i^2.$$

The function ϕ satisfies $-(\partial_r^2 + \frac{2}{r}\partial_r)\phi = \mathbf{1}_{r<1}$, and is given by

$$\phi(r) = \quad \frac{1}{2} - \frac{1}{6}r^2 \text{ for } r < 1, \qquad \frac{1}{3r} \text{ for } r > 1.$$

(Note $\partial_r \phi = -1/3$ at $r = 1^\pm$.) Thus

$$\Psi_i = \quad \frac{R_i^2}{2} - \frac{r^2}{6} \text{ for } r = |x - x_i| < R_i, \qquad \frac{R_i^3}{3r} \text{ for } r > R_i.$$

We notice that

$$\dot{\Psi}_i = \quad R_i \dot{R}_i \text{ for } r < R_i, \qquad \frac{R_i^2 \dot{R}_i}{r} \text{ for } r > R_i,$$

hence

$$\Delta \dot{\Psi} = 0 \text{ in } \mathbb{R} \setminus \Gamma, \quad \left[n \cdot \nabla \dot{\Psi}\right]_-^+ = -\dot{R}_i \text{ on } \Gamma_i, \quad \dot{\Psi}(x) \to 0 \text{ as } |x| \to \infty.$$

Comparing with (4.3) we see that $\dot{\Psi} = u = \mathcal{T}v$. By consequence, we get the *dissipation inequality* as follows:

$$(L\dot{L})^2 = \left(\int_{\mathbb{R}^3} \nabla \Psi \cdot \nabla \dot{\Psi}\right)^2 \le \int_{\mathbb{R}^3} |\nabla \Psi|^2 \int_{\mathbb{R}^3} |\nabla \dot{\Psi}|^2 = L^2 \int_{\mathbb{R}^3} |\nabla u|^2 = L^2(-\dot{E}).$$

Hence $(\dot{L})^2 \leq -\dot{E}$.

We prove the *interpolation inequality* $EL \geq 1$ by using an explicit expression for L^2 obtained as follows:

$$L^2 = \int_{\mathbb{R}^3} \Psi(-\Delta\Psi) = \int_{\cup B_i} \Psi = \sum_{i,j} \int_{B_i} \Psi_j$$

We evaluate the integrals using scaling for $i = j$ and the mean value property for the harmonic function Ψ_j on B_i for $i \neq j$:

$$\int_{B_i} \Psi_i = R_i^{2+3} \int_0^1 \left(\frac{1}{2} - \frac{r^2}{6} \right) 4\pi r^2 \, \mathrm{d}r = R_i^5 \frac{8\pi}{15},$$

$$\int_{B_i} \Psi_j = \frac{4\pi}{3} R_i^3 \Psi_j(x_i) = \frac{4\pi}{3} \frac{R_i^3 R_j^3}{3|x_i - x_j|}.$$

Hence

$$L^2 = \sum_i \frac{8\pi}{15} R_i^5 + \sum_{j \neq i} \frac{4\pi}{3} \frac{R_i^3 R_j^3}{3|x_i - x_j|} \geq \sum_i \frac{8\pi}{15} R_i^5. \tag{4.4}$$

Normalizing so that $\sum R_i^3 = 1$, we get

$$1 = \sum R_i^3 = \left(\sum R_i^{\frac{4}{3}} R_i^{\frac{5}{3}} \right)^{\frac{3}{2}} \leq \left(\sum R_i^2 \right) \left(\sum R_i^5 \right)^{\frac{1}{2}}$$

by Hölder's inequality with $p = \frac{3}{2}$, $q = 3$ $(p^{-1} + q^{-1} = 1)$. Hence $EL \geq 1$.

Now applying the ODE lemma yields

$$\int_0^T E(t)^2 \, \mathrm{d}t \geq \frac{1}{6} \int_0^T (t^{-1/3})^2 \, \mathrm{d}t \quad \text{for all } T \geq 16L_0^3.$$

This estimate provides a power-law bound on energy dissipation with the exponent one expects physically. We remark, however, that the inequality (4.4) is likely to be quite pessimistic in dense systems, and so one might hope for improvements regarding the prefactor or time of validity.

5. Smoluchowski's coagulation equations

5.1. *Introduction*

A simple mean-field model of coarsening by coalescence leads to Smoluchowski's coagulation equations. Particles or clusters of size x and size y combine at a rate proportional to the population of each and a rate kernel $K(x, y)$. Schematically, the number of particles of size x is affected by the processes

$$[x] + [y] \to [x + y], \qquad [x - y] + [y] \to [x].$$

One writes a mean-field rate equation for the number density $n(x, t)$ in the form

$$\partial_t n(x, t) = \frac{1}{2} \int_0^x K(x-y, y) n(x-y, t) n(y, t) \, \mathrm{d}y - \int_0^\infty K(x, y) n(x, t) n(y, t) \, \mathrm{d}y.$$

$$(5.1)$$

Perhaps due to its simplicity, this model has found an amazingly diverse set of applications over a vast range of scales. It has been used to study microdroplet formation (in clouds, ink fog, smoke, fuel, paint, etc.), the kinetics of polymerization, hashing algorithms, and the clustering of colloids, phytoplankton in "marine snow," planetesimals in stellar accretion disks, and stars themselves. Much scientific effort has gone into determining appropriate rate kernels $K(x, y)$ for different physical models. For simplicity, here we will only consider the constant kernel $K = 2$, for which we can get a solution formula via the Laplace transform.

Since particles only combine, one expects the size distribution to shift toward larger particles and the typical particle size to grow in time. So one must rescale size to observe nontrivial long-time dynamical behavior. Equation (5.1) has the scaling invariance that if $n(x, t)$ is a solution, then so is

$$\tilde{n}(x, t) = a \, n(bx, ct)$$

provided $a = bc$.

5.2. *A 'new' framework for dynamic scaling analysis*

An issue of considerable significance in applications is whether typically the size distribution will approach a scale-invariant form, for which $\tilde{n} = n$. This is a self-similar solution or *scaling* solution for short.

In these notes we will study this question within a larger framework for understanding dynamic scaling behavior that has been outlined in the paper [33]. The basic issues are:

(i) What scaling solutions exist?

(ii) What are their domains of attraction?

(iii) What are the most general scaling limit points?
(These comprise the "scaling attractor.")

(iv) How can we describe the dynamics on the set of limit points?
(This is arguably the "ultimate dynamics" of the system.)

(v) How complicated can the ultimate dynamics be?

Though evidently stated in dynamical terms, this framework is strongly motivated by classical results in probability theory that date back to the 1920s, involved with establishing necessary and sufficient conditions for convergence in the central limit theorem. The issue concerns scaling limits of sums $S_n = \sum_{j=1}^{n} X_j$ of independent and identically distributed random variables, as $n \to \infty$. The whole theory is beautifully exposed in Feller's great book [19], and provides complete answers to the questions in the framework above:

- The normal distribution is the unique scale-invariant distribution of finite variance. But more generally the class of scale-invariant distributions make up a two-parameter family of (heavy-tailed) distributions called the Lévy stable laws.

- The normal distribution attracts all distributions of finite variance. But in general, simple necessary and sufficient conditions for a scaling limit to exist are known in terms of the power-law behavior (*regular variation*, to be precise) of the second-moment distribution function.

- The most general scaling limits that can arise for some subsequence $n_j \to \infty$ are the *infinitely divisible distributions*. These form an infinite-dimensional family parametrized by the famous Lévy-Khintchine representation formula in terms of a measure satisfying certain finiteness conditions.

- There exist distributions (Doeblin's universal laws) for which *every possible* scaling limit is realized along some subsequence. This is a hallmark of chaos—sensitive dependence on initial conditions.

The analytical methods used to establish these classical limit theorems are not essentially probabilistic in nature. Rather, they appear to be natural tools for analyzing scaling dynamics in many problems. For Smoluchowski's equation with $K = 2$, the results that we will (mostly) prove here are stated informally as follows (see [34, 33]):

- We find a unique scale-invariant solution with finite number $\int_0^\infty n\,\mathrm{d}x$ and mass $\int_0^\infty xn\,\mathrm{d}x$. But there is a one-parameter familiy of infinite-mass self-similar solutions, with profiles given by *Mittag-Leffler* probability distribution functions.
- Solutions approach self-similar form as $t \to \infty$ if and only if initially $x \mapsto \int_0^x yn\,\mathrm{d}y$ is almost power-law—regularly varying at ∞.
- The scaling attractor (set of subsequential scaling limits) can be parametrized by measures satisfying certain finiteness conditions.
- The nonlinear dynamics on the scaling attractor is *linearized* in terms of this measure representation.
- This ultimate dynamics exhibits sensitive dependence on initial data.

These results are strikingly analogous to those of classical probability theory. For well-localized data (finite mass), there is one universal scaling behavior, analogous to the central limit theorem. For many physical applications this is the most relevant case. However, one can study scaling dynamics in a more general context, and there one finds a rich set of mathematical possibilities. These should not be dismissed as uninteresting, given the wide range of applications of Smoluchowski's model. Heavy-tailed distributions have come to be important in numerous applications of probability, for example.

5.3. *Solution by Laplace transform*

The rigorous study of solutions of the coagulation equation (5.1) begins with the general moment identity

$$\partial_t \int_0^\infty a(x)n(x,t)\,\mathrm{d}x = \frac{1}{2}\int_0^\infty \int_0^x a(x)n(x-y,t)n(y,t)K(x-y,y)\,\mathrm{d}y\,\mathrm{d}x$$
$$- \int_0^\infty \int_0^\infty a(x)n(x,t)n(y,t)K(x,y)\,\mathrm{d}y\,\mathrm{d}x$$
$$= \frac{1}{2}\int_0^\infty (a(x+y)-a(x)-a(y))n(x,t)n(y,t)K(x,y)\,\mathrm{d}x\,\mathrm{d}y. \quad (5.2)$$

Taking $a = x$ formally yields conservation of total mass:

$$\partial_t \int_0^\infty xn(x,t)\,\mathrm{d}x = 0.$$

For $K = 2$, the total number $N(t) = \int_0^\infty n(x,t)\,\mathrm{d}x$ satisfies $\partial_t N = -N^2$. We will normalize and scale x and t so that we always have

$$N(t) = \frac{1}{t}. \quad (5.3)$$

Next take $a(x) = 1 - e^{-qx}$. Since $a(x + y) - a(x) - a(y) = -a(x)a(y)$, with

$$\phi(q, t) = \int_0^\infty (1 - e^{-qx})n(x, t)\,dt, \tag{5.4}$$

a quantity related to the Laplace transform of n, we have the simple equation

$$\partial_t \phi = -\phi^2. \tag{5.5}$$

Note that

$$\phi(0, t) = 0, \quad \phi(\infty, t) = N(t) = \frac{1}{t}, \quad \partial_q \phi = \int_0^\infty e^{-qx} x n(x, t)\,dx. \tag{5.6}$$

Since $\partial_t(1/\phi) = 1$, for $t, t_0 > 0$ we obtain the solution formulae

$$\frac{1}{\phi(q, t)} - \frac{1}{\phi(q, t_0)} = t - t_0, \quad \phi(q, t) = \frac{\phi(q, t_0)}{1 + (t - t_0)\phi(q, t_0)}. \tag{5.7}$$

This solution formula serves as the basis for a theory of the initial-value problem for which the size distribution $n(x, t)\,dx = \nu_t(\,dx)$ is a finite measure on $(0, \infty)$ which initially can be completely arbitrary, subject to the normalization

$$\int_0^\infty \nu_{t_0}(\,dx) = \frac{1}{t_0}. \tag{5.8}$$

Let us sketch how this works. First, look for solutions that are lattice measures, of the form

$$\nu_t(\,dx) = \sum_{j=1}^\infty c_j(t)\delta(x - j\Delta x)$$

where $\delta(x - j\Delta x)$ is a Dirac mass at $j\Delta x$. With initially $\sum_{j=1}^\infty c_j(t_0) = 1/t_0$, solve the discrete equations

$$\partial_t c_j = \sum_{k=1}^{j-1} c_{j-k} c_k \Delta x - 2c_j N(t) \tag{5.9}$$

inductively for $j = 1, 2, \ldots$, with $N(t) = 1/t$. Then prove that

$$\sum_{j=1}^\infty c_j(t) \le \frac{1}{t} \quad \text{for all } t \ge t_0.$$

(Hint: $N_J(t) = \sum_{j=1}^J c_j(t)$ satisfies $\partial_t(N_J - N) \le (N_J - N)^2$, $N_J(t_0) - N(t_0) \le 0$.) Integrate (5.9) in t and apply the Laplace transform to deduce that the function $\phi(q, t) = \int_0^\infty (1 - e^{-qx})\nu_t(\,dx)$ satisfies

$$\phi(q, t) = \phi(q, t_0) + \int_{t_0}^t \phi(q, s)^2\,ds.$$

This implies (5.7), and we infer that $t \mapsto \nu_t$ is weakly continuous in the sense of measures by the continuity theorem for Laplace transforms. In general, we approximate a general measure ν_{t_0} by lattice measures as above. We pass to limits using the continuity theorem for Laplace transforms together with (5.7):

$$\nu_{t_0}^{\Delta x} \to \nu_{t_0} \Leftrightarrow \phi_0^{\Delta x}(q) \to \phi(q, t_0) \text{ for all } q > 0$$
$$\Leftrightarrow \phi^{\Delta x}(q, t) \to \phi(q, t) \text{ for all } q \geq 0, \quad t \geq t_0$$

We obtain existence and uniqueness, and solutions depend continuously on initial data with respect to weak convergence. Also, initial data depend continuously on the solution! See [34] for discussion of a precise sense in which this yields a weak solution of Smoluchowski's equation. The upshot is that for any measure ν_{t_0} on $(0, \infty)$ satisfying (5.8) with $t_0 > 0$, there is a unique measure solution defined for all $t \geq t_0$, meaning a weakly continuous map $t \mapsto \nu_t$ such that for $t \geq t_0$, ν_t is a finite measure on $(0, \infty)$ such that $\phi(q, t) = \int_0^\infty (1 - e^{-qx}) \nu_t(\,\mathrm{d}x)$ satisfies (5.7).

5.4. *Scaling solutions and domains of attraction*

Based upon the solution of the coagulation equation by Laplace transform, a complete classification of scaling solutions and their domains of attraction was worked out in [34] for $K = 2$, $x + y$ and xy. For $K = 2$, all nontrivial scaling limits can be classified as follows.

Theorem 5.1: *Take $t_0 = 1$ and suppose ν_t is a measure solution of Smoluchowski's equation, so that (5.7) holds, and introduce the probability distribution function*

$$F_t(x) = \int_0^x \nu_t(\,\mathrm{d}x) \Big/ \int_0^\infty \nu_t(\,\mathrm{d}x) \quad \left(= t \int_0^x n(y, t)\,\mathrm{d}y\right).$$

(i) Suppose that there exists $\lambda(t) \to \infty$ and a probability distribution F_* so that

$$F_t(\lambda(t)x) \to F_*(x) \quad \text{as } t \to \infty \tag{5.10}$$

at all points of continuity, where $F_*(x) < 1$ for some $x > 0$. Then,

$$\int_0^x y\nu_1(\,\mathrm{d}y) \sim x^{1-\rho}L(x) \quad \text{as } x \to \infty \tag{5.11}$$

for some $\rho \in (0, 1]$ and L slowly varying at ∞.

(ii) Conversely, suppose that (5.11) holds. Then (5.10) holds, with

$$F_* = F_\rho(x) = \sum_{k=1}^{\infty} \frac{(-1)^{k+1} x^{\rho k}}{\Gamma(1 + \rho k)},$$

a Mittag–Leffler distribution, whose Laplace transform is

$$\mathcal{L}F_\rho(q) = \int_0^\infty e^{-qx} F_\rho(\,\mathrm{d}x) = \frac{1}{1 + q^\rho}.$$

Remark 5.2: Finite mass

$$\int_0^\infty x\nu_1(\,\mathrm{d}x) < \infty$$

gives $\rho = 1$, and $F_1(x) = 1 - e^{-x}$ corresponding to

$$n(t, x) = \frac{1}{t^2} e^{-x/t}.$$

This is an analog of the central limit theorem in probability theory. The Mittag-Leffler distributions F_ρ for $0 < \rho < 1$ have infinite mass and are analogs of the (heavy-tailed) *Lévy stable laws* of probability theory.

Proof: The strategy of the proof is to use the rigidity property of scaling limits, and the Tauberian theorem (mentioned in the Background section above). Assume, as in the statement of the theorem, that there exists $\lambda(t) \to \infty$ and a nontrivial probability distribution F_* so that

$$F_t(\lambda(t)x) \to F_*(x) \qquad \text{as } t \to \infty,$$

at all points of continuity. By the continuity theorem for Laplace transforms,

$$\int_0^\infty e^{-qx} F_t(\lambda(x)\,\mathrm{d}x) \to \int_0^\infty e^{-qx} F_*(\,\mathrm{d}x) = \mathcal{L}F_*(q) \qquad \text{for all } q > 0.$$

In terms of ϕ and $\phi_1(q) := \phi(q, 1)$, this means that for all $q > 0$,

$$1 - t\phi(q/\lambda, t) = 1 - \frac{t\phi_1(q/\lambda)}{1 + (t - 1)\phi_1(q/\lambda)} \to \mathcal{L}F_*(q) \in (0, 1).$$

Therefore, for all $q > 0$,

$$t\phi_1(q/\lambda(t)) \to g(q) \in (0, \infty) \qquad \text{as } t \to \infty.$$

By the rigidity property for scaling limits, we must have $g(q) = cq^\rho$ for some $c > 0$, $0 \le \rho < \infty$. Then

$$\mathcal{L}F_*(q) = 1 - \frac{cq^\rho}{1 + cq^\rho} = \frac{1}{1 + cq^\rho}.$$

Since $F_*(\infty) = \mathcal{L}F_*(0+) = 1$, we have $\rho > 0$. Since $-\partial_q \mathcal{L}F_* = \int_0^\infty e^{-qx} x F_*(\,dx)$ is positive decreasing, we have $\rho \le 1$.

By scaling $\lambda(t)$, we can achieve $c = 1$. Moreover, the rigidity property implies

$$\phi_1(q) \sim q^\rho \tilde{L}(1/q) \quad \text{as } q \to 0^+,$$

for some $\tilde{L}$ slowly varying at ∞. By the Tauberian theorem, the result

$$\int_0^x y\nu_1(\,dy) \sim x^{1-\rho} L(x) \quad \text{as } x \to \infty$$

is equivalent to

$$\partial_q \phi_1(q) = \int_0^\infty e^{-qy} y\nu_1(\,dy) \sim q^{\rho-1} L(1/q)\Gamma(2-\rho), \text{ as } q \to 0^+.$$

The proof (in both directions) is finished with the use of the following lemma. $\qquad\square$

Lemma 5.3: *The following are equivalent.*

(1) $\phi_1(q) \sim q^\rho \tilde{L}(1/q)$ *as* $q \to 0^+$.
(2) $\phi_1'(q) \sim \rho q^{\rho-1} \tilde{L}(1/q)$ *as* $q \to 0^+$.

Proof: We first show that $(1) \Rightarrow (2)$. Since $\phi_1'' = -\int_0^\infty e^{-qy} y^2 \nu_1(\,dy) < 0$, we have that for fixed $a > 1$,

$$\phi_1'(q) \ge \frac{\phi_1(aq)\phi_1(q)}{aq - q} = \frac{(aq)^\rho \hat{L}(aq) - q^\rho \hat{L}(q)}{q(a-1)} = q^{\rho-1}\hat{L}(q)\frac{a^\rho \hat{L}(aq)/\hat{L}(q) - 1}{a - 1}$$

and $\hat{L}(q) = \phi_1(q)/q \sim \tilde{L}(1/q)$. Hence (take $a \to 1$)

$$\liminf_{q \to 0} \frac{\phi_1'(q)}{q^{\rho-1}\hat{L}(q)} \ge \frac{a^\rho - 1}{a - 1} \to \rho.$$

Similarly, for fixed $a < 1$, we get

$$\limsup_{q \to 0} \frac{\phi_1'(q)}{q^{\rho-1}\hat{L}(q)} \le \frac{a^\rho - 1}{a - 1} \to \rho.$$

In order to show that $(2) \Rightarrow (1)$, let $\hat{L}(q) = \phi_1'(q)/\rho q^{\rho-1} \sim \tilde{L}(1/q)$. Then

$$\phi_1(q) = \int_0^q \phi_1'(s)\,ds = \int_0^q \rho s^{\rho-1}\hat{L}(s)\,ds = q^\rho \hat{L}(q)\int_0^1 \rho x^{\rho-1}\frac{\hat{L}(qx)}{\hat{L}(q)}\,dx,$$

where in the last equality, we changed variables according to $s = qx$. We know

$$\frac{\hat{L}(qx)}{\hat{L}(q)} \to 1, \qquad \text{for all } x > 0,$$

and we also need the fact (not difficult to prove) that $\tilde{L}(x) \geq C_\varepsilon x^{-\varepsilon}$ for all $\varepsilon > 0$. For fixed ε we use dominated convergence to conclude that

$$\frac{\phi_1(q)}{q^\rho \hat{L}(q)} \to \int_0^1 \rho x^{\rho-1}\, \mathrm{d}x = 1, \text{ as } q \to 0. \qquad \square$$

5.5. *The scaling attractor*

In systems with complicated dynamics, a fundamental notion aimed at capturing all long-time behavior, not only limiting states as $t \to \infty$, is that of the attractor. In finite-dimensional systems, one definition describes the attractor in terms of all possible limit points of bounded sequences of solutions. Modulo rescaling in size, this is precisely what we aim to describe here, following [33]. The resulting object, which we call the *scaling attractor*, turns out to have a remarkable characterization analogous to the Levy-Khintchine representation of infinitely divisible laws in probability theory.

Definition 5.4: Suppose $\hat{F}$ is a probability distribution function such that there exists a sequence of solutions $\nu_t^{(n)}$ defined for $t \geq t_0$ and numbers t_n, $\beta_n \to \infty$ such that

$$F_{t_n}^{(n)}(\beta_n x) \to \hat{F}(x) \qquad \text{as } n \to \infty$$

at each point of continuity. Then we say that $\hat{F}$ belongs to the *(proper) scaling attractor $\mathcal{A}$*.

One property enjoyed by the attractor in a finite system is that it is an invariant set forward and backward in time. A related property holds for the scaling attractor. First we note the following scaling property for measure solutions of Smoluchowski's equation with $K = 2$: Let $a > 0, b > 0$ be given, and let ν_t be a solution on $[t_0, \infty)$. Now, let

$$\tilde{\nu}_t(\, \mathrm{d}x) = a\nu_{at}(b\, \mathrm{d}x),$$

with $\tilde{F}_t(x) = F_{at}(bx)$. Then, $\tilde{\nu}$ is again a solution, on $[t_0/a, \infty)$, because of the fact that

$$\tilde{\phi}(q, t) = \int_0^\infty (1 - \mathrm{e}^{-qx})\nu_{at}(b\, \mathrm{d}x) = a\phi(q/b, at)$$

satisfies $\partial_t \tilde{\phi} = a^2 (\partial_t \phi)(q/b, at) = -a^2 \phi^2 = -\tilde{\phi}^2$.

Now, suppose we have a sequence as in the definition above. Put

$$\tilde{F}_t^{(n)}(x) = F_{t_n t}^{(n)}(\beta_n x).$$

Then

$$\tilde{F}_1^{(n)}(x) \to \hat{F}(x) \text{ as } n \to \infty.$$

Correspondingly, $\tilde{\phi}^{(n)}(q, 1) \to \hat{\phi}(q)$ by the continuity theorem, and

$$\tilde{\phi}^{(n)}(q, t) = \frac{\tilde{\phi}^{(n)}(q, 1)}{1 + (t - 1)\tilde{\phi}^{(n)}(q, 1)} \to \frac{\hat{\phi}(q)}{1 + (t - 1)\hat{\phi}(q)} =: \phi(q, t).$$

We have

$$\phi(q, t) = \int_0^\infty (1 - e^{-qx}) \nu_t(\,\mathrm{d}x),$$

where ν_t is a solution on $[t_1, \infty)$ for all $t_1 > 0$. Starting from any $t_0 > 0$, such a solution is defined *backwards in time* as far as it is meaningful.

Definition 5.5: A solution with $K = 2$ defined for all $t > 0$ is an *eternal solution*.

This analysis proves the following:

Theorem 5.6: *Points in the scaling attractor correspond one-to-one with eternal solutions. That is, $\hat{F} \in \mathcal{A}$ if and only if $\hat{F} = F_1$ for some eternal solution ν_t.*

We get an interesting characterization of the scaling attractor by studying limits as $t \downarrow 0$. Observe that

$$\phi(q, t) = \frac{\hat{\phi}(q)}{1 + (t - 1)\hat{\phi}(q)} \to \frac{\hat{\phi}(q)}{1 - \hat{\phi}(q)} =: \Phi(q), \text{ as } t \to 0^+.$$

This raises the question: What does this mean in terms of weak convergence of measures? Note that $\int_0^\infty \nu_t(\,\mathrm{d}x) = 1/t \to \infty$ as $t \to \infty$, and $\hat{\phi}(q) \to 1$ as $q \to 0^+$, so $\Phi(\infty) = \infty$. Also $t\phi(q, t) \to 0$, so the limit $F_t \to \delta(x - 0)$ is trivial.

The answer turns out to be to look at $G_t(\,\mathrm{d}x) = x\nu_t(\,\mathrm{d}x)$. One has

$$\phi(q, t) = \int_0^\infty \frac{1 - e^{-qx}}{x} G_t(\,\mathrm{d}x), \quad \partial_q \phi(q, t) = \int_0^\infty e^{-qx} G_t(\,\mathrm{d}x) = \mathcal{L}G_t(q).$$

Since

$$\frac{1}{\phi} = \frac{1}{\hat{\phi}} + (t - 1),$$

we get that $\partial_q \phi / \phi^2 = \partial_q \hat{\phi} / \hat{\phi}(q)^2$, and hence as $t \to 0$,

$$\partial_q \phi(q, t) = \mathcal{L} G_t(q) \to \frac{\Phi^2}{\hat{\phi}^2} = \frac{\partial_q \hat{\phi}}{(1 - \hat{\phi})^2} = \partial_q \Phi(q).$$

By the extended continuity theorem, there exists a measure H on $[0, \infty)$ with

$$G_t \to H \quad \text{as } t \to 0^+.$$

From

$$\phi(q, t) = \phi(\varepsilon, t) + \int_\varepsilon^q \partial_q \phi(q', t) \mathrm{d}q',$$

taking $t \to 0$ and $\varepsilon \downarrow 0$, we get

$$\Phi(q) = \int_0^q \mathcal{L} H(q') \mathrm{d}q' = \int_0^\infty \frac{1 - \mathrm{e}^{-qx}}{x} H(\mathrm{d}x).$$

Since $\Phi(\infty) = \infty$,

$$\text{either} \quad H(0) > 0 \quad \text{or} \quad \int_0^\infty x^{-1} H(\mathrm{d}x) = \infty. \tag{5.12}$$

Definition 5.7: A measure G on $[0, \infty)$ is a *generating measure* if

$$\int_{[0,x]} G(\mathrm{d}y) + \int_{[x,\infty)} y^{-1} G(\mathrm{d}y) < \infty \quad \text{for all } x > 0,$$

i.e.,

$$\int_{[0,\infty)} (1 \wedge x^{-1}) G(\mathrm{d}x) < \infty.$$

G is *divergent* if either $G(0) > 0$ or $\int_{(0,\infty)} y^{-1} G(\mathrm{d}y) = +\infty$.

Theorem 5.8: *To each non-divergent generating measure $\hat{G}$ with*

$$\int_{(0,\infty)} y^{-1} \hat{G}(\mathrm{d}y) = \frac{1}{t_0}, \quad t_0 > 0,$$

there corresponds a unique solution ν_t on $[t_0, \infty)$ with $\nu_{t_0}(\mathrm{d}x) = x^{-1} \hat{G}(\mathrm{d}x)$, and conversely. Furthermore, to each eternal solution ν_t on $(0, \infty)$ there corresponds a divergent generating measure H, such that

$$G_t \to H \quad \text{as } t \to 0.$$

Conversely, to each divergent generating measure H corresponds a unique eternal solution ν_t as above, determined by

$$\phi(q, t) = \frac{\Phi(q)}{1 + t\Phi(q)}, \quad \Phi(q) = \int_{[0,\infty)} \frac{1 - \mathrm{e}^{-qx}}{x} H(\mathrm{d}x).$$

Proof: We will prove the converse result. For small $\varepsilon > 0$ put

$$\hat{G}_\varepsilon = H|_{[\varepsilon,\infty)} + H(0)\delta(x - \varepsilon).$$

Define $t_0(\varepsilon)$ by

$$\frac{1}{t_0(\varepsilon)} = \int_{(0,\infty)} x^{-1}\hat{G}_\varepsilon(\,\mathrm{d}x) = \int_{[\varepsilon,\infty)} x^{-1}H(\,\mathrm{d}x) + \frac{1}{\varepsilon}H(0).$$

Then $t_0(\varepsilon) \to 0$ as $\varepsilon \to 0$, and $\hat{G}_\varepsilon \to H$ on $[0,\infty)$. $\hat{G}_\varepsilon$ determines a solution ν_t^ε on $[t_0(\varepsilon), \infty)$ with

$$\phi_0^\varepsilon(q) = \phi^\varepsilon(q, t_0(\varepsilon)) = \int_0^\infty \frac{1 - \mathrm{e}^{-qx}}{x}\hat{G}_\varepsilon(\,\mathrm{d}x) \to \int_0^\infty \frac{1 - \mathrm{e}^{-qx}}{x}H(\,\mathrm{d}x) = \Phi(q),$$

as $\varepsilon \to 0$. Then, for $t > t_0(\varepsilon)$,

$$\phi^\varepsilon(q, t) = \frac{\phi_0^\varepsilon(q)}{1 + (t - t_0(\varepsilon))\phi_0^\varepsilon(q)} \to \frac{\Phi(q)}{1 + t\Phi(q)} =: \phi(q, t),$$

corresponding to a unique eternal solution ν_t. $\qquad\square$

The result of this theorem is that *each point on the scaling attractor $\mathcal{A}$ corresponds to a unique divergent generating measure H*:

$$F_1 \in \mathcal{A} \longleftrightarrow \nu_t \text{ is eternal } \longleftrightarrow H \text{ is a divergent generating measure.}$$

5.6. *Linearization of dynamics on the scaling attractor*

Understanding the dynamics on the scaling attractor in terms of the measures H turns out to be a simple consequence of a scaling property of solutions. The upshot is that nonlinear clustering dynamics governed by the coagulation equation (5.1) with $K = 2$ becomes *linear* in terms of H.

Suppose ν_t is an eternal solution. Given $a, b > 0$, let $\tilde{\nu}_t(\,\mathrm{d}x) = a\nu_{at}(b\,\mathrm{d}x)$ and $\tilde{F}_t(x) = F_{at}(bx)$. Then, since ν_t is eternal, $\tilde{\nu}_t$ is also eternal and furthermore, $\tilde{F}_1(x) = F_a(bx)$. Also, we have

$$\tilde{G}_t \to \tilde{H} \qquad \text{weakly as } t \to 0.$$

But

$$\tilde{G}_t(\,\mathrm{d}x) = x\tilde{\nu}_t(\,\mathrm{d}x) = ax\nu_{at}(b\,\mathrm{d}x) = \frac{a}{b}G_{at}(b\,\mathrm{d}x) \to \frac{a}{b}H(b\,\mathrm{d}x), \text{ as } t \to 0.$$

Hence

$$\tilde{H}(x) = \frac{a}{b}H(bx).$$

Theorem 5.9: *Under the correspondence $\mathcal{G}$ that maps the scaling attractor $\mathcal{A}$ to divergent generating measures, given by*

$$\mathcal{G}(F_1) = H_1 = H,$$

the scaling dynamics on $\mathcal{A}$ given by $F_1(x) \mapsto F_t(bx)$ is represented via the map

$$H(\,\mathrm{d}x) \mapsto H_t(\,\mathrm{d}x) = \mathcal{G}(F_t(b\,\mathrm{d}x)) = \frac{t}{b} H_1(b\,\mathrm{d}x).$$

Remark 5.10: In greatest generality, the scaling dynamics is complicated! One can show that there exists an H_1 so that the trajectory $t \mapsto F_t$ is *dense* in $\mathcal{A}$. This means that solutions exhibit sensitive dependence on initial data, the hallmark of chaos. To show this, basically we need to show that for every divergent generating measure H, there exist $t_n, b_n \to \infty$ such that

$$\frac{t_n}{b_n} H_1(b_n\,\mathrm{d}x) \to \hat{H},$$

in an appropriate topology. One arranges this by carefully "packing the tail" of the measure H_1 in a way similar to the construction of Doeblin's universal laws in probability [19]. Details will appear in [33].

    Self-similar solutions. These correspond to solutions invariant under continuous rescaling with $b(t) \to \infty$ as $t \to \infty$, so that $F_t(b(t)x) = F_1(x)$, i.e.,

$$H(x) = \frac{t}{b(t)} H(b(t)x) \qquad \text{for all } x > 0.$$

Take $t \to \infty$ and apply the rigidity lemma. Then H must be a pure power:

$$H(x) = \tilde{c} x^p \qquad \text{for some } \tilde{c} > 0, \, 0 \le p < \infty.$$

Then

$$\partial_q \Phi(q) = \int_0^\infty \mathrm{e}^{-qx} H(\,\mathrm{d}x) = \tilde{c} \int_0^\infty \mathrm{e}^{-qx}\,\mathrm{d}(x^p) = \tilde{\tilde{c}} q^{-p},$$

so $\Phi(q) = cq^{1-p}$. Note that $p < 1$, since $\int_1^\infty x^{-1} H(\,\mathrm{d}x) < \infty$. Then,

$$t\phi(q,t) = \int_0^\infty (q - \mathrm{e}^{-qx}) F_t(\,\mathrm{d}x) = \frac{tcq^{1-p}}{1 + tcq^{1-p}}$$

gives

$$\mathcal{L}F_t(q) = 1 - t\phi = \frac{1}{1 + tcq^{1-p}} = \mathcal{L}\hat{F}_1((ct)^{\frac{1}{1-p}}q),$$

with

$$\mathcal{L}\hat{F}_1(q) = \frac{1}{1 + q^{1-p}} = \frac{1}{1 + q^\rho}, \qquad \text{for } \rho = 1 - p \in (0,1].$$

Acknowledgements

I especially thank Barbara Niethammer for her hospitality and many discussions during my visit to Berlin. This work was supported by the Deutsche Forschungsgemeinschaft in the form of a Mercator professorship at Humboldt University, by the Max Planck Institute for Mathematics in the Sciences in Leipzig, and by the Institute for Mathematical Sciences at the National University of Singapore. This material is based upon work supported by the National Science Foundation under grant DMS 03-05985.

References

1. N. D. ALIKAKOS, P. W. BATES, AND X. CHEN, *Convergence of the Cahn-Hilliard equation to the Hele-Shaw model*, Arch. Rational Mech. Anal., 128 (1994), pp. 165–205.

2. N. D. ALIKAKOS, G. FUSCO, AND G. KARALI, *Ostwald ripening in two dimensions—the rigorous derivation of the equations from the Mullins-Sekerka dynamics*, J. Differential Equations, 205 (2004), pp. 1–49.

3. S. ALLEN AND J. CAHN, *Microscopic theory for antiphase boundary motion and its application to antiphase domain coarsening*, Acta Metallurgica, 27 (1979), pp. 1085–1095.

4. G. ANDREWS AND J. M. BALL, *Asymptotic behaviour and changes of phase in one-dimensional nonlinear viscoelasticity*, J. Differential Equations, 44 (1982), pp. 306–341. Special issue dedicated to J. P. LaSalle.

5. G. S. BALES AND A. ZANGWILL, *Morphological instability of a terrace edge during step-flow growth*, Phys. Rev. B, 41 (1990), pp. 5500–5508.

6. N. H. BINGHAM, C. M. GOLDIE, AND J. L. TEUGELS, *Regular variation*, vol. 27 of Encyclopedia of Mathematics and its Applications, Cambridge University Press, Cambridge, 1987.

7. W. K. BURTON, N. CABRERA, AND F. C. FRANK, *The growth of crystals and the equilibrium structure of their surfaces*, Philos. Trans. Roy. Soc. London. Ser. A., 243 (1951), pp. 299–358.

8. G. CAGINALP, *Stefan and Hele-Shaw type models as asymptotic limits of the phase-field equations*, Phys. Rev. A (3), 39 (1989), pp. 5887–5896.

9. J. CARR AND R. PEGO, *Invariant manifolds for metastable patterns in $u_t = \epsilon^2 u_{xx} - f(u)$*, Proc. Roy. Soc. Edinburgh Sect. A, 116 (1990), pp. 133–160.

10. ———, *Self-similarity in a coarsening model in one dimension*, Proc. Roy. Soc. London Ser. A, 436 (1992), pp. 569–583.

11. J. CARR AND R. L. PEGO, *Metastable patterns in solutions of $u_t = \epsilon^2 u_{xx} - f(u)$*, Comm. Pure Appl. Math., 42 (1989), pp. 523–576.

12. J. CARR AND O. PENROSE, *Asymptotic behaviour of solutions to a simplified Lifshitz-Slyozov equation*, Phys. D, 124 (1998), pp. 166–176.

13. X. CHEN, *Generation, propagation, and annihilation of metastable patterns*, J. Differential Equations, 206 (2004), pp. 399–437.

14. S. DAI, *Universal bounds on coarsening rates for some models of phase transitions*, PhD thesis, University of Maryland, College Park, 2005.

15. S. DAI AND R. L. PEGO, *Universal bounds on coarsening rates for mean-field models of phase transitions*, SIAM J. Math. Anal., 37 (2005), pp. 347–371 (electronic).

16. J. DAVIS, *The new diamond age*, Wired, 11.09 (2003). http://www.wired.com/wired/archive/11.09/diamond.html.

17. B. DERRIDA, *Coarsening phenomena in one dimension*, in Complex systems and binary networks (Guanajuato, 1995), vol. 461 of Lecture Notes in Phys., Springer, Berlin, 1995, pp. 164–182.

18. J.-P. ECKMANN AND J. ROUGEMONT, *Coarsening by Ginzburg-Landau dynamics*, Comm. Math. Phys., 199 (1998), pp. 441–470.

19. W. FELLER, *An introduction to probability theory and its applications. Vol. II.*, Second edition, John Wiley & Sons Inc., New York, 1971.

20. P. C. FIFE, *Dynamics of internal layers and diffusive interfaces*, vol. 53 of CBMS-NSF Regional Conference Series in Applied Mathematics, Society for Industrial and Applied Mathematics (SIAM), Philadelphia, PA, 1988.

21. G. FUSCO AND J. K. HALE, *Slow-motion manifolds, dormant instability, and singular perturbations*, J. Dynam. Differential Equations, 1 (1989), pp. 75–94.

22. T. GALLAY AND A. MIELKE, *Convergence results for a coarsening model using global linearization*, J. Nonlinear Sci., 13 (2003), pp. 311–346.

23. B. GIRON, B. MEERSON, AND P. V. SASOROV, *Weak selection and stability of localized distributions in ostwald ripening*, Phys. Rev. E, 58 (1998), pp. 4213–4216.

24. K. GLASNER, *A diffuse interface approach to Hele-Shaw flow*, Nonlinearity, 16 (2003), pp. 49–66.

25. J. K. HALE AND P. MASSAT, *Asymptotic behavior of gradient-like systems*, in Dynamical systems II (Proc. 2nd Intl. Symp., Gainesville, Fla., 1981), A. R. Bednarek and L. Cesari, eds., New York, 1982, Academic Press Inc., pp. 85–101.

26. H.-C. JEONG AND E. WILLIAMS, *Steps on surfaces: experiment and theory*, Surf. Sci. Reports, 34 (1999), pp. 175–294.

27. R. V. KOHN AND F. OTTO, *Upper bounds on coarsening rates*, Comm. Math. Phys., 229 (2002), pp. 375–395.

28. R. V. KOHN AND X. YAN, *Upper bound on the coarsening rate for an epitaxial growth model*, Comm. Pure Appl. Math, 56 (2003), pp. 1549–1564.

29. R. V. KOHN AND X. YAN, *Coarsening rates for models of multicomponent phase separation*, Interfaces Free Bound., 6 (2004), pp. 135–149.

30. B. LI AND J.-G. LIU, *Epitaxial growth without slope selection: energetics, coarsening, and dynamic scaling*, J. Nonlinear Sci., 14 (2004), pp. 429–451 (2005).

31. I. M. LIFSHITZ AND V. V. SLYOZOV, *The kinetics of precipitation from supersaturated solid solutions*, J. Phys. Chem. Solids, 19 (1961), pp. 35–50.

32. R. MENDOZA, J. ALKEMPER, AND P. W. VOORHEES, *The morphological evolution of dendritic microstructures during coarsening*, Metall. and Mater. Trans., 34A (2003), pp. 481–489.

33. G. MENON AND R. L. PEGO, *The scaling attractor and ultimate dynamics for Smoluchowski's coagulation equation*, J. Nonlinear Sci., to appear.

34. G. MENON AND R. L. PEGO, *Approach to self-similarity in Smoluchowski's coagulation equations*, Comm. Pure Appl. Math., 57 (2004), pp. 1197–1232.

35. D. MOLDOVAN AND L. GOLUBOVIC, *Interfacial coarsening dynamics in epitaxial growth with slope selection*, Phys. Rev. E, 61 (2000), pp. 6190–6214.

36. T. NAGAI AND K. KAWASAKI, *Statistical dynamics of interacting kinks II*, Physica A, 134 (1986), pp. 483–521.

37. B. NIETHAMMER, *Derivation of the LSW-theory for Ostwald ripening by homogenization methods*, Arch. Ration. Mech. Anal., 147 (1999), pp. 119–178.

38. ——, *The LSW model for Ostwald ripening with kinetic undercooling*, Proc. Roy. Soc. Edinburgh Sect. A, 130 (2000), pp. 1337–1361.

39. B. NIETHAMMER AND F. OTTO, *Ostwald ripening: the screening length revisited*, Calc. Var. Partial Differential Equations, 13 (2001), pp. 33–68.

40. B. NIETHAMMER AND R. L. PEGO, *Non-self-similar behavior in the LSW theory of Ostwald ripening*, J. Statist. Phys., 95 (1999), pp. 867–902.

41. ——, *On the initial-value problem in the Lifshitz-Slyozov-Wagner theory of Ostwald ripening*, SIAM J. Math. Anal., 31 (2000), pp. 467–485 (electronic).

42. ——, *Well-posedness for measure transport in a family of nonlocal domain coarsening models*, Indiana Univ. Math. J., 54 (2005), pp. 499–530.

43. A. NOVICK-COHEN AND R. L. PEGO, *Stable patterns in a viscous diffusion equation*, Trans. Amer. Math. Soc., 324 (1991), pp. 331–351.

44. F. OTTO, *Dynamics of labrinthine pattern formation in magnetic fluids: a mean-field theory*, Arch. Rational Mech. Anal., 14 (1998), pp. 63–103.

45. F. OTTO, P. PENZLER, A. RÄTZ, T. RUMP, AND A. VOIGT, *A diffuse-interface approximation for step flow in epitaxial growth*, Nonlinearity, 17 (2004), pp. 477–491.

46. R. L. PEGO, *Front migration in the nonlinear Cahn-Hilliard equation*, Proc. Roy. Soc. London Ser. A, 422 (1989), pp. 261–278.

47. R. L. PEGO, *Stabilization in a gradient system with a conservation law*, Proc. Amer. Math. Soc., 114 (1992), pp. 1017–1024.

48. J. RUBINSTEIN, P. STERNBERG, AND J. B. KELLER, *Fast reaction, slow diffusion, and curve shortening*, SIAM J. Appl. Math., 49 (1989), pp. 116–133.

49. E. SENETA, *Regularly varying functions*, Springer-Verlag, Berlin, 1976. Lecture Notes in Mathematics, Vol. 508.

50. C. WAGNER, *Theorie der alterung von niederschlägen durch umlösen*, Z. Elektrochemie, 65 (1961), pp. 581–594.

51. S. J. WATSON, *Coarsening dynamics of growing facetted crystal surfaces: the annealing to growth transtion*. submitted.

Quantized Vortices in Superfluids — A Mathematical and Computational Study

Qiang Du

Department of Mathematics,
Penn State University,
University Park, PA 16802, USA
Email: qdu@math.psu.edu
Url: http://www.math.psu.edu/qdu

The appearance of quantized vortices is a typical signature of superfluidity. It has received a lot of attention in the studies of superfluid Helium, superconductivity and more recently the Bose-Einstein condensation. The significance of the research on the quantized vortex phenomena was recently highlighted by the recent Nobel Physics Prizes in the years 2001 and 2003.

Throughout the last few decades, both theoretical and computational studies have shed light on the characteristics of the quantized vortex nucleation and dynamics. In this short lecture notes, we intend to provide a concise description of the physical background, several relevant mathematical models, and the numerical methods developed for the study of the motion and interaction of quantized vortices in various contexts. In particular, we emphasize on issues related to the celebrated Ginzburg-Landau models of superconductivity and the mean field Gross-Pitaevskii equations. Much of the discussions given here are taken from our earlier works in the field.

Contents

1. Introduction

Quantized vortices have a long history that begins with the studies of liquid Helium and superconductors. Their appearance is viewed as a typical signature of superfluidity which describes a phase of matter characterized by the complete absence of viscosity. In other words, if placed in a closed loop, superfluids can flow endlessly without friction. The examples of superfluidity can be found in superconductivity, liquid Helium and in Bose-Einstein condensation. In 1955, Feynman made the prediction that a superfluid rotation may be subject to an array of quantized singularities, namely, the quantized vortices [83]. The seminal work of Abrikosov in 1957 already made predictions of the vortex lattice in superconductors a decade before the experimental confirmation. The research on the quantized vortex phenomena has since flourished and it was recently highlighted by the Nobel Physics Prizes awarded to Cornell, Weimann and Ketterle in 2001 and to Ginzburg, Abrikosov and Leggett in 2003, who have made decisive contributions to Bose-Einstein condensation, superconductivity and superfluidity and to the understanding of the quantized vortex state.

In recent years, there have been many works on the mathematical analysis and numerical simulations of quantized vortices. It is truly remarkable that many of the phenomenological properties of quantized vortices have been well captured by relatively simple mathematical models, for example, the Ginzburg-Landau equations and the Gross-Pitaesvkii equations. The phenomenological model of Ginzburg and Landau, the center piece of 2003 Nobel physics prize winning work, has been widely used in the study of superconductivity, and it is also the focus of our study here.

The structures of quantized vortices have been studied through various approaches ranging from asymptotic analysis, numerical simulations and rigorous mathematical analysis. For instance, in the context of superconductivity, the nucleation mechanism of quantized vortices due to the presence of the applied magnetic field has become more and more mathematically rigorous through the analysis of the phenomenological Ginzburg-Landau models. Moreover, vortices may be set in motion due to vortex interactions, thermal fluctuations and applied voltages and currents. Those effects may be again studied through the Ginzburg-Landau models and their variants. The vortex motion, unfortunately, induces electrical resistance and causes the loss of superconductivity. In the Bose Einstein condensates, such motion also causes energy dissipation and leads to the loss of superfluidity. Understanding the dynamics of quantized vortices thus bears tremendous importance both scientifically and technologically.

Despite the much progress made in the last decade, it should be pointed out that the rigorous mathematical study of a large part of the subject on vortex dynamics remains nearly non-existent. Indeed, what become available in the literature are primarily studies of the various dynamical laws of well separated vortices deduced from the Ginzburg Landau models and mathematically justified in the case of a static constant applied magnetic field and a small coherence length of the superconductors (or a small healing length for BECs). Much of these studies, however, have not incorporated issues like the effect of applied current and thermal fluctuations. On the other hand, numerical simulations have become useful tools that could help providing a more clear picture on the exotic vortex dynamics driven by various forces, even though there are also challenging computational issues to be tackled.

In this chapter, we give some physical background to both problems in superconductivity and in Bose-Einstein condensation and the associated vortex phenomena. We also present some related mathematical models and describe the mathematical and computational studies of these models. Some computational results given in the literature as well as open questions are also provided here. Much of our discussions are taken from our earlier works in the field, thus, they may have limited scope. Nevertheless, they touch most of the important issues related to the vortex phenomena such as their basic structure, nucleation pattern, interaction and dynamics, and pinning effects. We refer to the list of the references given at the end of the chapter for more detailed and more complete studies on the subject.

2. Superconductivity and mathematical models

We begin with a brief account of the basic phenomena in superconductivity and an introduction to some basic terminologies.

2.1. *What is superconductivity?*

Superconductivity was traditionally defined as a phenomenon occurring in certain materials at low temperatures, characterized by the complete absence of electrical resistance and the damping of the interior magnetic field.

The superconductivity of certain metals, such as mercury, lead and tin, at very low temperatures was discovered by H. Kamerlingh-Onnes in 1911 (see [132] for a historical account). Since then, superconductivity has become one of the most fascinating subject of modern science. *"the fascination with superconductivity is associated with the words : perfect, infinite, and zero!"*, according to B. Maple, a physicist at UCSB. Indeed, what Kamerlingh-Onnes first observed was that electrical resistance became zero below some critical temperature, thus leading to infinite conductance and perfect conductivity. According to him, *"Mercury has passed into a new state, which on account of its extraordinary electrical properties may be called the superconductive state"*. The discovery of superconductivity was awarded the Nobel Prize in 1913.

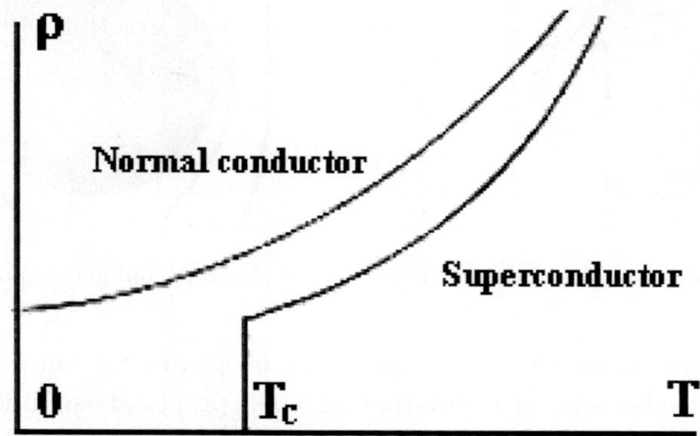

Fig. 1.   The loss of electrical resistance ($\rho = 0$) for superconductors.

Kamerlingh-Onnes also recognized the importance of superconductivity both scientifically and commercially. Closed currents in a ring of superconducting material have been observed to flow without decay for over two years, and the resistivity of some of these materials has been estimated to be no greater than 10^{-23} ohm/cm, thus giving rise to the phenomenon referred as the existence of persistent current.

In addition, superconductors are also characterized by the property of perfect diamagnetism, a phenomenon known now as the Meissner effect which is first discovered in 1933 by W. Meissner and R. Ochsenfeld. What they observed is that not only a magnetic field is excluded from a superconductor, i.e., if a magnetic field is applied to a superconducting material at a temperature below the critical temperature, it does not penetrate into the material, but also that a magnetic field is expelled from a superconductor, i.e., if a superconductor subject to a magnetic field is cooled through the critical temperature, the magnetic field is expelled from the material. Of course, sufficiently large magnetic fields cannot be excluded from the material, so that the Meissner effect also predicts the existence of a critical magnetic field above which the material ceases to be superconducting even at temperatures below the critical temperature. Furthermore, passage through the critical temperature is reversible which leads to the fact that superconductivity is a true thermodynamic state.

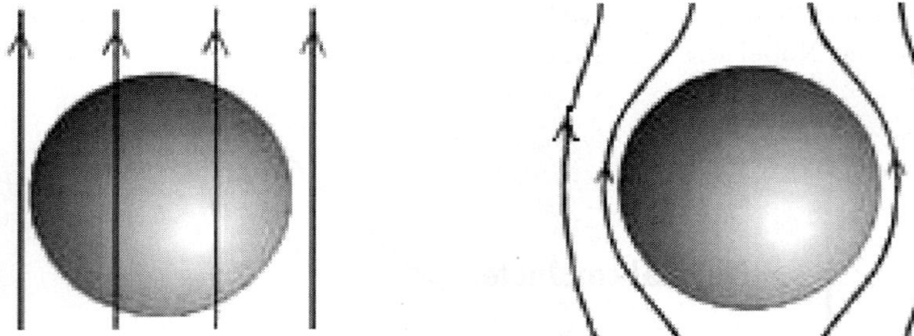

Fig. 2. The magnetic field penetrates the sample above T_c and gets expelled below T_c

Due to the extremely low temperature necessary for known materials, e. g. , metals to become superconducting, their practical usefulness was limited and therefore general interest in superconductivity waned for a number of years. However, in 1986, Bednorz and Müller made a Nobel-prize winning discovery of high-temperature superconductors (HTS). Their discovery of a family of Copper-Oxide compounds with high superconducting transition temperatures at 35 Kelvin was quickly followed by the discovery of $YBa_2Cu_3O_{7-x}$, a compound that becomes superconducting at 93K, a temperature that can be achieved with liquid Nitrogen (whose boiling point is 77 Kelvin). The promise of exciting new applications of high-T_c superconductivity has naturally brought a resurgence in interest of superconductivity. The quest to find samples with ever higher superconducting transition temperatures has since continued with the present record being around 135 Kelvin, almost halfway to room temperature.

In light of the discoveries made in the last twenty years, the definition of superconductivity is in fact rewritten. It is now referring to an electronic state of matter characterized by zero resistance, perfect diamagnetism, and long-range quantum mechanical order.

2.2. *Type-II superconductors and the vortex state*

The superconducting phenomena were first thought only appearing as a sharp transition between the normal state and the superconducting state, it was however discovered that there is in fact a gradual transition in many superconductors which are named as type-II superconductors. Type-II superconductivity was first recognized by deHaas and Voogd in 1930, but their work became accepted only after the discovery of the Meissner effect . It is understood now that, below the critical temperature T_c, the response of a superconducting material to an externally imposed magnetic field is most conveniently described by the diagrams given in Figure 3, which shows the minimum energy state of the superconductor as a function of H, the applied magnetic field, according to different regimes of the dimensionless material parameter κ (known as the Ginzburg-Landau parameter). This parameter κ is given by the ratio of the penetration depth and the coherence length, and its value determines the type of the superconducting material [65, 148]. For $\kappa < \frac{1}{\sqrt{2}}$, it is a type-I superconductor, while for $\kappa > \frac{1}{\sqrt{2}}$, it is a type-II superconductor. In the former case, there is a critical magnetic field H_C below which the material favors the superconducting Meissner state, but above which it favors the normal state. Such a thermodynamic critical field is then naturally characterized as the field for which the energy of the Meissner state and the energy of the normal state equal each other. It is easy to see that $H_C = O(T_c - T)$, though in practice, the determination of the state in a type-I superconductor is also affected by other factors such as the sample geometry.

Since the pioneering work of Abrikosov [1], it is well known that in a macroscopic type-II superconductor, that is, a material with $\kappa > \frac{1}{\sqrt{2}}$, the magnetic field, in the so-called mixed vortex state, penetrates the sample in the form of flux lines or vortices (normal filaments) indexvortex filament embedded in a superconducting matrix. Each of these filaments carries with it a quantized amount of magnetic flux, and is circled by a vortex of superconducting current. Thus these filaments are often know as vortex lines. Their two dimensional planar cross-sections are referred as two dimensional vortices.

One of the most challenging problems to mathematicians working on the superconductivity models is the understanding of vortex phenomena in type-II superconductors, which include the recently discovered high-temperature superconductors.

The transition from the normal state to the vortex state takes place by a bifurcation as the magnetic field is lowered through some critical value H_{C2}, the so-called upper critical field. The lower critical field H_{C1}, on the other hand, is calculated so that at this field the energy of the wholly superconducting solution becomes equal to the energy of the single vortex filament solution for an infinite superconductor.

Quantized vortex structures have been studied extensively on the mezoscale using the well-known Ginzburg-Landau models of superconduc-

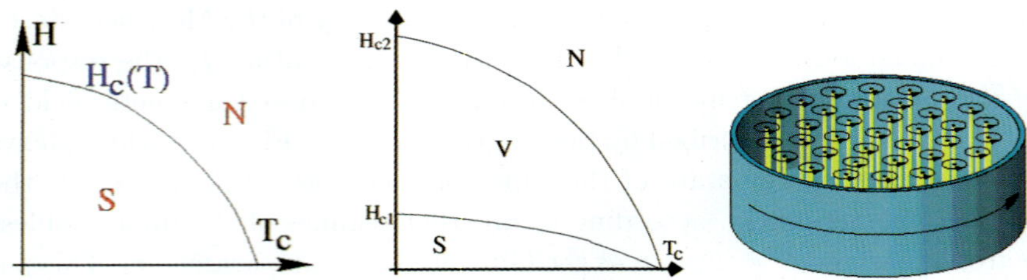

Fig. 3.   Equilibrium phase diagram for superconductors in an applied field and an illustration of quantized vortices.

tivity [65, 148]. The existence of vortex like solutions for the full nonlinear Ginzburg-Landau equations has been investigated by researchers using methods ranging from asymptotic analysis to numerical simulations. Much progress has been made in the recent years on establishing a mathematical framework for a rigorous description of both static and dynamic properties of the vortex solutions, in particular, as the coherence length tends to zero (κ goes to infinity), various results have been obtained. From a technological point of view, this is of interest since the recently discovered high critical temperature superconductors are known to have large values of κ, say κ in excess of 50.

Vortex lines may move due to internal interactions between these filaments and external forces (due to applied fields or thermal fluctuations) acting on them. Unfortunately, such vortex motion in an applied magnetic field induces an effective electrical resistance in the material, and thus a loss of superconductivity. Therefore, it is crucial to understand the dynamics of these vortex lines. At the same time, one is interested in studying mechanisms that can pin the vortices at fixed locations, i.e., preventing their motion. Various such vortex pinning mechanisms have been advanced by physicists, engineers, and material scientists. For example, normal (non-superconducting) impurities in an otherwise superconducting sample are believed to provide sites at which vortices are pinned. Likewise, anisotropy and other material inhomogeneities such as the thickness variations in thin samples are also believed to provide pinning sites. These mechanisms have been introduced into the general Ginzburg-Landau framework to derive various variants of the original Ginzburg-Landau models of superconductivity. Numerical simulations based on such models clearly suggest the pinning effect and some of these findings have been rigorously established in the mathematics literature in the last decade.

2.3. *Applications of superconductivity*

Ever since the discovery of superconductivity, various practical applications have been under consideration: powerful superconducting magnets could be made much smaller than a resistive magnet, generators made of superconductors could generate the same amount of electricity with smaller equipment and less energy, and the electricity could be stored in superconductors and distributed through superconducting power lines without significant loss.

As of today, applications of superconductivity that are currently being used include magnetic shielding devices, medical imaging systems, superconducting cyclotron, superconducting quantum interference devices (SQUIDS), infrared sensors, analog signal processing devices, and microwave devices. Applications that are being explored for the future include power transmission, superconducting magnets in generators, energy storage devices, particle accelerators, and magnetic levitated vehicle transportation. More recently, the concept of quantum computing has become important research directions in superconducting electronics [22].

In electronics, the miniaturization and the processing speed of computer chips are limited by the generation of heat and the charging time of capacitors due to the resistance of the interconnecting metal films. The use of new superconducting films may result in more densely packed chips which could transmit information at speeds that are faster by several orders of magnitude. The superconducting current controller (Current Fault Limiter or CFL) are being developed to control the reduction of fault currents (surges) in transmission lines.

Superconducting magnets are already crucial components of several technologies. Magnetic resonance imaging (MRI) is playing an ever increasing role in diagnostic medicine. The intense magnetic fields that are needed for these instruments are a perfect application of superconductors. Magnetic Source Imaging or MSI uses superconductors to provide the measurement of magnetic fields produced by biological systems such the human body. It is different from magnetobiology, which is the study of magnetic field effects on biological systems.

Magnetically levitated (MAGLEV) trains are considered as a future application of HTS development. The idea of MAGLEV transportation has been around since the early 1900s. The benefit of eliminating the wheel/rail friction to obtain higher speeds and lower maintenance costs has great appeal. The basic idea of a MAGLEV train is based on the levitation effect of superconductors in a strong magnetic field so that no friction is produced due to the lack of physical contact between the train and the rails (guideways).

In short, research in superconductivity is not only an intriguing scientific endeavor, but it also leads to enormous economical value and has significant impact on the human society. There have been estimates that predict the growth in the superconductor market in US alone may reach nearly \$5 billion in value by 2010 and as high as \$38 billion by 2020.

2.4. *Superconductivity models and mathematical problems*

Along with the historical development of experimental discoveries in superconductivity, there have been numerous theoretical studies trying to decipher the mystery of this intriguing phenomena. However, a good theoretical understanding of the low-temperature superconductivity was not arrived at until the 1950s. Indeed, a completely acceptable microscopic theory did not exist until Bardeen-Cooper-Schriefer (BCS) published their landmark paper in 1957. This work was awarded the Nobel physics prize in 1972. According to the BCS theory, superconductivity is due to an effective attraction between conducting electrons. Near T_c, two electrons not only experience a repulsive Coulomb force, but also an attractive force through electron-phonon interactions, thus forming the so called Cooper pairs. Such electron pairs are coherent structures that can pass through the conductor in unison. Briefly, due to the screening effect of the phonons, the electrons are separated by some distance. According to the BCS theory, as a negatively charged electron passes by positively charged ions in the lattice of the superconductor, the lattice distorts. That is, the attraction between the negative electron and the positive ion causes a lattice vibration from ion to ion until the other electron of the pair absorbs the vibration. This causes phonons to be emitted, thus forming a trough of positive charges around the electron. Before the electron passes by and before the lattice springs back to its normal position, a second electron is drawn into the trough. Effectively, it is as if the forces exerted by the phonons overcome the natural repulsion of electrons. The exchange of phonon keeps the Cooper pairs together, though the pairs are constantly breaking and reforming. Such a scenario gives the basic mechanism of low T_c superconductivity. The BCS theory not only leads to fundamental understanding of the low T_c superconductivity, but also gives the so-called BCS gap equations that can be used to determine the critical transition temperature. We refer to [75] for some works on the numerical solution of the BCS gap equations. Historically, however, various mezoscopic theories had been proposed even earlier than the BCS theory, most notably the London theory in 1935, the non-local theory of Pippard and the theory of Ginzburg and Landau in 1957. The Ginzburg-Landau theory was based on the Landau's general theory of second order phase transitions, but it was not immediately widely accepted, largely due to its phenomenological natural. In 1959, Gorkov showed that in some appropriate limit, the mezoscopic Ginzburg-Landau theory can be derived from the microscopic BCS theory, putting the Ginzburg-Landau theory on a firm ground. Moreover, the seminal work by Abrikosov in 1957 on the type-II vortex state based on the G-L model was a theoretical prediction that preceded experimental confirmation by almost a decade. From a technological standpoint, type II superconductors are the ones of greatest interest, mainly because they can retain superconductivity properties in the presence of large applied magnetic fields. In 2003, Ginzburg-Landau model was cited as the center piece of the Nobel prize winning work of Ginzburg, Abrikosov and Legget, showing its wide acceptance and significant impact.

There are many types of mathematical equations and problems appearing in the models of superconductivity, for instance, the BCS gap equation gives an integral equation of the first kind [75], the equilibrium Ginzburg-Landau models are a non-linear, non-convex variational problems, the time-dependent Ginzburg-Landau models give rise to a degenerate parabolic system, the mean field vortex density models lead to hyperbolic conservation laws, the studies of thermal fluctuation are linked to stochastic differential equations. Despite of the recent progress on the mathematical analysis and numerical simulations of the various superconductivity models, there is still a diverse class of many interesting and challenging questions to be answered in the future.

3. The mathematical theory of Ginzburg-Landau models

The phenomenological model of Ginzburg and Landau [65, 148], and its various generalizations have been widely used in the mathematical and numerical studies of the vortex phenomena in superconductivity. The Nobel prize in 2003 awarded to Ginzburg, Abrikosov and Legget highlighted its wide acceptance and significant impact. Recent mathematical studies on the Ginzburg-Landau (G-L) models have provided many rigorous justifications concerning the vortex state in type-II superconductors which were previously understood only based on physical heuristics and observations.

3.1. *The free energy postulated by Ginzburg and Landau*

Benefited from the theory of Gorter and Casimir, that of London, and the non-local model of Pippard, Ginzburg and Landau proposed their phenomenological model of superconductivity in 1950, seven years before the BCS theory. Since then, it has been formally derived from the microscopic BCS, generalized to many variations for high Tc superconductors. To this day, it remained as a standard initial approach to study problems in superconductivity. We refer to [65] for an earlier review on the basic mathematical theory. Numerical approximations of the Ginzburg-Landau have also been developed there systematicaly, see also [59] for a brief survey on the computational aspect.

The Ginzburg-Landau theory is based on Landau's general theory for second order phase transition. It was originally given to describe the equilibrium state and later generalized to the time-dependent cases. Let $\Omega \subset R^d$ be a region occupied by the superconducting sample. The primary variables used in the time dependent Ginzburg-Landau model are the complex scalar-valued *order parameter* ψ, the real vector-valued *magnetic potential* $\mathbf{A}$, and the real scalar-valued electric potential $\bar{\Phi}$. In a non-dimensionalized form, these variables are related to the physically observable variables by the following:

$$
\begin{array}{rl}
\text{density of superconducting charge carriers} & |\psi|^2 \\
\text{induced magnetic field} & \text{curl}\,\mathbf{A} \\
\text{current} & \mathbf{j} = \mathbf{curl}\,\text{curl}\,\mathbf{A} \\
\text{electric field} & \frac{\partial}{\partial t}\mathbf{A} + \nabla\bar{\Phi}.
\end{array}
$$

For an applied magnetic field H, the conventional Ginzburg-Landau free energy density function is given by

$$
f = f_n + \alpha|\psi|^2 + \frac{\beta}{2}|\psi|^4 + \frac{|h|^2}{8\pi} + \frac{1}{2m_s}\left|\left(\hbar\nabla - \frac{ie_s\mathbf{A}}{c}\right)\psi\right|^2
$$

where f_n is the free energy density of the normal state in the absence of magnetic fields, $h = curl\,\mathbf{A}$ is the induced magnetic field and α and β are constants in space, whose values depend on the temperature, c is the speed of light, e_s and m_s are the charge and mass, respectively, of the superconducting carriers (the Cooper pairs). For temperature T near the critical temperature T_c, we have $\alpha = \alpha_c(T - T_c)$ while $\beta = \beta_c$ for two positive constants α_c and β_c.

A few comments about the free energy used in the Ginzburg-Landau models are in order. First, the choice of complex order parameter, though not entirely obvious, is by the analogy with liquid Helium, another superfluid. When modeling liquid Helium, $|\psi|^2$ is known to represent the superfluid density. Second, the physical intuition of Ginzburg and Landau led to a clean formulation of the coupling with electro-magnetic fields even though there was no obvious analog in superfluid Helium. By exploiting the similarity of the formalism to ordinary quantum mechanics, the fields were coupled in the usual way to *charges* e_s associated with *particles* of mass m_s, and thus leading to the above density form. Moreover, with the reversal of signs in $\alpha = \alpha_c(T - T_c)$ near the critical temperature T_c, the phase transition from the normal state to superconducting state became transparent.

The Ginzburg-Landau model produced many interesting and valid results. One important consequence is its prediction of the existence of two characteristic lengths in a superconductor. The first is a coherence length ξ which describes the scale of spatial variation of the order parameter and the second is the penetration depth which denotes the depth to which an external magnetic field can penetrate the superconductor. For most materials, we have λ to be in the range $450 - 900A$ while ξ can range from $1000 - 2000A$ in conventional superconductors to $2 - 3A$ in high T_c materials. With the Ginzburg-Landau model, the Meissner effects and the flux quantization phenomenon can all be readily explained.

Through a proper nondimensionalization, the free energy functional of a superconducting sample occupying the domain Ω, subject to a constant applied magnetic field H, may be specified as follows:

$$\mathcal{G}(\psi, \mathbf{A}) = \int_\Omega \left(\frac{1}{2} \left| \left(\frac{i}{\kappa}\nabla + \mathbf{A} \right) \psi \right|^2 + \frac{1}{4}(1 - |\psi|^2)^2 + \frac{1}{2}|\mathrm{curl}\,\mathbf{A} - H|^2 \right) d\Omega,$$

where the Ginzburg-Landau parameter κ, a material constant, is defined as the ratio of the penetration depth λ and the coherence length ξ.

3.2. *The equilibrium Ginzburg-Landau models*

The minimizers of the functional satisfies the basic Ginzburg-Landau equations:

$$\left(\frac{i}{\kappa}\nabla + \mathbf{A} \right)^2 \psi - \psi + |\psi|^2\psi = 0, \tag{3.1}$$

$$\mathbf{curl}\,\mathrm{curl}\,\mathbf{A} + \frac{i}{2\kappa}(\psi^*\nabla\psi - \psi\nabla\psi^*) + |\psi|^2\mathbf{A} = 0. \tag{3.2}$$

ψ^* is the complex conjugate of ψ.

On the boundary $\Gamma = \partial\Omega$, the natural boundary conditions are

$$(\frac{i}{\kappa}\nabla\psi + \mathbf{A}\psi) \cdot \mathbf{n} = 0 \,, \tag{3.3}$$

$$\operatorname{curl}\mathbf{A} = H \,. \tag{3.4}$$

The above equations and boundary conditions are often called the equilibrium Ginzburg-Landau models, or simply the G-L models. Note that for the general three dimensional problems, the interactions between the fields inside the superconducting sample and the external field are important. Various measurement of critical fields are also affected by the geometric shape of the sample [113, 118, 121, 135]. To apply the Ginzburg-Landau theory in such situations, a coupled system of equations must be solved in both the sample and its exterior. Energetically speaking, in the case where Ω is a bounded domain in 3d, it may be necessary to reformulate the free energy as follows:

$$\mathcal{G}(\psi, \mathbf{A}) = \int_\Omega \left(\frac{1}{2}\left|\left(\frac{i}{\kappa}\nabla + \mathbf{A}\right)\psi\right|^2 + \frac{1}{4}(1 - |\psi|^2)^2 \right) d\Omega$$
$$+ \frac{1}{2}\int_{R^3} |\operatorname{curl}\mathbf{A} - H|^2 \, dR^3 \,.$$

This set of nonlinear Ginzburg-Landau equations in the (bounded) interior of Ω is coupled with the linear Maxwell equations in the (unbounded) exterior $\Omega^e = R^3 \setminus \bar{\Omega}$ with far field conditions at infinity and interface conditions on $\Gamma = \partial\Omega$. If we only consider the interior problem (typically valid for a two dimensional cross-section of a long three dimensional cylinder with the applied magnetic field perpendicular to the cross-section), then on Γ, it is customary to use the natural boundary conditions.

Note that for the zero applied magnetic field, the pure superconducting state $|\psi| = 1$ and $\operatorname{curl}\mathbf{A} = 0$ is a solution while for any applied fields, there is always a normal state solution given by $\psi = 0$ and $\operatorname{curl}\mathbf{A} = H$. Moreover, for very large applied fields, it is known that the normal solution is the only solution while for smaller fields, there are other solutions which display different types of behaviors. In particular, for large values of κ, it is known that there are solutions containing isolated vortices which are zeros of the order parameter ψ with nonzero topological degrees.

In the literature, more general boundary conditions have also been studied. For instance, to study the proximity effect, one may use

$$(\frac{i}{\kappa}\nabla\psi + \mathbf{A}\psi) \cdot \mathbf{n} = -i\gamma\psi \,.$$

We refer to [25, 65, 148] for more discussions.

3.3. *Time dependent Ginzburg-Landau equations*

Let η_1 and η_2 be given relaxation parameters, the conventional time-dependent Ginzburg-Landau (TDGL) model is given by

$$\eta_1\left(\frac{\partial\psi}{\partial t} + i\kappa\bar{\Phi}\psi\right) + \left(\frac{i}{\kappa}\nabla + \mathbf{A}\right)^2\psi - \psi + |\psi|^2\psi = 0, \quad (3.5)$$

$$\eta_2\left(\frac{\partial\mathbf{A}}{\partial t} + \nabla\bar{\Phi}\right) + \mathbf{curl}\,\mathrm{curl}\,\mathbf{A} + \frac{i}{2\kappa}(\psi^*\nabla\psi - \psi\nabla\psi^*) + |\psi|^2\mathbf{A} = 0, \quad (3.6)$$

with boundary conditions

$$\left(\frac{i}{\kappa}\nabla\psi + \mathbf{A}\psi\right)\cdot\mathbf{n} = 0, \tag{3.7}$$

$$\mathrm{curl}\,\mathbf{A} = H, \tag{3.8}$$

$$\eta_2\left(\frac{\partial\mathbf{A}}{\partial t} + \nabla\bar{\Phi}\right)\cdot\mathbf{n} = \mathbf{J}\cdot\mathbf{n} \tag{3.9}$$

where $\mathbf{J}$ is an applied current. The initial conditions are:

$$\psi(\mathbf{x}, 0) = \psi_0(\mathbf{x}) \quad \text{and} \quad \mathbf{A}(\mathbf{x}, 0) = \mathbf{A}_0(\mathbf{x}) \quad \text{in} \quad \Omega.$$

Note that we only focus on the interior problem for now and we assume that η_1 and η_2 are both positive real numbers (dynamics with complex valued η_1 have been studied [48]).

First, it is convenient to introduce an auxiliary variable $\Phi_a(\mathbf{x}, t) = (\mathbf{J} \cdot \mathbf{x})/\eta_2$ and define $\Phi = \bar{\Phi} - \Phi_a$. The triple $(\psi, \mathbf{A}, \Phi)$ is often used as the primary variables for the TDGL equations which are related to the energy functional by

$$\eta_1\left(\frac{\partial\psi}{\partial t} + i\kappa\bar{\Phi}\psi\right) = -\frac{\partial\mathcal{G}}{\partial\psi}(\psi, \mathbf{A}), \tag{3.10}$$

$$\eta_2\left(\frac{\partial\mathbf{A}}{\partial t} + \nabla\Phi\right) = -\frac{\partial\mathcal{G}}{\partial\mathbf{A}}(\psi, \mathbf{A}). \tag{3.11}$$

Another interpretation of the equation (3.11) is to relate with the Maxwell equation:

$$\mathbf{curl}\,H = \mathbf{curl}\,\mathrm{curl}\,\mathbf{A} = \mathbf{J}_{total}$$

and the total current $\mathbf{J}_{total}$ is made of the normal current $\mathbf{J}_n$ given by the Ohm's Law:

$$\mathbf{J}_n = \eta_2 E = -\eta_2\left(\frac{\partial\mathbf{A}}{\partial t} + \nabla\Phi\right)$$

with E being the electric field, and the supercurrent $\mathbf{J}_s$ which is given by

$$\mathbf{J}_s = |\psi|^2 \mathbf{v}_s = |\psi|^2 (\frac{1}{\kappa}\nabla\phi - \mathbf{A})$$

with $\mathbf{v}_s$ being the particle (Cooper-pairs) velocity and ϕ being the phase variable of the complex order parameter ψ.

3.4. *Gauge invariance and some basic theory*

Both the G-L and the TDGL equations enjoy the *gauge invariance* property. For instance, in the static G-L model, the standard gauge transformation is given by:

$$(\psi, \mathbf{A}) \rightarrow (\psi\exp(i\kappa\eta),\ \mathbf{A} + \nabla\eta)\ . \tag{3.12}$$

for any smoothly defined real-valued scalar function $\eta = \eta(\mathbf{x})$. The energy functional and all the physically observable quantities are obviously invariant under the gauge transformation. In the TDGL, the transformation is extended to:

$$(\psi, \mathbf{A}, \Phi) \rightarrow (\psi\exp(i\kappa\eta),\ \mathbf{A} + \nabla\eta,\ \Phi - \partial_t\eta) \tag{3.13}$$

for any smoothly defined real-valued scalar function $\eta = \eta(\mathbf{x}, t)$, see [52, 113] for more detailed discussions.

Numerical minimization of the free energy functional is made difficult due to the gauge invariance. However, nice remedy has been developed to avoid such pitfalls [65]. Define:

$$\mathcal{F}(\psi, \mathbf{A}) = \mathcal{G}(\psi, \mathbf{A}) + \int_\Omega |\mathrm{div}\,\mathbf{A}|^2 dx\ . \tag{3.14}$$

By choosing proper gauge transformation, it can be shown that:

Theorem 3.1: *The following minimization problems are equivalent:*

$$\left\{ \begin{array}{l} \textbf{\textit{Min}}\,\mathcal{G}(\psi, \mathbf{A}) \\ s.t.\ \psi \in \mathcal{H}^1(\Omega), \\ \mathbf{A} \in \mathbf{H}^1(\Omega). \end{array} \right. \Leftrightarrow \left\{ \begin{array}{l} \textbf{\textit{Min}}\,\mathcal{G}(\psi, \mathbf{A}) \\ s.t.\ \psi \in \mathcal{H}^1(\Omega), \\ \mathbf{A} \in \mathbf{H}^1_n(div,\Omega). \end{array} \right. \Leftrightarrow \left\{ \begin{array}{l} \textbf{\textit{Min}}\,\mathcal{F}(\psi, \mathbf{A}) \\ s.t.\ \psi \in \mathcal{H}^1(\Omega), \\ \mathbf{A} \in \mathbf{H}^1_n(\Omega). \end{array} \right.$$

where $\mathbf{H}^1_n(\Omega)$ *is a subspace of* $\mathbf{H}^1(\Omega)$ *with vanishing normal component on the boundary while* $\mathbf{H}^1_n(div,\Omega)$ *is a subspace of divergence free functions in* $\mathbf{H}^1_n(\Omega)$.

With the equivalent formulation, one can simply enforce the Coulomb gauge implicitly by solving for the variational problems with respect to $\mathcal{F}$

in $\mathcal{H}^1(\Omega) \times \mathbf{H}_n^1(\Omega)$. The penalty term $\|\operatorname{div}\mathbf{A}\|^2$ serves as a null Lagrangian which vanishes at the energy minimizer.

For the equilibrium G-L equation, one can prove the following maximum principle for the complex order parameter ψ [65]:

Theorem 3.2: *Let $(\psi, \mathbf{A})$ be the solution of the G-L equations, then*

$$|\psi(\mathbf{x})| \leq 1$$

in Ω.

In other words, the magnitude of the order parameter is bounded above by 1, the value of the superconducting state in our nondimensionalized setting. The same conclusion holds also for the solution of the TDGL if the initial condition satisfies the same property [52].

The TDGL equations (3.10-3.11) may be viewed as the gauge invariant gradient flow of the free energy. The well-posedness of the nonlinear TDGL system was first reported in 1992 at the first world congress of nonlinear analysts [50, 52] through a careful examination of the various gauge choices. Indeed, the popular gauge choices include [52]:

- the London gauge: as specified by $\operatorname{div}\mathbf{A} = 0$ and appropriate boundary conditions, in the steady state case, this is also called the Coulomb gauge;
- the zero electric potential gauge: $\Phi = 0$;
- the Lorentz gauge: as specified by a coupled potential constraint $\operatorname{div}\mathbf{A} = -\Phi$.

In the case of zero applied current (voltage), it it easy to see that with the choice of the zero electric potential gauge, we have

Theorem 3.3: *For the solution of TDGL equations with $\Phi = \bar{\Phi} = 0$, we have*

$$\eta_1\left\|\frac{\partial\psi}{\partial t}\right\|^2 + \eta_2\left\|\frac{\partial\mathbf{A}}{\partial t}\right\|^2 + \frac{1}{2}\frac{d\mathcal{G}}{dt}(\psi, \mathbf{A}) = 0 \, .$$

The energy laws like the above lead immediately to the decay of energy in time [50, 52] and the bounds on the time derivatives.

As the energy only gives the control on the curl $\mathbf{A}$, a regularization was introduced in [52]:

$$\eta_1\left(\frac{\partial\psi}{\partial t} + i\kappa\bar{\Phi}\psi\right) + \left(\frac{i}{\kappa}\nabla + \mathbf{A}\right)^2\psi - \psi + |\psi|^2\psi = 0\,, \qquad (3.15)$$

$$\eta_2\left(\frac{\partial\mathbf{A}}{\partial t} + \nabla\bar{\Phi}\right) + \mathbf{curl}\,\mathrm{curl}\,\mathbf{A} - \epsilon\mathrm{div}\,\mathrm{div}\,\mathbf{A}$$

$$+ \frac{i}{2\kappa}(\psi^*\nabla\psi - \psi\nabla\psi^*) + |\psi|^2\mathbf{A} = 0\,. \qquad (3.16)$$

It follows that for the modified system, we have

$$\eta_1\|\frac{\partial\psi}{\partial t}\|^2 + \eta_2\|\frac{\partial\mathbf{A}}{\partial t}\|^2 + \frac{1}{2}\frac{d}{dt}(\mathcal{G}(\psi,\mathbf{A}) + \epsilon\|\mathrm{div}\,\mathbf{A}\|^2) = 0\,.$$

The well-posedness of the regularized problem thus follows. By taking the divergence of the second equation in the regularization, one may then derive additional uniform estimates on the div $\mathbf{A}$ with respect to ϵ, which would allow us to pass to the limit as $\epsilon \to 0$. Hence, we get the existence and uniqueness of strong solutions of the original system without the regularization. Sharper results based on some better energy estimates and the long time solution behavior have later been studied, for instance, in [113, 147] and [85]. In particular, by utilizing the gauge invariance, it has been shown that

Lemma 3.4: *Let* $(\psi, \mathbf{A})$ *satisfy* $curl\,\mathbf{A} \times \mathbf{n} = \mathbf{H}_0 \times \mathbf{n}$, $\mathbf{A} \cdot \mathbf{n} = 0$ *and* $\nabla\psi \cdot \mathbf{n} = 0$ *on* $\partial\Omega$. *Let* $(\psi_*, \mathbf{A}_*)$ *be a steady state solution of the G-L equations in the gauge* $\mathrm{div}\,\mathbf{A}_* = 0$ *in* Ω *and* $\mathbf{A}_* \cdot \mathbf{n} = 0$ *on* $\partial\Omega$, *then there exist constants* $\theta_0, \sigma_0 \in (0, 1)$ *such that*

$$|\mathcal{G}(\psi, \mathbf{A}) - \mathcal{G}(\psi_*, \mathbf{A}_*)|^{\theta_0} \leq \|grad\,\mathcal{G}(\psi, \mathbf{A})\|_{L^2(\Omega)}$$

for any $\|(\psi, \mathbf{A}) - (\psi_*, \mathbf{A}_*)\|_{C^2(\Omega)} \leq \sigma_0$.

The above lemma, coupled with the theory of L. Simon, leads to the conclusion that, if there is no applied current, then the time dependent solutions of the TDGL converges to the steady state solutions as $t \to \infty$ [113].

Naturally, the most interesting dynamics is when there is an applied current. More detailed discussions along this direction will be given later.

4. Numerical algorithms for Ginzburg-Landau models

The development of approximation methods of the Ginzburg-Landau model goes back to the 1950s shortly after the inception of the model [89]. Particularly notable works include the seminal paper by Abrikosov [1] on the vortex state in type-II superconductors based on the linearization of G-L equations near the upper critical field.

The systematic studies of the G-L models from the numerical analysis point of view, to our knowledge, have not been seriously developed until the publication of [65]. In [65], both rigorous mathematical theory on the well-posedness of the equilibrium G-L models and their physical background were presented, along with the systematic development of finite element approximation methods. The work in [65] was partly motivated by [2, 47] of casting the equilibrium models into a variational framework. Extensions to the dynamic models, i.e., the time-dependent Ginzburg-Landau (TDGL) equations were subsequently made [52, 51]. Since then, many other works have appeared in the literature, including the development of different types of numerical approximations of Ginzburg-Landau type models, their rigorous theoretical analysis as well as extensive simulations such as finite difference [2, 37, 42, 47, 54, 82, 87, 91, 93, 105], finite element [3, 32, 33, 35, 65, 66, 67, 96, 151, 154] and finite volume methods [68, 69, 71] for spatial discretizations and different time-stepping methods [4, 11, 51, 122, 123]. By now, almost all aspects of modern numerical analysis have been utilized by people working on the numerical solutions of the G-L models, ranging from the design and applications of various discretization methods and fast algorithms, domain decomposition and parallelization techniques, and adaptive computation strategies. Let us add that there are also mis-conceptions related to the computation of vortices based on the G-L models. At the end of this lecture notes, we will point out that the numerical resolution of large number of vortices within the G-L framework is a computational challenge: the level of difficulty is similar to that of resolving the high frequency oscillations in the Helmholtz equations, and the quantization effect is a signature and yet peculiar phenomenon that leads to dimension mismatches in the resolution needs. For a recent review, we refer to [59].

Here, we first discuss the spatial discretization schemes, then we also discuss various time-stepping schemes for time dependent models. As there have been a large amount of works on the numerical simulations of the G-L models in the last twenty years, we make no attempt to provide a comprehensive survey on all existing works on the subject due to limited space. In particular, our review of the works appeared in the vast physics literature is very much limited to those that have also received much attention in the numerical analysis community or have been examined more rigorously in the mathematics literature.

4.1. *Finite element approximations*

The basic theory of conforming finite element approximations for the steady state G-L equations in a bounded domain has been presented in [65]. Let us choose a pair of conforming finite element spaces $\mathcal{V}_k^h \times := \mathbf{V}_k^h \subset \mathcal{H}^1(\Omega) \times \mathbf{H}_n^1(\Omega)$ where h being a mesh parameter, and assume that they satisfy the approximation properties:

$$\inf_{g^h \in \mathcal{V}_k^h} \|\psi - g^h\|_1 \le ch^r \|\psi\|_{r+1}$$

$$\inf_{\mathbf{B^h} \in \mathbf{V_k^h}} \|\mathbf{A} - \mathbf{B^h}\|_1 \le \mathbf{ch^r} \|\mathbf{A}\|_{r+1}$$

for functions ψ and $\mathbf{A}$ of sufficient regularity.

Then, the discrete Galerkin finite element approximation can be formulated as follows [65]:

$$\mathbf{Min} \quad \mathcal{F}(\psi^h, \mathbf{A}^h)$$
$$\text{s.t. } (\psi^h, \mathbf{A}^h) \in \mathcal{V}_k^h \times \mathbf{V}_k^h \,.$$

It has been shown that the above problems generate a sequence of convergent approximate solutions as $h \to 0$ under the minimal regularity condition. Moreover, optimal order of error estimates of the following type have been derived for nonsingular solution branches:

$$\|\psi - \psi^h\|_1 + \|\mathbf{A} - \mathbf{A}^h\|_1 \le ch^k \left\{ \|\psi\|_{k+1} + \|\mathbf{A}\|_{k+1} \right\} \,.$$

The techniques used in [65] follow the idea of obtaining similar error estimates for the associated decoupled linear problems:

$$\mathbf{Min} \quad \left\{ \|\nabla\psi^h\|^2 + \|\psi^h\|^2 + \|\operatorname{curl}\mathbf{A}^h\|^2 + \|\operatorname{div}\mathbf{A}^h\|^2 + \|\mathbf{A}^h\|^2 \right\}$$
$$\text{s.t. } (\psi^h, \mathbf{A}^h) \in \mathcal{V}_k^h \times \mathbf{V}_k^h \,.$$

Then the nonlinear problems are shown to be the compact perturbations in the appropriate Sobolev spaces, so that on the nonsingular solution branches, the optimal order error estimates also hold.

It is worthwhile to note that no inf-sup condition is required for the finite element spaces. The gauge condition is not strictly imposed for finite mesh parameter h, but it is shown to be valid in the limit as $h \to 0$.

The finite element methods have later been generalized to other related models, such as the d-wave G-L equations [151], optimal control of G-L models [92] and the Lawrence-Doniach models for layered superconductors [25, 62, 96]. In [66, 67], various numerical simulations have been conducted based on the finite element approximations.

The application of finite element method to TDGL has been considered in [51]. Besides the basic convergence results for semi-discrete scheme, both first order and second in time discretization schemes have also been presented. Let Δt_n be the step size and $(\psi_n^h, \mathbf{A}_n^h)$ the numerical solution at time t_n. When the applied current is absent, the first order backward Euler scheme can be given a variational form: find $(\psi_{n+1}^h, \mathbf{A}_{n+1}^h)$ such that it solves the problem

$$\mathbf{Min}\, \mathcal{G}_{\Delta t_n}(\psi, \mathbf{A}) = \mathcal{G}(\psi, \mathbf{A}) + \frac{1}{\Delta t_n} \int_\Omega \left[\eta_1 |\psi - \psi_n^h|^2 + \eta_2 |\mathbf{A} - \mathbf{A}_n^h|^2 \right] d\Omega$$

s.t. $(\psi, \mathbf{A}) \in \mathcal{V}_k^h \times \mathbf{V}_k^h$.

Consequently, we have the energy decreasing property in the discrete approximation:

$$\mathcal{G}(\psi_{n+1}^h, \mathbf{A}_{n+1}^h) \leq \mathcal{G}(\psi_n^h, \mathbf{A}_n^h) .$$

It turns out that a second order in time scheme can also be similarly formulated: first find $(\psi_*^h, \mathbf{A}_*^h)$ that minimizes

$$\mathcal{G}(\psi, \mathbf{A}) + \frac{1}{2\Delta t_n} \int_\Omega \left[\eta_1 |\psi - \psi_n^h|^2 + \eta_2 |\mathbf{A} - \mathbf{A}_n^h|^2 \right] d\Omega$$

then we let

$$(\psi_{n+1}^h, \mathbf{A}_{n+1}^h) = 2(\psi_*^h, \mathbf{A}_*^h) - (\psi_n^h, \mathbf{A}_n^h).$$

We note that the derivation of error estimates of the fully discrete scheme was, however, not rigorously provided there. By using a mixed formulation, [32] has presented a more complete theory for the approximation of the TDGL along with optimal order error estimates in two space dimension. Later, [35] considered approximations to a related optimal control problem. Generalizations to the time-dependent Lawrence-Doniach model have been presented in [96].

4.2. *Finite difference approximations*

The finite difference approximations of the G-L models have been the most widely used approach. Though conventional difference schemes have been studied in [82, 103], much of the focus has been on the gauge invariant difference approximation. The motivation has come from the fact that the underlying physical model enjoys the gauge invariance property. In [2], a gauge invariant finite difference scheme was proposed for the steady-state G-L equations on a uniform rectangular grid, in the spirit of discrete gauge field theory. Many subsequent works have followed up on such an approach via an introduction of the so called link and bond variables, and various extensions have also been made [37, 38, 82, 87, 91, 93, 104, 105]. For the approximation of the magnetic vector potential, such an approach is naturally related to the idea of staggered grid (Marker-and-Cell) used in computational electro-magnetics and fluid dynamics.

For simplicity, let us consider the two dimensional setting with a uniform rectangular mesh of grid size h. Following the notation in [54], the discrete gauge invariant energy functional is given by:

$$\mathcal{G}^h(\vec{\psi}, \vec{A}) = \frac{1}{2} \sum_{jk} \frac{1}{\kappa^2} |\psi_k \exp(-i\kappa a_{jk} h) - \psi_j|^2$$

$$+ \sum_j \frac{h^2}{4}(1 - |\psi_j|^2)^2 + \frac{1}{2} \sum_{jklm} (a_{jk} + a_{kl} + a_{lm} + a_{mj} - Hh)^2 \quad (4.1)$$

where the first sum is over all neighboring edges, the second over all vertices and the third over all square cells. $\vec{\psi} = \{\psi_j\}$ are the approximations of the order parameter at the cell vertices and $\vec{A} = \{a_{jk}\}$ are the approximations of the signed tangential component of the magnetic vector potential at the mid-point of the cell edges. See Figure4 for an illustration.

The gauge invariant backward Euler scheme is given by [54]:

$$\eta_1 \frac{\psi_j^n - \psi_j^{n-1} \exp\left(-i\kappa \bar{\Phi}_j^n \Delta t\right)}{\Delta t} = -\frac{1}{h^2} \frac{\partial \mathcal{G}^h}{\partial \psi_j^n}(\vec{\psi}^n, \vec{A}^n) ,$$

and

$$\eta_2 \left(\frac{a_{jk}^n - a_{jk}^{n-1}}{\Delta t} + \frac{\Phi_k^n - \Phi_j^n}{h} \right) = -\frac{1}{h^2} \frac{\partial \mathcal{G}^h}{\partial a_{jk}^n}(\vec{\psi}^n, \vec{A}^n))$$

for $n = 1, 2, ..., N = T/\Delta t$. Here, $\{\bar{\Phi}_j^n\}$ are the approximations of the scalar electric potential. Note that the equations at the nodes on the boundary may require slight modifications. The approximation preserves the variational structure at the discrete level:

Theorem 4.1: *The solution of gauge invariant difference approximation to the TDGL equations is the minimizer of*

$$\mathcal{G}^h(\vec{\psi}^n, \vec{A}^n) + \frac{\eta_1 h^2}{\Delta t_n} \sum_j |\psi_j^n - \psi_j^{n-1} \exp\left(-i\kappa \bar{\Phi}_j^n \Delta t\right)|^2$$

$$+ \frac{\eta_2 h^2}{\Delta t_n} \sum_{jk} |a_{jk}^n - a_{jk}^{n-1} + \frac{\Phi_k^n - \Phi_j^n}{h}|^2 .$$

A complete and rigorous analysis for such a discretization has been provided in [54], including the proof of the discrete maximum principle. The optimal order error estimates have been obtained through some discrete comparison results. At $t_n = n\Delta t$, let the components of $\vec{\varphi}^n$ be defined by

$$\varphi_j^n := \psi(\mathbf{x}_j, t_n)$$

and the components of $\vec{\underline{A}}^n$ be defined by

$$\alpha_{jk}^n := \int_0^1 \mathbf{A}((1-s)\mathbf{x}_j + s\mathbf{x}_k, t_n) \cdot \vec{\mathbf{t}}_{jk} \, ds \ .$$

Then,

Theorem 4.2: *Given the parameters* $\kappa, \eta_1, \eta_2, J, H$ *and the final time* T, *there exist some constant* $c > 0$, *independent of* $h, \Delta t$ *and* n, *such that for* $h, \Delta t$ *small and* $1 \le n \le N = T/\Delta t$, *we have*

$$\|\vec{\psi}^n - \vec{\varphi}^n\|_{U,2} + \|\vec{A}^n - \vec{\underline{A}}^n\|_{W,2} \le c(\Delta t + h^2)$$

where $\|\cdot\|_{U,2}$ *and* $\|\cdot\|_{W,2}$ *are some discrete norms as defined in* [54].

It remains to see if a general higher order in time fully discrete gauge invariant scheme can be developed for the TDGL equations. We note here also that, in practical numerical simulations, explicit or semi-implicit in time difference schemes have been mostly employed. In general, these schemes are only conditionally stable at best.

4.3. *Finite volume approximations*

As more and more attention are being paid to the study of the effect of the sample geometry and topology on the superconductivity phenomena, methods based on unstructured grids become more competitive in such cases. Besides the finite element methods we have discussed, finite volume methods have also been developed which have the combined advantage of being able to work with an unstructured grid while preserving the discrete gauge invariance [68, 69, 71].

A standard extension to the staggered grid used in the gauge invariant difference approximation is the Voronoi-Delaunay pair. Given a set of distinct points $\{\mathbf{x}_j\}_{j=1}^n \subset R^2$, we can define for each point $\mathbf{x}_j$, $j = 1, \ldots, n$, the corresponding Voronoi region V_j, $j = 1, \ldots, n$, by

$$V_j = \left\{ \mathbf{y} \in R^2 \mid |\mathbf{x}_j - \mathbf{y}| < |\mathbf{x}_k - \mathbf{y}| \text{ for } k = 1, \cdots, n \text{ and } k \neq j \right\} .$$

We refer to $\{V_j\}_{j=1}^n$ as the *Voronoi tessellation* corresponding to the generators $\{\mathbf{x}_j\}_{j=1}^n$. The dual tessellation to a Voronoi tessellation consisting of triangles is referred to as a *Delaunay triangulation*. Given a discrete vector field $\vec{A}$ tangentially defined at the mid-points of the triangle edges of a Delaunay triangulation, it is easily seen that such vectors are normal to the edges of the Voronoi regions. Thus, discrete calculus can be defined for the curl operator on the Delaunay triangles and for the div operator on the Voronoi cells [128].

Then, the central idea is to find suitable discretization of the G-L energy functional is to construct a gauge invariant approximation to $|\nabla\psi - i\kappa\mathbf{A}\psi|$. It was noticed in [71] that one may first project the vector $\mathbf{v} = \nabla\psi - i\kappa\mathbf{A}\psi$ into the tangential components along the triangle edges and obtain the following simple identity:

$$|\tau_{ijk}||\mathbf{v}|^2 = \sum \cot\theta_i |v_{jk}|^2$$

with $v_{jk} = \mathbf{v} \cdot (\mathbf{x}_j - \mathbf{x}_k)$ and the sum is over all three edges and θ_i is the opposite angle. Notice that if we let θ_{i1} and θ_{i2} be the two opposing angles corresponding to the same edge $\mathbf{x}_j\mathbf{x}_k$, then

$$\cot\theta_{i1} + \cot\theta_{i2} = \frac{|\Gamma_{jk}|}{|\mathbf{x}_j - \mathbf{x}_k|}$$

where Γ_{jk} is the common edge between the two adjacent Voronoi regions V_{i1} and V_{i2}. Now, $|v_{jk}|$ can be approximated by

$$|\psi_k \exp(-i\kappa a_{jk}|\mathbf{x}_j - \mathbf{x}_k|) - \psi_j| .$$

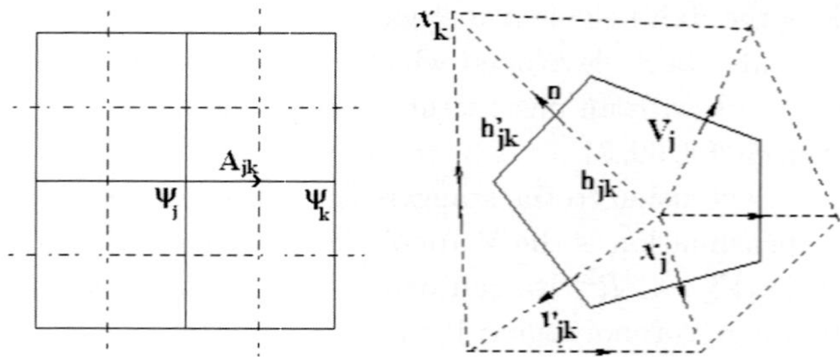

Fig. 4.   A staggered uniform grid and a Voronoi-Delaunay mesh.

Thus, the Ginzburg-Landau functional is discretized as follows [71]:

$$
\mathcal{G}^h(\vec{\psi}^h, \vec{A}^h) = \sum_{j=1}^{n} \frac{1}{4|V_j|} (1 - |\psi_j|^2)^2
$$

$$
+ \sum_{j=1}^{n} \left\{ \sum_{k \in \chi_j} \frac{|\Gamma_{jk}|}{2\kappa^2 |\mathbf{x}_j - \mathbf{x}_k|} |\psi_k \exp(-i\kappa a_{jk} |\mathbf{x}_j - \mathbf{x}_k|) - \psi_j|^2 \right\} \quad (4.2)
$$

$$
+ \sum_{\tau_{jkl}} \frac{1}{2\tau_{jkl}} (a_{jk} h_{jk} + a_{kl} h_{kl} + a_{lj} h_{lj} - H \tau_{jkl})^2
$$

where for any j, the index set χ_j denotes the indices of all vertices which are adjacent to the vertex $\mathbf{x}_j$.

The discrete gauge invariance is understood in the sense that

$$
\mathcal{G}^h(\vec{\psi}^h, \vec{A}^h) = \mathcal{G}^h(T_\phi^h(\vec{\psi}^h, \vec{A}^h)) \,,
$$

where the transformation T^h is defined by the map:

$$
\begin{cases}
\psi_j \to \psi_j e^{i\kappa\phi_j} \\
a_{jk} \to a_{jk} + \frac{\phi_k - \phi_j}{|\mathbf{x}_j - \mathbf{x}_k|} \,,
\end{cases}
$$

corresponding to any scale field $\vec{\phi}^h$.

The gauge invariant difference approximation on a rectangular grid discussed earlier is in fact a special case of the above finite volume scheme. This can be seen by making the equivalence of the rectangular cells with the Voronoi cells and the equivalence of the dual cells with pairs of right Delaunay triangles sharing a common edge opposing the right angles.

Similar to the technique introduced in [65] and in the the finite difference setting, a modified functional can be defined to enforce the gauge choice implicitly. Let us define

$$
\mathcal{F}^h(\vec{\psi}^h, \vec{A}^h) = \mathcal{G}^h(\vec{\psi}^h, \vec{A}^h) + \sum_j \frac{1}{|V_j|} \left(\sum_{k \in \chi_j} a_{kj} |\Gamma_{kj}| \right)^2 \,.
$$

Then, it is easy to show that the minimizer of $\mathcal{F}^h$ is also a minimizer of $\mathcal{G}^h$ and it is also divergence free in the discrete sense:

$$
\sum_{k \in \chi_j} a_{kj} |\Gamma_{kj}| = 0 \,, \quad \forall j \,.
$$

Moreover, it has been shown that the minimizer of $\mathcal{F}^h$ satisfies the discrete maximum principle:

Theorem 4.3: *Let $(\vec{\psi}^h, \vec{A}^h)$ be the minimizer of the discrete G-L functional (4.2), then*

$$|\psi_j| \leq 1$$

for all j.

This, coupled with suitable energy estimates, leads to the convergence of the discrete approximations as the mesh size goes to zero, see [71] for details.

In [68, 69], such finite volume scheme was extended to solve a reduced set of TDGL models defined on a thin spherical shell. In addition to the basic convergence properties, a novel feature of the discussions in [68, 69] is the consideration of a special Voronoi-Delaunay pair, namely, the spherical centroidal Voronoi tessellations and the corresponding Delaunay triangulations [63]. The SCVTs are naturally extensions of the centroidal Voronoi tessellations in Euclidean spaces [60]. For the standard CVTs in the Euclidean spaces, the generators of the Voronoi regions all coincide with the mass centers of the corresponding regions. CVTs may be viewed as the optimal tessellation SCVTs, on the other hand, are those Voronoi tessellations defined by the standard Euclidean distance whose generators are the constrained centers of mass. Note that for points on the sphere, the Voronoi tessellations defined by the standard Euclidean distance are the same as the ones defined via the geodesic distance. It has been shown that by using the SCVTs, the discrete approximations exhibit a higher order convergence comparing with the conventional Voronoi-Delaunay pair. Moreover, super-convergent gradient recovery schemes for the SCVT meshes are also developed to obtain more accurate estimation of the induced magnetic field and the super-current.

Detailed numerical simulations have also been performed in [68, 69] using the SCVT based gauge invariant finite volume scheme. Both static vortex configurations and vortex dynamics under applied current have been studied.

4.4. *Artificial boundary conditions*

For a full three dimensional simulation of the G-L model, taking into account the effect of the induced magnetic field of the superconducting sample on the field exterior to the sample, the numerical solution of the Ginzburg-Landau equations in the superconducting sample needs to be solved in conjunction with the solution of the Maxwell equations in the exterior.

To overcome the unboundedness of the exterior domain, approximations have to be introduced or alternative formulations have to be considered. The crudest but often very effective approximation is simply to ignore the effect in the exterior completely. Such approximations are particularly valid for high-kappa materials (in such cases, some reduced G-L models have been proposed and used in numerical simulations of vortex lines [27, 41, 61].

As an alternative, a rigorous theory based on artificial boundary conditions has been presented in [74] to transform the computation domain to a finite ball enclosing the superconducting sample.

The key step is to note that the magnetic energy $\|\operatorname{curl}\mathbf{A}\|^2_{L^2(R^3)}$, coupled with the null-Lagrangian term $\|\operatorname{div}\mathbf{A}\|^2_{L^2(R^3)}$ used to enforce the Coulomb gauge, is equivalent to $\|\nabla\mathbf{A}\|^2_{L^2(R^3)}$ which can be decomposed as $\|\nabla\mathbf{A}\|^2_{L^2(B_r)} + \|\nabla\mathbf{A}\|^2_{L^2(B^e_r)}$. Here B_r denotes a ball of radius r and $B^e_r = R^3 \setminus B_r$ denotes the exterior.

For $\mathbf{A}$'s that are harmonic in B^e_r, we have

$$\|\nabla\mathbf{A}\|^2_{B^e_r} = \int_{S_r} \mathbf{A}\frac{\partial\mathbf{A}}{\partial\mathbf{n}}dS = \int_{S_r} \mathbf{A}J(\mathbf{A})dS$$

where S_r denotes the sphere of radius r, J is the Steklov-Poincare operator (or the Dirichlet to Neumann map). Since J can be explicitly expressed with the help of Legendre functions, various orders of approximations can be constructed.

Briefly, let the Legendre polynomial and Legendre function be given by

$$P^0_n(t) = P_n(t) = \frac{1}{2^n n!}\frac{d^n(t^2-1)^n}{dt^n}, \quad P^m_n(t) = (1-t^2)^{\frac{m}{2}}\frac{d^m}{dt^m}P_n(t) ,$$

and let

$$G(\gamma) = -\frac{1}{4\pi R^3} - \sum_{n=1}^{\infty}\frac{(n+1)(2n+1)}{4\pi R^3}P_n(\cos\gamma) ,$$

for γ defined by $\cos\gamma = \cos\theta\cos\theta' + \sin\theta\sin\theta'\cos(\varphi-\varphi')$ for a pair of points $x = (r,\theta,\varphi)$ and $y = (r,\theta',\varphi')$. Then, with $\mathbf{A}_0$ satisfying $\operatorname{curl}\mathbf{A}_0 = H$, we have the following equivalent form of the energy functional [74]

$$\mathcal{F}(\psi,\mathbf{A}) = \int_\Omega \frac{1}{4}(1-|\psi|^2)^2 d\Omega + \frac{1}{2}\int_\Omega \left|\left(\frac{i}{\kappa}\nabla + \mathbf{A} + \mathbf{A}_0\right)\psi\right|^2 d\Omega$$

$$+ \frac{1}{2}\int_{B_r} |\nabla\mathbf{A}|^2 dx + \int_{S_r}\int_{S_r} \mathbf{A}(x)\cdot G(\gamma)I\cdot\mathbf{A}(y)dSdS .$$

In a nut-shell, the exterior energy is now effectively transformed into an energy defined on the boundary of the ball.

Various approximations of the boundary energy can be made, for instance, using a finite element discretization in the interior of B_r, the following discrete problem has been considered in [74]:

$$\mathcal{F}(\psi^h, \mathbf{A}^h) = \frac{1}{2} \int_\Omega \left| \left(\frac{i}{\kappa} \nabla + \mathbf{A}^h + \mathbf{A}_0 \right) \psi^h \right|^2 d\Omega$$

$$+ \int_\Omega \frac{1}{4} (1 - |\psi^h|^2)^2 d\Omega + \frac{1}{2} \int_{B_r} |\nabla \mathbf{A}^h|^2 dx$$

$$+ \sum_{n=1}^N \frac{(n+1)(2n+1)}{4\pi r^3} \int_{S_r} \int_{S_r} \mathbf{A}^h(x)\, P_n(\cos\gamma)\, \mathbf{A}^h(y)\, dSdS \, .$$

Under the approximation assumptions on the finite element spaces made earlier, it has been proved that

Theorem 4.4: *For a smooth exact solution on a nonsingular solution branch, the approximate solutions satisfy the following error estimates:*

$$\|\psi - \psi^h\|_{1,\Omega} + \|\mathbf{A} - \mathbf{A}^h\|_{1,B_r}$$

$$\leq c_1 h^k (\|\psi\|_{k+1,\Omega} + \|\mathbf{A}\|_{k+1,B_r}) + c_2(\Omega, r) \left(\frac{diam(\Omega)}{r} \right)^N \|\mathbf{A}\|_{1,B_r} \, ,$$

for some constants c_1 and $c_2 = c_2(\Omega, r)$ independent of h and N.

It has been seen in numerical computation that in practice with $N = 6$, it is sufficient to choose the computational domain with r no more than twice of the diameter of Ω in order to obtain good accuracy of the approximation in the exterior domain [74]. For other works on the use of artificial boundary conditions, see [94, 153] for detailed references for works developed in China and also [90] for works in the West.

4.5. *More on time-discretization*

For TDGL models, while most of the rigorous mathematical analysis have been focused on the fully implicit in time discretizations, explicit marching schemes and semi-implicit marching schemes [152] have also been frequently used in numerical simulations due to their simplicity in implementation.

Theoretically, to make the time-marching more efficient, other useful ideas have also been considered in the literature. For example, a linearized Crank-Nicolson scheme has been considered in [122], similar to the semi-implicit approach. Analytical studies of an alternating marching scheme have also been made in [123] where for the order parameter and magnetic potential are solved in alternating steps and thus reducing the size of the implicit nonlinear system by half.

To effectively solving the nonlinear and linear systems employed in the implicit schemes, constructions of suitable preconditioners can be very helpful. In this regard, the Sobolev gradient methods studied in [125, 126] fit into such a framework. Essentially, the gradient flows in the H^{-1} space considered there are equivalent to employing the inverse of Laplace operator Δ^{-1} as the preconditioner for the standard gradient flow in the L^2 space.

For high values of κ, the Ginzburg-Landau parameter, the original G-L models can often be simplified. A particular simplification corresponding to high applied magnetic field has been considered in [27, 61]. The reduced equations are very much similar to the so-called Gross-Pitaevskii equations used to model the BEC superfluid [4]. In [11], some efficient splitting schemes for computing the ground state solutions of the BEC condensate based on the normalized gradient flow has been studied which may be readily applied to the solution of the reduced G-L models.

4.6. *Multi-level, adaptive and parallel algorithms*

The numerical simulations of the vortex state in type-II superconductors based on the G-L models become computationally challenging when there is a need to resolve a large number of vortices. More efficient implementations of the numerical schemes thus become necessary. There have been a lot of interesting attempts made along this direction. For example, multilevel finite element methods have been analyzed in [97] for a d-wave G-L model, following earlier works of [55, 151]. Posterior error estimates and adaptive finite element methods have been studied in [33, 98] with both rigorous analysis and numerical examples.

Parallelization is naturally another important avenue for greater computational efficiency. In [91], parallelized MPI-based implementation of the explicit finite difference discretization schemes has been presented along with many large scale simulations. Using a natural domain decomposition strategy, a number of parallel algorithms for the simulation of layered superconductors based on the Lawrence-Doniach model have been studied in [62]. The implementation has been made first using PVM, with an MPI version developed subsequently. Numerical results indicated significant speed-ups and good scalability. Moreover, interesting numerical simulations of three dimensional vortex tubes, their dynamics and pinning effect have been made there as well.

4.7. *Other methods*

For the time-discretization, there are also studies on various time-splitting methods. A class of splitting schemes for computing the ground state solutions of the BEC condensate based on the normalized gradient flow [11] can be readily generalized to the solution of G-L equations.

With the time-splitting scheme, it has become a popular approach to use spectral Galerkin or spectral collocation methods for the spatial discretization. For example, with a Fourier type spectral approximation, the nonlinear ODEs can be solved in real space while the linear PDEs can be solved in the spectral/frequency space [11, 12]. One can employ FFT to perform the transformation between the real space and frequency space representations.

5. Vortex configurations – Analysis and simulation

In type-II superconductors, when the applied magnetic field H increases to the vicinity of the so-called lower critical field H_{C1}, the solutions start to nucleate the quantized vortices, that is, the order parameter has isolated zeros with non-trivial topological degrees [113, 145, 148]. Such a vortex state persists until the applied field H reaches H_{C2}. The detailed analysis and simulations are presented here.

5.1. *Phase diagrams and equilibrium solution branch*

Without an applied magnetic field, the global minimum of the G-L energy functional is given by the superconducting state solution.

As the applied magnetic field increases, penetration of the magnetic field is first limited to the boundary region on the scale of the penetration depth. Once the field is larger than the lower-critical field H_{C1}, it starts to penetrate the material sample through the cores of the so called quantized vortices which are described as the isolated zeros of the order parameter with nontrivial topological degrees (with changes of phases in integer multiples of 2π (hence the flux quantization) . The phenomena of quantized vortices are well-known features of superconductivity. For type-II superconductors, the study of Abrikosov on the vortex lattices based on the G-L models has become a seminal work (Nobel Physics Prize in 2003) that exhibits the great predictive power of the G-L theory.

When H becomes larger than H_{C2}, the vortex structures (lattices) would generally be destroyed and the solution becomes one that corresponding to the normal state (see later discussion for the field H_{C3} corresponding to the surface superconductivity).

For a two dimensional geometry, the bifurcation diagrams of the G-L models are quite complex depending on the parameters involved. For large fields, the only solution is the normal solution while for smaller fields, the normal solution always exists but there are other solutions which display different types of behaviors, depending on the values of κ and d (the diameter of the sample domain). What distinguishes them are features like the existence (or the lack of existence) of vortex solutions, the global and local stability of solutions, and the hysteresis phenomena. We refer to [3, 113] for detailed calculations and some rigorous analysis.

5.2. *Vortex solutions*

In Fig.5, we present a few typical plots for the numerical solutions of the steady state G-L equations which include a surface plot of the magnitude of the order parameter, a surface plot of the induced magnetic field given by curl **A** and a vector plot of the superconducting current. The solution corresponds to a vortex profile with a single vortex at the center of a rectangular superconducting sample [3].

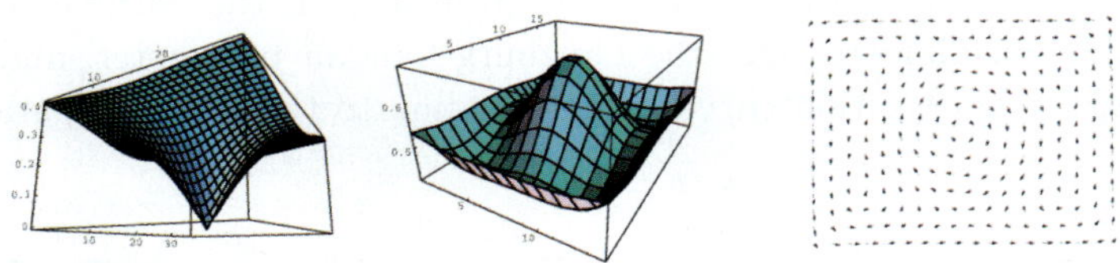

Fig. 5.   Order parameter, induced field and super-current in a 2d rectangular domain.

In general, a typical two dimensional vortex profile can be obtained by taking the special ansatz in R^2:

$$\psi(\mathbf{x}) = f(r)e^{in\theta} \ , \ \mathbf{A}(\mathbf{x}) = A(r)\vec{e}_\theta \ \text{ for } \ \mathbf{x} = (r\cos(\theta), r\sin(\theta))$$

where $\vec{e}_\theta$ is the unit vector in the azimuthal direction, the magnitude f and the phase variable θ are given by the solution of the equations for $r > 0$:

$$\frac{1}{r}\frac{d}{dr}\left(r\frac{d}{dr}f\right) - (\kappa A - \frac{n}{r})^2 f + (1 - f^2)f = 0$$

$$\frac{1}{r}\frac{d}{dr}\left(r\frac{d}{dr}(rA)\right) = f^2\left(A - \frac{n}{\kappa r}\right)$$

with A, f bounded near the origin and

$$A \to 0, \quad f \to 1, \quad \text{as} \quad r \to \infty.$$

The integer n gives the topological degree of the vortex. A solution to the above problem typically is illustrated in figure 6. The stability of such vortex solutions has been carried out both numerically and analytically in recent years [3, 65, 146].

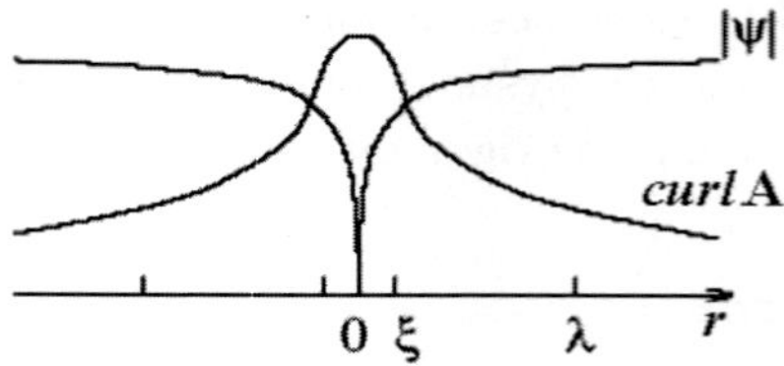

Fig. 6.   An illustration of the order parameter magnitude and the induced field.

More systematic analysis and simulations of the phase transitions, vortex nucleation and critical fields can be found in, for instance, [3, 9, 10, 17, 20, 99, 113, 135, 143, 144] and the references cited therein. Figure 7 contains a computed phase diagram given in[3] in the parameter space of the domain diameter and the Ginzburg-Landau parameter, and some associated magnetization curves corresponding to the different parameter regimes.

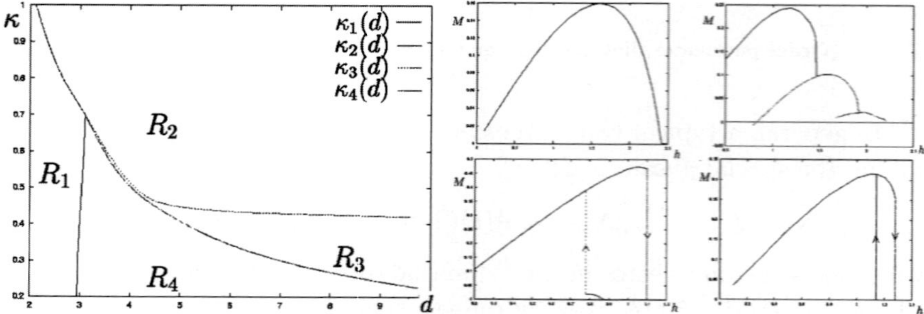

Fig. 7.   A phase diagram (left) with respect to the sample size (horizontal axis) and G-L parameter (vertical axis); associated magnetization curves in four parameter regimes (right).

Given a bounded domain, it has been well understood both physically and in more recent years mathematically that, for sufficiently large κ, there are vortex solutions which are the global minimizers of the free energy for a certain range of fields. The number of vortices depends on the strength of the applied field. The maximum number of vortices increases with d and κ which exhibits typical type-II behavior.

For very small fields, the global minimizer is the superconducting solution. As the field is increased, the superconducting solution loses its global stability and for even larger fields loses its local stability. Then the global minimizer starts to nucleate vortices.

The lower critical field H_{C1} may be characterized as the value of H for which the energy of the Meissner solution (superconducting solution) becomes equal to the energy of the solution with a single-vortex. The upper critical field H_{C2} is the field at which the densely packed vortex solutions disappear into the normal state.

The expected behavior of the minimizers of G-L energy functional are as follows: when $H = 0$, the trivial superconducting solution is a minimizer; when $H = O(H_{C1})$, a few vortices of degree one appear, and their positions are determined by the sample geometry; if $H >> H_{C1}$, the density of the vortices of the minimizers are expected to be uniform and proportional to H, and they repel one another through the Coulombian interaction; Then, if $H \rightarrow H_{C2}$, the density of vortices increases such that the vortices are separated by a distance shorter than the coherence length thus the solution becomes normal almost everywhere.

As the applied magnetic field gets bigger, the transition from normal to the last vortex solution is of second order since the vortex solution is stable, which occurs at a field that is usually called H_{C3} in the literature. The onset of superconductivity in decreasing fields (instability of normal solutions and computation of the fields of nucleation) has been analyzed in [20, 10]. Other works concerning the linearized problem include [118, 135]. Their works provide, as d and κ tends to ∞, an asymptotic estimate of H_{C3}, the field at which the normal solution bifurcates to a vortex solution. This is what is called surface superconductivity. A linearization of the Ginzburg-Landau equation has been done near the normal solution which is consistent with the work of Saint James and de Gennes [142]. In [135], the state of the material has been rigorously analyzed when the magnetic field is further decreased from the nucleation.

5.3. *A rigorous result on vortex nucleation near H_{C1}*

For high value of κ, the vortex core becomes small and thus allow the concentration of the vortex energy. In this limit, some rigorous mathematical results can be established. We outline one of the such result here.

Rescaling the G-L energy as

$$\mathcal{G}(\psi, \mathbf{A}) = \frac{1}{2} \int_\Omega \left(\frac{1}{2\epsilon^2}(1 - |\psi|^2)^2 + |(\nabla - i\mathbf{A})\,\psi|^2 + |\operatorname{curl} \mathbf{A} - H|^2 \right) d\Omega \,,$$

with $\epsilon = O(1/\kappa)$, and define a renormalized energy for $\mathbf{b} = \{\mathbf{b}_j\}_1^k$ by

$$W_\Omega(\mathbf{b}, H) = \frac{1}{2} H^2 C(\Omega) + 2\pi H \sum_{j=1}^{k} \zeta_1(\mathbf{b}_j) + \tilde{g}_\Omega(\mathbf{b}) \,. \tag{5.1}$$

where $C(\Omega)$ is a suitably defined constant, ζ_1 is given by

$$-\Delta^2 \zeta_1 + \Delta \zeta_1 = 0 \quad \text{in } \Omega \,,$$

$$\zeta_1 = 0 \quad \text{and} \quad \Delta \zeta_1 = 1 \quad \text{on } \partial\Omega \,,$$

and $\tilde{g}_\Omega(\mathbf{b})$ has the property

$$\tilde{g}_\Omega(\mathbf{b}) = \begin{cases} +\infty \,, & \mathbf{b}_i = \mathbf{b}_j \text{ for some } i \neq j \,, \\ -\infty \,, & \mathbf{b} \in \partial\Omega^k \end{cases}$$

and otherwise, it is a smooth function in Ω^k.

We now give one of the early rigorous derived result on the vortex solutions in the vicinity of the lower critical field H_{C1} [113]:

Theorem 5.1: *There exists some $H_0(\Omega)$ such that for $H \geq H_0(\Omega)$, $W_\Omega(\mathbf{b}, H)$ has a local minimum for some $\mathbf{b} \in \Omega^d$.*
Moreover, for small ϵ, there are solutions $(\psi^\epsilon, \mathbf{A}^\epsilon)$ to the full steady state G-L equations (which are critical points of the rescaled G-L energy in the above) with the gauge choice $\mathbf{A}_\epsilon = \nabla^\perp \zeta_\epsilon$ in Ω, $\mathbf{A}_\epsilon \cdot \mathbf{n} = 0$ on $\partial\Omega$, such that:

$$\psi_\epsilon(\mathbf{x}) \to \psi^*(\mathbf{x}) \quad in \quad C^{1,\alpha}_{\mathrm{loc}}\left(\overline{\Omega}/\{\mathbf{b}_1, \mathbf{b}_2, \ldots, \mathbf{b}_k\}\right)$$

$$\zeta_\epsilon(\mathbf{x}) \to \zeta^*(\mathbf{x}) \quad in \quad H^2(\Omega)$$

where

$$-\Delta^2 \zeta^* + \Delta \zeta^* = 2\pi \sum_{j=1}^{k} \delta(\mathbf{b}_j) \quad in \quad \Omega \,,$$

with

$$\zeta^* = 0 \quad and \quad \Delta \zeta^* = H_0 \quad on \quad \partial\Omega$$

and

$$\psi^*(\mathbf{x}) = \prod_{j=1}^{k} \frac{\mathbf{x} - \mathbf{b}_j}{|\mathbf{x} - \mathbf{b}_j|} e^{ih^*(\mathbf{x})}, \quad in \ \ \Omega \ .$$

In addition, if we write $\psi^*(\mathbf{x}) = e^{i\theta_b(\mathbf{x})+ih^*(\mathbf{x})} = e^{i\phi_b}$, *then* ϕ_b *is a multi-valued harmonic function on* $\Omega \setminus \{\mathbf{b}_1, \mathbf{b}_2, \ldots, \mathbf{b}_k\}$ *with* $\frac{\partial \phi_b}{\partial n} = 0$ *on* $\partial\Omega$.

The result above has been used in [113] to give rigorous justifications of hysteresis phenomena. In [143, 145, 146], more results on the existence of stable vortex solutions and more precise characterizations of the critical magnetic fields have been obtained.

5.4. *Effect due to spatial inhomogeneities*

Spatial inhomogeneities play important roles in the study of vortex state, especially in relation to the so called vortex pinning. When modeling spatial inhomogeneities, studies on the thin films of variable thickness [24] and SNS junctions [111, 72, 73] have been given particular attention.

A sample in which a layer of normal material is sandwiched between two layers of superconducting material is called an SNS junction or Josephson junction and is well known for its quantum mechanical Josephson effects [148], which is another Nobel prizing winning work and it is named after Josephson who predicted its existence in 1962.

Quantum tunneling is a process arising from the wave nature of the electron. It is observed that a flow of electric current of Cooper Pairs may tunnel through two superconducting materials that are separated by an extremely thin insulator or normal layer. This arrangement is called a Josephson junction. The key here is that the two superconductors act to preserve their long-range order across the insulating barrier. Rapid alternating currents occur within the insulator when a steady voltage is applied across the superconductors. The current flow is known as the Josephson Current and the quantum tunneling of the insulator by the Cooper Pairs is the Josephson Effect.

Studies of SNS junctions are particularly useful in many applications including the design of microwave devices using high-T_c superconductors. The Josephson junction can act as a super-fast switching devise that can perform switching functions such as switching voltages approximately ten times faster than ordinary semi-conducting circuits. This is a distinct advantage in a computer, which depends on short, on-off electrical pulses. The so called superconducting Qubits are made of Josephson junctions which play important roles in the research on quantum computing.

The nonlinear Josephson equations proposed by Josephson was partly motivated by the Ginzburg-Landau theory. Since then, various studies have been made to derive or simplify Ginzburg-Landau type phenomenological models to fit the junction setting. For example, the G-L type models for SNS junctions was derived in [25] to account for both the superconducting layers and the normal layer. A discussion of supercurrent across a one-dimensional junction is also presented there.

A particular form of the energy functional adopted in [25] is given by

$$\mathcal{G}(\psi, \mathbf{A}) = \frac{1}{2} \int_\Omega \left(\frac{1}{2} (a(\mathbf{x}) - |\psi|^2)^2 + \left| \left(\frac{i}{m(\mathbf{x})\kappa} \nabla + \mathbf{A} \right) \psi \right|^2 \right) d\Omega$$
$$+ \int_\Omega |\mathrm{curl}\, \mathbf{A} - H|^2 d\Omega ,$$

where $a(\mathbf{x}) = 1$ in superconductors but $a(\mathbf{x}) \leq 0$ in normal or insulating samples. $m = m(\mathbf{x})$ is a function used to denote the variation in the mass of superconducting carriers and the normal electron.

In order to study the pinning effect in three dimensional thin films while reducing the complexity of the coupled system of nonlinear PDEs, attempts have been made to reduce the three-dimensional system to a planar system when the normal layer is very thin. A thin normal layer limit was given in [72] but it did not incorporate the effects of weak links between the superconducting layers in the leading order equations. The generations in two limiting regimes were made in [73]: the high-κ, high-field limit as well as the limiting model as the thickness of the middle normal layer approaches zero. In particular, it has been shown that the effects of weak links may appear in the leading order equations. Other related mathematical works include [139, 141].

For a sufficiently thin superconducting film, it was shown in [24] that the three-dimensional Ginzburg-Landau model of superconductivity [65, 148] may be reduced to a two-dimensional thin film model given by the minimization of the following functional in $H^1(\Omega, R^2)$:

$$J_a(u) = \frac{1}{2} \int_\Omega a(x) \left[|\nabla_{\mathbf{A}_0} u|^2 + \frac{1}{2\epsilon^2} (1 - |u|^2)^2 \right], \tag{5.2}$$

where Ω is a bounded smooth domain in R^2, $a \in C^\infty(\bar{\Omega})$ is a given function measuring the variation in the thickness of the film, and $\mathbf{A}_0(x)$, the in-plane component of the magnetic potential, is determined from the vertical component H of the external applied magnetic field by

$$\begin{cases} \operatorname{div}(a(x)\mathbf{A}_0) = 0, & \operatorname{curl}\mathbf{A}_0 = H, \quad \text{in } \Omega, \\ \mathbf{A}_0 \cdot \mathbf{n} = 0, & \text{on } \partial\Omega. \end{cases}$$

Let us assume that $a(\mathbf{x}) \geq a_0 > 0$ for all $\mathbf{x} \in \bar{\Omega}$, $\mathbf{n}$ denotes the outward normal to $\partial\Omega$, $\nabla_{\mathbf{A}_0} u = \nabla u - i\mathbf{A}_0 u$, and let $\xi = H\xi_0 \in H^2(\Omega, R)$ satisfy

$$a(\mathbf{x})\mathbf{A}_0(\mathbf{x}) = \nabla^{\perp}\xi = (-\xi_{x_2}, \xi_{x_1}), \quad \text{in } \Omega, \tag{5.3}$$

with ξ_0 solving

$$\begin{cases} -\operatorname{div}(\frac{1}{a(\mathbf{x})}\nabla\xi_0) = -1, & \text{in } \Omega, \\ \xi_0 = 0, & \text{on } \partial\Omega. \end{cases} \tag{5.4}$$

We may easily see that $-C \leq \xi_0 < 0$ for some constant $C > 0$ only depending on Ω and $a = a(\mathbf{x})$.

In [43], the minimizers of the free energy functional (5.2) for the thin film in the set

$$D_M^a = \left\{ u \in H^1(\Omega, R^2) : \ F_a(u) < M|\ln\epsilon| \right\}, \tag{5.5}$$

has been studied under the conditions:

Assumption 5.1: Define

$$\Lambda = \left\{ x \in \Omega, |\xi_0(x)/a(x)| = \max\lim_{y\in\Omega} |\xi_0(y)/a(y)| \right\}, \tag{5.6}$$

and assume that the constant M in (5.5) is so chosen that there is a positive integer n such that

$$\left[\frac{M}{\pi \max_\Lambda a(\mathbf{x})}, \frac{M}{\pi \min_\Lambda a(\mathbf{x})} \right] \subset (n, n+1). \tag{5.7}$$

The above assumption on n with the property (5.7) is needed in proving that the minimizer of J_a in the constrained set is in fact a solution of the reduced Ginzburg-Landau equation (5.8):

$$\begin{cases} -(\nabla - i\mathbf{A}_0) \cdot a(\mathbf{x})(\nabla u - i\mathbf{A}_0 u) = \dfrac{a(\mathbf{x})}{\epsilon^2} u(1 - |u|^2), & \text{in } \Omega \\ \partial_{\mathbf{n}} u = 0, & \text{on } \partial\Omega \end{cases} \tag{5.8}$$

The assumption is equivalent to assume that $\max_\Lambda a(\mathbf{x}) < 2\min_\Lambda a(\mathbf{x})$.

Concerning the vortex free solutions of (5.8), we have the following uniqueness results [44].

Theorem 5.2: *The stable vortex-less solution of (5.8) is unique for sufficiently small ϵ and*

$$H \leq C\epsilon^{-\alpha}$$

where $C > 0$ and $\alpha > 0$ are positive constants.

The more interesting issues are naturally related to the vortex solutions. In this regard, estimates on the critical magnetic field have also been proved in [43] which also give characterization of the vortex configuration. A couple of sample results are provided below.

Theorem 5.3: *Let $k_a = 1/(2\max\lim_\Omega |\xi_0(\mathbf{x})/a(\mathbf{x})|)$, there exist $k_2^\epsilon = O(1)$, $k_3^\epsilon = o(1)$, and $\epsilon_0 = \epsilon_0(M) > 0$ such that*

$$H_{C1} = k_a|\ln\epsilon| + k_2^\epsilon \tag{5.9}$$

for $\epsilon < \epsilon_0$. Moreover, the following holds
(i) if $H \leq H_{C1}$, there exists a solution u_ϵ of (5.8) which minimizes $J_a(u)$ in D_M^a, and it satisfies $\frac{1}{2} \leq |u_\epsilon| \leq 1$;
(ii) if $H_{C1} + k_3^\epsilon \leq H \leq H_{C1} + O(1)$, there exists a solution u_ϵ of (5.8) that minimizes $J_a(u)$ in D_M^a. The solution has a bounded positive number of vortices b_i^ϵ of degree one, such that

$$\mathrm{dist}(b_i^\epsilon, \Lambda) \to 0, \quad as\ \epsilon \to 0, \tag{5.10}$$

and there exists a constant $\alpha > 0$ such that $\mathrm{dist}(b_i^\epsilon, a_j^\epsilon) \geq \alpha$ for $i \neq j$.

Theorem 5.4: *Considering a sequence $u_n = u_{\epsilon_n}$ of solutions of (5.8) given by the above theorem, then up to subsequence, there exist d points $c_i \in \Lambda$ such that $u_n \to u_*$ weakly in $W^{1,p}$ $(p < 2)$ and strongly in $H^1_{\mathrm{loc}}(\Omega \setminus \cup_{i=1}^d \{c_i\})$, where u_* is a solution of*

$$\begin{cases} -\nabla \cdot (a(\mathbf{x})\nabla u_*) = a(\mathbf{x})u_*|\nabla u_*|^2, & in\ \Omega \setminus \cup_{i=1}^d\{c_i\} \\ \frac{\partial u_*}{\partial \mathbf{n}} = 0, & on\ \partial\Omega \\ |u_*| = 1, & a.e.\ on\ \Omega \end{cases} \tag{5.11}$$

As the set Λ is determined by the coefficient function $a = a(\mathbf{x})$, the above results clear indicate the pinning effect due to the variation of the thickness.

Similarly, mathematical analysis on the effect of normal inclusions have also been studied, for instance, in [6].

In essence, the variation or generalization to incorporate either variable thickness or normal inclusions in the G-L framework lies in the following change of the free energy density:

$$\frac{1}{2\epsilon^2}(1 - |\psi|^2)^2 + |(\nabla - i\mathbf{A})\,\psi|^2$$

$$\Rightarrow \frac{a(\mathbf{x})}{2\epsilon^2}(b(\mathbf{x}) - |\psi|^2)^2 + a(\mathbf{x})\left|\left(M^{-1}(\mathbf{x})\nabla - i\mathbf{A}\right)\psi\right|^2 .$$

The function $a = a(\mathbf{x})$ appears in the thin film model to model the variable thickness. The function $b = b(\mathbf{x})$ is used to model the normal inclusion so that in the superconductor $b(\mathbf{x}) = 1$ while in the normal region $b(\mathbf{x}) < 0$. The matrix M may be used to model the spatial anisotropy [41].

6. Dynamics of quantized vortices

The motion of vortices has played a central role in the study of superconductivity, Vortices may be set in motion due to vortex interactions, thermal fluctuations and applied voltages and currents. An applied current exerts the Lorentz force on the vortices, and the motion of vortices unfortunately, induces electrical resistance and cause the loss of superconductivity. Thus, understanding the dynamics of quantized vortices thus bears tremendous importance in the practical application of superconductivity.

When the coherence length of the superconductor is very small, represented by a large Ginzburg-Landau parameter κ, dynamical vortex motion laws of well separated vortices have been deduced and mathematically justified in the case of a static constant applied magnetic field. The effect of spatial inhomogeneities and pinning forces have also been included in the studies. In spite of the much progress made in the last decade, it should be pointed out that the rigorous mathematical study of a large part of the subject on vortex dynamics remains nearly non-existent, in particular, much of the studies on the vortex dynamic laws have not incorporated issues like the effect of applied current and thermal fluctuations. On the other hand, numerical simulations have become useful tools that could help providing a more clear picture on the exotic vortex dynamics driven by various forces.

Here, we briefly discuss the vortex dynamics described by the time-dependent Ginzburg-Landau models. We should note that other types of dynamic equations are also available, such as the wave dynamics (which is believed to be valid near 0K), and the more general Glauber dynamics.

6.1. *Dynamics of vortex nucleation*

Typical snapshots of the solutions of TDGL for a two dimensional square domain are given in the Fig. 8. An initial superconducting state is used with an applied field $H > H_{C1}$. As time goes on, the magnitude of the order parameter $|\psi|$ first decreases near along the edges of the square sample and correspondingly the induced magnetic field increases in such regions, creating a front propagating into the sample . Later, the front becomes unstable, and individual vortices start to nucleate near the edges, and move into the interior of sample. Due to the repulsion among the vortices, they finally settle into an equilibrium configuration. In the plots, contour lines for $|\psi|$ are drawn. $|\psi|^2$ is proportional to the density of superconducting charge carriers, thus, $|\psi| = 0$ corresponds to the normal state and, in our nondimensionalization, $|\psi| = 1$ corresponds to the superconducting state. For more numerical simulations, see $[41, 66, 67, 87, 91, 117]$.

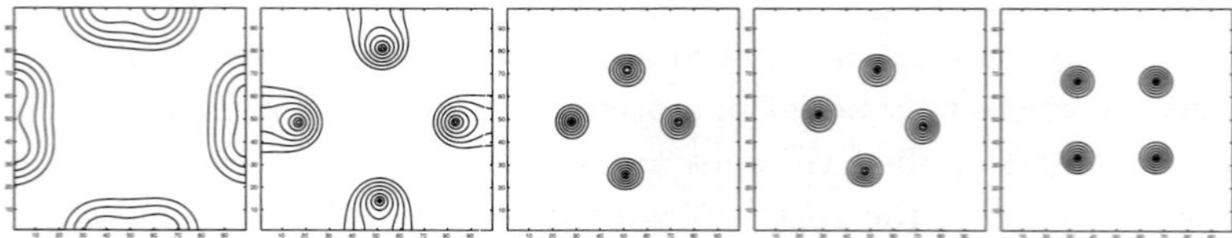

Fig. 8.   Contour plots of $|\psi|$ in a 2-d square at t= 5, 10, 20, 25, ∞.

For high-T_c superconductors, the value of the G-L parameter κ is very large, which is a fact that has been utilized to derive various reductions of the original G-L equations.

6.2. *Dynamics of individual vortices*

The rigorous study of vortex dynamical laws started with the case of $\mathbf{H} = 0$, i.e., the following model problems were first considered:

$$\frac{1}{\lambda_\epsilon} \frac{\partial u}{\partial t} = \Delta u + \frac{1}{\epsilon^2} u \left(1 - |u|^2\right) \quad \text{in} \quad \Omega \times R_+ , \tag{6.1}$$

$$u(x, 0) = u_\epsilon^0(x), \quad x \in \Omega, \tag{6.2}$$

$$u(x, t) = g(x), \quad x \in \partial\Omega . \tag{6.3}$$

Here, ϵ is inversely proportional to κ. The dynamics of the vortices in the limit $\epsilon \to 0$ can be considered within the framework of a general program initiated by J. Neu, and later extended, and improved by many others (see [133, 134] and [31, 30, 77, 78]) using the method of matched asymptotic expansions. Such formal expansions have now been largely rigorously proved [112, 115, 100].

For $\Omega \subset R^2$, assuming that u_ϵ^0 consists of d vortices,

$$u_\epsilon^0(x) \to u_0(x) = \prod_{j=1}^{d} \frac{(x - b_j)}{|x - b_j|} e^{i\,h(x)} \,,$$

it has been shown that to leading order, the dynamic vortex motion laws are given by [112]:

$$m_i \frac{d\,b_i\,(t)}{dt} = -\nabla_{b_i} W_n(b), \qquad i = 1, 2, \ldots d,$$

where $b = (b_1, \ldots, b_d)$ are positions of vortices, the constants $\{m_i\}$ are called mobilities of the vortices, and W_n is the so-called renormalized energy [21, 112]. An example of such an renormalized energy (incorporating the effect of applied magnetic field) is given by (5.1). By examining the particular forms of the renormalized energy, it has been shown that isolated vortices of the same signs tend to repel while isolated vortices of opposite signs attract.

The case of motion laws derived as the limit of the full time-dependent Ginzburg-Landau models has also been studied, for instance, in [30, 48].

6.3. *High-κ, high field dynamics*

In the high-κ limit considered in [27, 61], the applied field is assumed to be high so that it penetrates the sample completely to the leading order, i.e., the induced field is equal to the applied field. With a proper scaling and invoking a gauge choice $\Phi = 0$, and $\mathbf{A} \cdot \mathbf{n} = 0$, the time-independent magnetic potential $\mathbf{A}_0$ can be solved for separately from the Maxwell's equations, i.e.,

$$\nabla \cdot \mathbf{A}_0 = 0 \,, \quad \text{in} \quad \Omega \,,$$
$$\nabla \times \mathbf{A}_0 = H \,, \quad \text{in} \quad \Omega \,,$$
$$\mathbf{A}_0 \cdot \mathbf{n} = 0 \,, \quad \text{on} \quad \partial\Omega \,.$$

Ignoring the proximity effect, the resulting simplified leading-order system for the order parameter ψ_0 is given by:

$$\frac{\partial \psi_0}{\partial t} + i\Phi_a \psi + (i\nabla + \mathbf{A}_0)^2 \psi_0 - \psi_0 + |\psi_0|^2 \psi_0 = 0 \quad \text{in } \Omega \tag{6.4}$$

$$(i\nabla + \mathbf{A}_0)\psi_0 \cdot \mathbf{n} = 0 \quad \text{on } \Gamma \,, \tag{6.5}$$

$$\psi_0(\mathbf{x}, 0) = \varphi(\mathbf{x}) \quad \text{in } \Omega \,. \tag{6.6}$$

Interestingly, with the presence of Φ_a and $\mathbf{A}_0$, the above high-κ high-field (HKHF) model preserves much of the physics described by the original TDGL equations. The dynamics can also be much more diverse than the case $\mathbf{A}_0 = 0$, $\Phi_a = 0$, illustrated by the examples given later.

6.4. *Dynamics involving spatial inhomogeneities*

Various modified dynamic equations have been considered when dealing with spatial inhomogeneities. For instance, when modeling normal inclusions, we may consider

$$\frac{\partial u}{\partial t} = \Delta u + \frac{1}{\epsilon^2} u(a(x) - |u|^2), \ \text{in } \Omega \times (0, +\infty)$$

for some function $a > 0$. Under proper scalings, the dynamic laws would change dramatically with a non-degenerate coefficient a, namely, the equation of motion becomes effectively [46, 101]:

$$\frac{db_j}{dt} = -\nabla \log a(b_j) \, ,$$

in the limit $\epsilon \to 0$. The same dynamic law applies to the model for the variable thickness thin film:

$$\frac{\partial u}{\partial t} = a \left(\Delta u + \frac{1}{\epsilon^2} (1 - |u|^2)u \right), \ \text{in } \Omega \times (0, +\infty)$$

where $a > 0$ is a function measuring the relative thickness of the three dimensional thin film [28, 43, 44]. Physically, this is associated to the phenomena that the vortices turn to move to regions where the film is thin. Extensions to the cases with the presence of the applied magnetic field have also been obtained [43, 45, 46, 143].

The G-L models have been used to study the pinning effect from many different angles, for instance, pinning due to variable thickness in thin films [24, 113], spatial inhomogeneities and normal inclusions and anisotropy [25, 41, 72, 73].

More general version involving the effect of applied magnetic field can also be considered [73]:

$$\frac{\partial u}{\partial t} = (\nabla - i\mathbf{A}_0)^2 u + \frac{1}{\epsilon^2}(a(x) - |u|^2)u, \ \text{in } \Omega \times (0, +\infty) \, .$$

The corresponding free energy also shares remarkable resemblance with the Gross-Pitaevskii free energy used to model the vortex state of a BEC under the rotating magnetic trap. In such case, $a(x)$ is a conventional trapping potential while curl $\mathbf{A}_0$ represents the angular velocity of the rotating trap [4].

6.5. *Dynamics driven by the applied current*

Since there is no loss in electrical energy when superconductors carry electrical current, relatively narrow wires made of superconducting materials can be used to carry huge currents. However, there is a certain maximum current (the so called critical current) that these materials can be made to carry, above which they stop being superconductors. In general, the value of critical current density is a function of temperature; i.e., the colder the temperature of the superconductor is, the more current it can carry.

To give more details, an applied current J generally exerts a Lorentz force $F = J \times B$ on each vortex, analog to the action of the Magnus force on a spinning tennis ball. The motion of vortices due to the Lorentz force induces an electric field, and thus produces electrical resistance. Thus, in superconductivity, it is important to understand the interaction of the vortices with the applied current and study the critical values of the applied current which will dislodge the vortices from their equilibrium positions.

The TDGL equations may be used as a prototype model for the study of critical current. In fact, it is equally enlightening to study an even simpler system [58], the high-κ high-field (HKHF) model given in (6.4-6.6). Assume that Ω is the unit square (or disk) in R^2, and the applied current is applied along the x_2 direction, then, we may set $\Phi_a(x_1, x_2) = Jx_2$ for some constant applied current J, and $\mathbf{A}_0(x_1, x_2) = H(x_2, -x_1)/2$ for some constant applied field H.

Beyond the basic well-posedness [61], there exists very little mathematical analysis on the HKHF equation. In fact, even many questions on its steady state solutions remain largely unanswered. Possible different parameter regimes are roughly outlines in [58]. In [70], a perturbative technique was used to show that for H near H_{C1}, for sufficiently small ϵ, and for small enough J, the steady state has a solution which possesses a single vortex, see Fig. 9.

In Fig. 10, an incomplete phase diagram describing the dynamic solutions using H and J as the parameters is given though numerical evidence have indicated various possibilities that we now outline. For detailed numerical simulations substantiating the discussion here, we refer to [61] and also more recent computations in [68, 69].

Consider for simplicity the case H is near H_{C1}, there are several possible regimes in which the time-dependent solutions may behave differently as observed in [58].

First of all, it is trivial to see that the solution beginning with the nor-

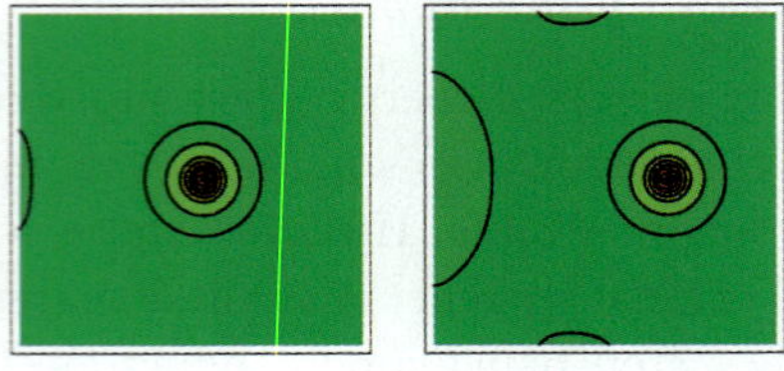

Fig. 9.   Steady states with perturbed vortex locations for $J = 0.1, 0.5$.

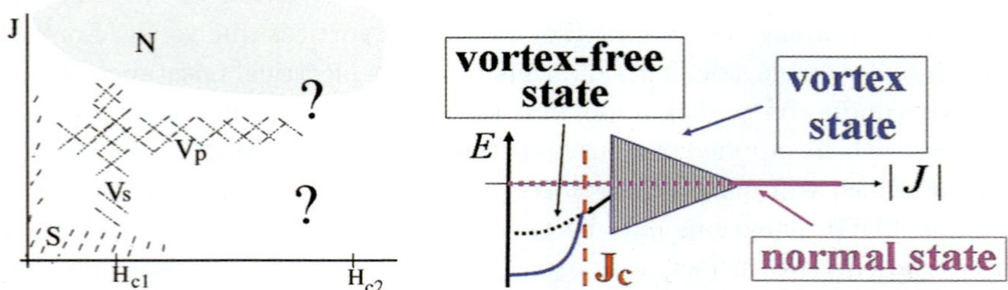

Fig. 10.   Left: An incomplete phase diagram (left) and an energy bifurcation diagram (right) for the HKHF model driven by current.

mal state ($\psi = 0$ everywhere) will remain normal in time, this is hardly of any interest dynamically. For small J, in accordance with the result of [70], due to the hysteresis effect, the solution starting at the superconducting state (no vortex) would approach a steady state which is vortex-free. Meanwhile, the solution starting with a vortex would undergo a gentle shift of the vortex position due to the Lorentz force, but eventually comes to a complete rest and thus provides a steady state solution with an off-center vortex. Numerically we find that as the applied current J gets stronger, the vortex would shift more from their original equilibrium positions. For larger J, however, the superconducting solution would start to nucleate vortices because the Lorentz force is now strong enough to overcome the boundary barrier of the applied magnetic field. These vortices will move through the sample, and the Lorentz force is again strong enough to push them over the barrier on the other side of the boundary, thus lead to possibly time-periodic motion of single vortex or vortex arrays, see Fig. 11 and 12. Such a current J is thought to be above the critical current.

The plots of the free energy in time for two different values of current are given in Fig. 13, supporting the existence of time-periodic solutions with

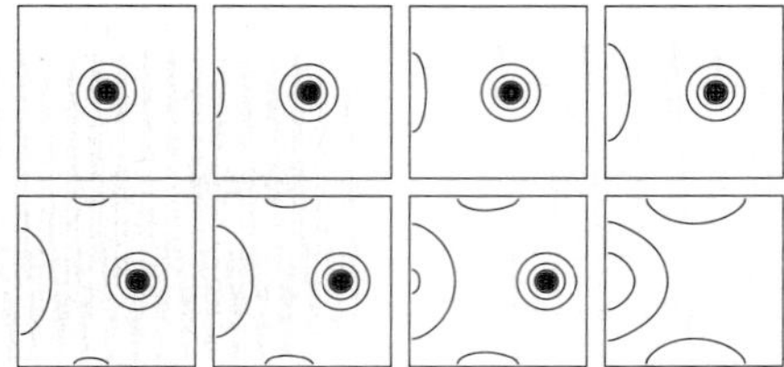

Fig. 11.   Motion of a single vortex in the presence of an applied current.

non-trivial spatial structures.

If J is exceedingly large, the time dependent solution would eventually collapse to the normal state regardless of its initial state.

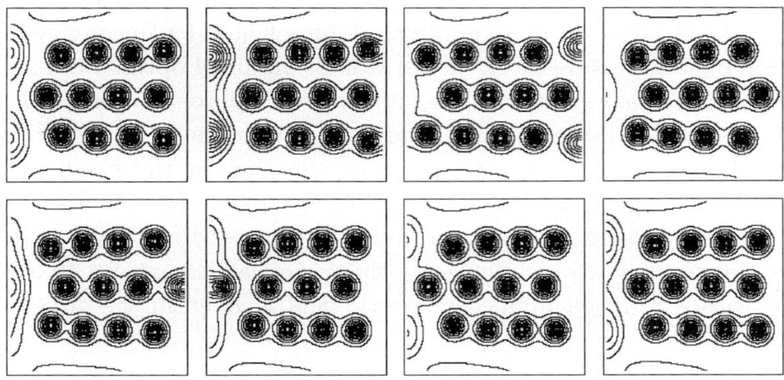

Fig. 12.   Motion of vortex array in the presence of an applied current.

We note that similar discussions for the full TDGL also remain to be carried out. As the size of the sample domain also affects the dynamics of the equations and their long time behavior, a much more complex and multi-dimensional diagram than that in Fig. 13 should be constructed. Numerical simulations can again be very helpful.

6.6. *Vortex state in a thin superconducting spherical shell*

The geometry of spherical shell is not only used in superconductivity applications, but also provides an ideal setting for one to examine the vortex state. Numerical simulations of the vortex configurations on the spherical shell have been done in [68, 69] based on a high resolution finite volume

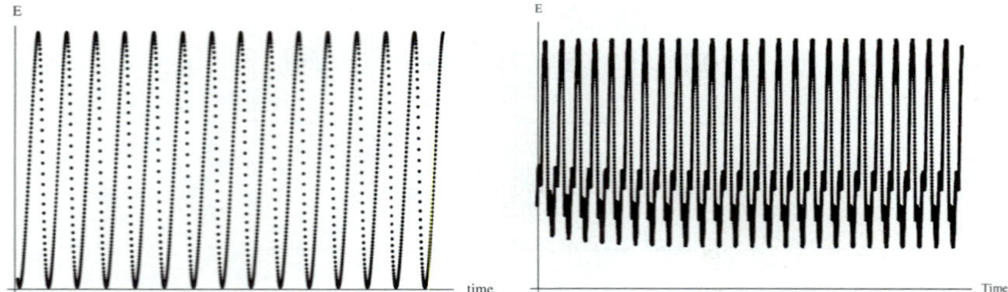

Fig. 13.   Plots of energy in time: evidence of time-periodic solution.

approximation for a simplified G-L model that extended earlier studies by us as well as by other physicists on superconducting thin films [24].

An illustration of SCVTs with almost perfect uniformity is presented in figure 14. Analysis of these approximation schemes for linear convection diffusion equations were given in [69] which contained optimal order error estimates and numerical tests. The finite volume approximation of the G-L model enjoys the discrete gauge invariance, and leads to a high order recovery of the physically interesting quantities such as the magnetic field and the supercurrent. In figure 14, two density plots of a solution $|\psi|$, the magnitude of the order parameter, are also presented as they are viewed from different angles, which illustrate the vortex lattice pattern on a hemisphere with a constant applied magnetic field is given in the direction of the north-pole.

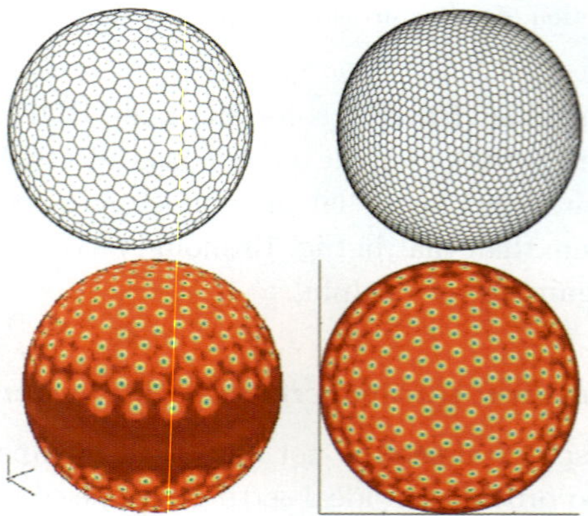

Fig. 14.   SCVT meshes (top) and a computed vortex lattice.

With increasing values of the applied magnetic field, some of the energy minimizing vortex configurations are given in Fig. 6.6. We refer to [68] for more detailed descriptions of the corresponding parameter values.

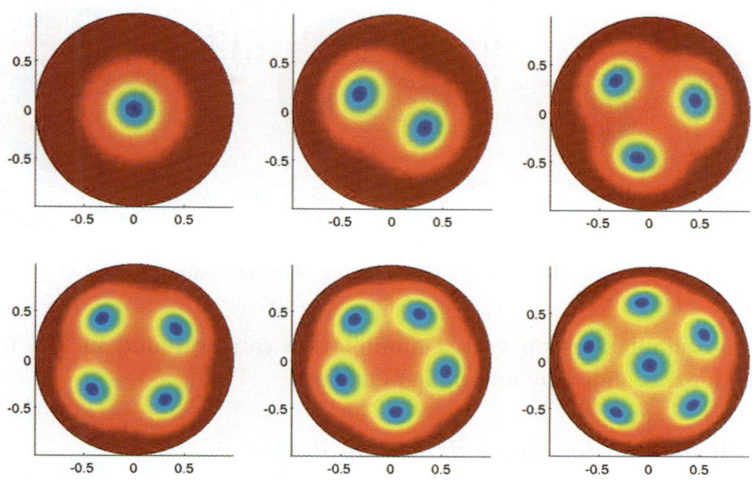

Fig. 15.   Energy minimizing vortex configurations on the upper hemisphere.

In Fig.16, minimum energy values $\mathcal{G}(H)$ corresponding to different external magnetic field strengths H are plotted, along with the magnetization curve given by the derivatives of minimum energy $\mathcal{G}(H)$.

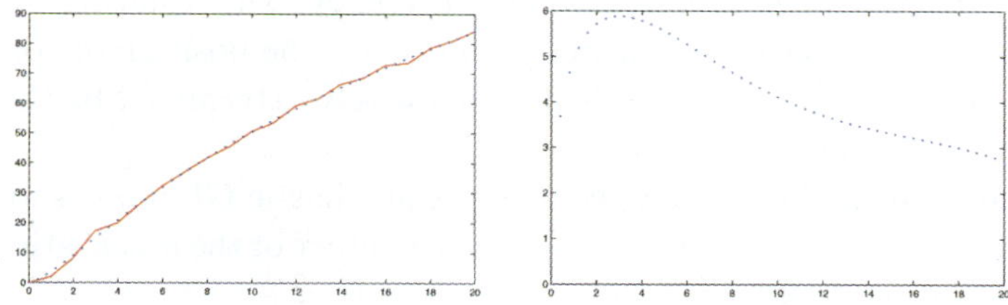

Fig. 16.   Minimum energy with its polynomial fitting and the magnetization curve.

The spatially homogeneous applied magnetic field naturally produces vortices of opposite signs on the two hemispherical shells, thus providing a window for us to see the details of vortex nucleation and vortex annihilation (see [16] for some analysis in a simpler setting). In Fig.17, with the applied magnetic field aligned along the poles, the density plots of order parameter show that a pair of vortices of opposite signs first nucleate near the equator, then later split and move into the interior of the hemispheres [68].

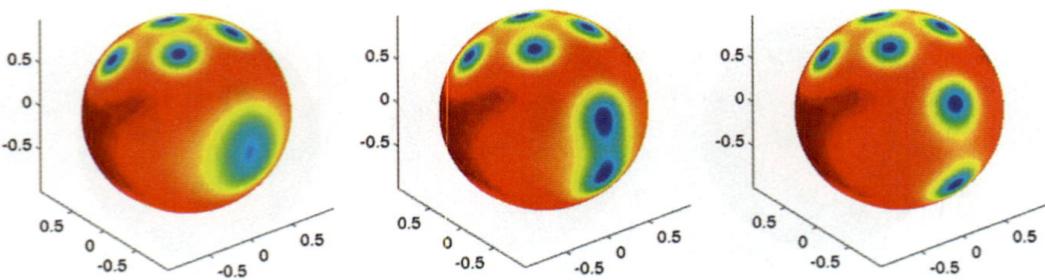

Fig. 17. The nucleation and the splitting of vortex pairs near equator.

Other simulations on vortex annihilation can be found in [61] in the planar domain, aided by an applied current.

6.7. *Stochastic dynamics driven by noises*

The conventional Ginzburg-Landau theory is applicable only to highly idealized physical contexts that do not take into account factors like inhomogeneities, thermal fluctuations, and random applied fields. For example, it is well known that thermal fluctuations and material defects play a central role in the pinning of vortices in type-II superconductors [107, 148]. In the study of stochastic vortex dynamics given in [95], the average position of vortices is determined by averaging the equations that determine the movement of vortices, i.e. equations involving ψ and $\mathbf{A}$. The resulting equations are necessarily stochastic in nature meaning that the position of vortices in a sample at a particular time, t, is not completely determined by the initial condition of the sample at $t = 0$.

To model such phenomena, we may use stochastic GL models following some Langevin dynamics [86, 48, 49]. In the context of the stochastic HKHF model, we have [39, 40]

$$\frac{\partial \psi}{\partial t} + (i\nabla + \mathbf{A})^2 \psi - \psi + |\psi|^2 \psi = \eta \quad \text{in } \Omega \tag{6.7}$$

$$(i\nabla + \mathbf{A})\psi \cdot \mathbf{n} = 0 \text{ on } \Gamma, \tag{6.8}$$

$$\psi(\mathbf{x}, 0) = \varphi(\mathbf{x}) \text{ in } \Omega, \tag{6.9}$$

where $\eta = \sigma dW/dt$ is a random complex-valued field in time and space and $W(t), t \geq 0$ is a H^1-valued Wiener process with covariance operator Q which is a Hilbert-Schmidt operator. For the sake of theoretical investigations, it is assumed to be of finite trace. The above equations belong to the class of stochastic G-L models with an additive noise.

When modeling the effects of thermal fluctuations, the variance of η (or σ) is on the order of $K(1 - T/T_c)^{-2}$, where T and T_c are the temperature and the critical transition temperature, respectively, and K is a constant. One may consider also the stochastic dynamics based on a multiplicative noise

$$\frac{\partial \psi}{\partial t} + (i\nabla + \mathbf{A})^2 \psi - \psi + |\psi|^2 \psi = \eta \psi \quad \text{in } \Omega$$

with similar boundary and initial conditions. In figure18, snapshots of solutions of the multiplicative model with $\sigma = 4$ are provided, see [39, 40] for details.

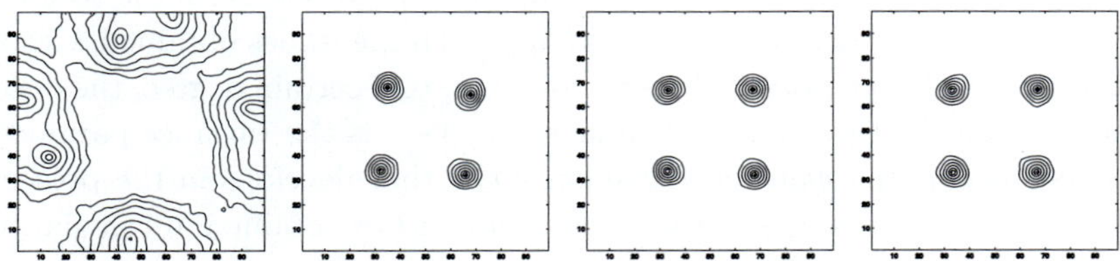

Fig. 18.   Plots of $|\psi|$ at different time with multiplicative noise.

In Fig.19, isosurface plots of quasi-static vortex tubes are presented corresponding to different values of variance of the additive noise term using for the simulation of Langevin dynamics [39, 40]. The snapshots are taken at the same time instant and the vortex lattice melting effect can be observed in the process.

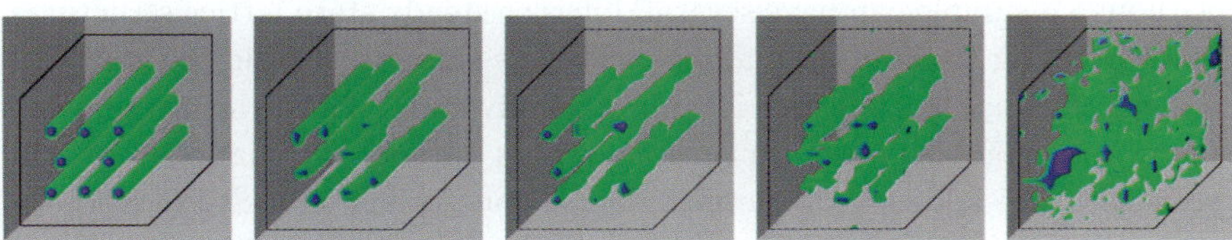

Fig. 19.   Snapshots of the vortex tubes in a cubic sample with increasing additive noises.

Theoretically, the relations between the time evolution of the deterministic model and the model with the noise effects have also been established, for example, it is shown that as the variance approaches zero, the solution of the stochastic high-κ time-dependent Ginzburg-Landau equations approaches the solution of the deterministic high-κ time-dependent Ginzburg-Landau equations.

Theorem 6.1: *Let $\psi_\sigma(t)$ be the solution to the stochastic differential equation (6.7) corresponding to variance σ. Then,*

$$\sup_{t\in[0,T]} E\|\psi_\sigma(t) - \psi_0(t)\|^2 \leq C\sigma^2 tr(Q) \quad \text{for some constant } C$$

and, consequently,

$$\sup_{t\in[0,T]} E\|\psi_\sigma(t) - \psi_0(t)\|^2 \to 0 \quad \text{as } \sigma \to 0.$$

Intuitively, when the variance decreases to zero, the level of fluctuations introduced to the stochastic time-dependent Ginzburg-Landau equation goes to zero, and as the level of fluctuations goes to zero, the resulting values of ψ at the steady state should approach the values obtained without fluctuations. This theoretical result confirms, to a certain degree, the observations made in the numerical simulations, i.e., as the variance parameter σ gets smaller, the *quasi-steady state lattice* that develops in the presence of fluctuations converges to the *steady state lattice* obtained in calculations without fluctuations [39, 40].

Computationally, thermal fluctuations are evident in contour plots of the density of superconducting electrons since vortices are in constant motion, the vortices are distorted, and the resulting vortex lattice is also distorted. With small variances and at large times, the vortices seems to be at an equilibrium where each vortex is on the *average* at a fixed position in the sample. When fluctuations are subsequently stopped at this *quasi-steady state*, the resulting vortex lattice evolves into a steady state configuration for the sample with respect to the number of vortices present in the material. Although a sample can have several different steady state lattice structures corresponding to different realizations of the same stochastic process with a given variance, each lattice is a quasi-steady state. Thus, the lattice symmetry and regularity seen in pure materials without fluctuations are still seen with the addition of thermal fluctuations in our model. We also observe that while the additive noise tends to make the positions of vortices vibrate, multiplicative noises tends to alter the core sizes of the vortices which is even more apparent when discrete lattice random functions are used.

The main thread needed to connect the computational studies with experimental studies of thermal fluctuations in superconductors is the exact relationship between the magnitude of the variance and the ambient temperature of the sample. It is known that thermal fluctuations are *small* random forces, but we have not precisely connected the temperature (or the critical temperature) to the variance in the non-dimensional time-dependent Ginzburg-Landau equations that are used in the computations.

6.8. *Variants of G-L models: Lawrence-Doniach and d-wave models*

The great success of the Ginzburg-Landau models for low T_c superconductivity generated tremendous interests in extending them to other settings including layered materials and high T_c superconductors.

The Lawrence-Doniach (LD) model is a derivative of the basic G-L model for a layered superconductor with G-L energy given in individual layers and the Josephson-like coupling between the layers [108]. The LD model provides an effective description of a number of interesting problems such as the appearance of the so called pancake vortices and the 2d and 3d cross-over [36]. Such pancake vortices may interact both magnetically and through Josephson coupling. They may align under certain conditions into elastic vortex lines similar to those in three dimensional bulk isotropic superconductors. Under different conditions, however, the pancake vortices in each layer may move independently of the other layers, and the system acts like a stack of independent thin film superconductors. Thus in layered superconductors two different effective dimensionalities of the vortex pancake lattice may appear, each with different characteristic properties [130].

Mathematically, in order to simplify the equations, the LD model has been examined in the setting where the superconducting material possesses a high Ginzburg–Landau parameter κ [26, 62]:

$$\frac{\partial \psi_n}{\partial t} + i\Phi_n^a \psi_n - \psi_n + |\psi_n|^2 \psi_n + (i\nabla + \mathbf{A}_n^a)^2 \psi_n$$

$$+\rho\left[2\psi_n - \psi_{(n+1)}\exp(-i\phi_n^{n+1}) - \psi_{(n-1)}\exp(i\phi_{(n-1)}^n)\right] = 0$$

for layers labeled by $n = 0, 1, \ldots, N$, where $\mathbf{A}^a$ and Φ^a are prescribed magnetic and electric potentials,

$$\phi_n^{n+1} = \int_{ns}^{(n+1)s} \mathbf{A}_z^a \quad \text{for } n = 0, \ldots, N-1,$$

where s being the layer thickness and

$$(i\nabla\psi_n + \mathbf{A}_n^a \psi_n) \cdot \mathbf{n} = 0$$

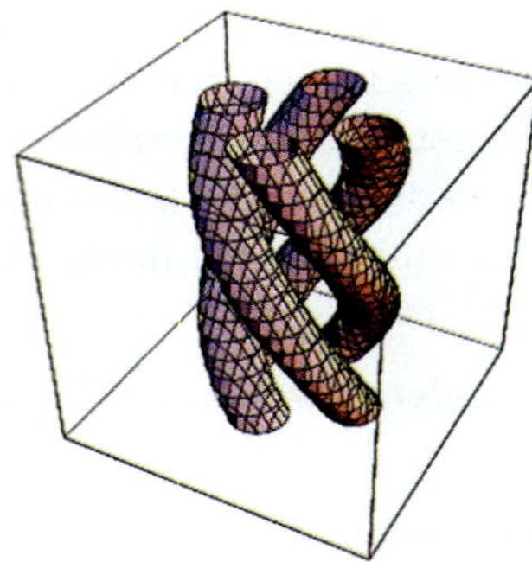

Fig. 20. Vortex torsion due to the pinning of the normal inclusions in a three dimensional superconductor sample.

on the boundary of each layer.

The numerical algorithms provided in [62] for the LD model were based on codes developed for the two-dimensional Ginzburg-Landau equations. For the LD model, simplifications were sought which address the complexity introduced by the coupling of the variables between the layers. This interdependence of the variables between layers may be solved for iteratively. The results of [62] show that with a judiciously-chosen iterative scheme, the coupling of the variables between the layers may be broken which allows for a straight-forward parallelization of the solution. The convergence of those iterative schemes can be studied rigorously. Numerical tests on various computing platforms such as the DEC-Alpha clusters Paragon have shown significant speed-up of the parallel algorithms. Three dimensional vortex tubes in a three dimensional layered sample pinned due to spatially distributed inhomogeneities were computed via the Lawrence-Doniach [62]. By adding various pinning sites which provides the collective pinning force, the vortex tubes may no longer be strictly aligned in the direction of the applied magnetic field.

Other numerical studies can be found, for example, in [109].

Recently, in contrast to the conventional s-wave pairing mechanism first suggested by Frolich in 1950, and confirmed by the discovery of the in low-T_c materials, both theoretical and experimental studies on d-wave pairing symmetry in some high-T_c superconductors have carried out. In particular, Ginzburg-Landau type models for the d-wave pairing and $s + d$ wave pairing have been studied and some of these were derived from the phenomenological Gorkov equations. The basic feature of these models is the use of multi-component order parameters. The free energy densities are expansions of terms invariant under the tetragonal symmetry (for the $s + d$ pairing) and the orthorhombic symmetry (for the pure d wave pairing), and are thus more complex than the conventional Ginzburg-Landau energy. In [55], both asymptotic analysis and numerical approximations were made for a Ginzburg-Landau model for d-wave superconductors that was first derived in [138]. The generalized G-L free energy for both the s and d wave order parameters is given by:

$$\mathcal{G}_{sd}(\psi_s, \psi_d, \mathbf{A}) = \int_\Omega \left\{ \kappa^2 |\text{curl } \mathbf{A} - H|^2 - 2\beta |\psi_s|^2 \right.$$
$$+ \frac{1}{2}(1 - |\psi_d|^2)^2 + \frac{4}{3}|\psi_s|^4 + \frac{8}{3}|\psi_s|^2|\psi_d|^2 + \frac{2}{3}\left(\psi_s^2 \psi_d^{*2} + \psi_d^2 \psi_s^{*2}\right)$$
$$\left. + 2|\Pi\psi_s|^2 + |\Pi\psi_d|^2 + 2\Re\{\Pi_x^*\psi_s \Pi_x \psi_d^* - \Pi_y^*\psi_s \Pi_y \psi_d^*\} \right\} d\Omega,$$

where $\Pi = i\nabla - \mathbf{A}$.

A rigorous mathematical framework was established and comparison with the conventional Ginzburg-Landau model for the low-T_c superconductors were made. Various simplifications and reductions of the model were also studied. The numerical results provided various new and exotic structures in the vortex solutions of the d wave Ginzburg-Landau model. The simulations in [55] illustrated that the vortex solutions there typically display a four fold symmetry. In certain parameter regimes, the basic stability properties of single and multiple vortices deviated significantly from the single component counter-part, see Fig.21 for an illustration of a stable double vortex profile.

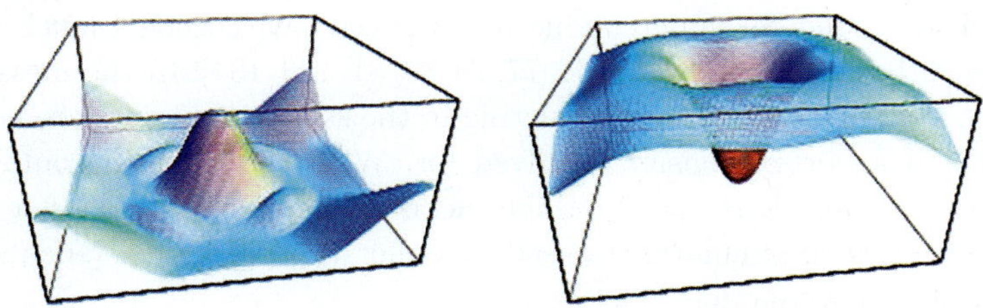

Fig. 21.   Surface plots of s and d wave densities for a double vortex profile.

In a limiting regime, the d-wave model may be viewed as a perturbation of the original GL model [55] with a modified free energy:

$$\int_\Omega \left(\frac{1}{2\epsilon^2}(1 - |\psi|^2)^2 + |\Pi\psi|^2 + \delta^2 \, |\Box\psi|^2 + |\mathrm{curl}\,\mathbf{A} - H|^2 \right) d\Omega \,,$$

where $\Box = (\partial_x + i\mathbf{A}_1)^2 - (\partial_y + i\mathbf{A}_2)^2$ and δ is some small parameter. The four-fold symmetry can then be easily seen from the perturbation term $|\Box\psi|^2$. This simplification could make the rigorous analysis of the vortex stability properties easier to carry out. More analytical and numerical studies on the d-wave G-L models can be found in [114, 138, 151].

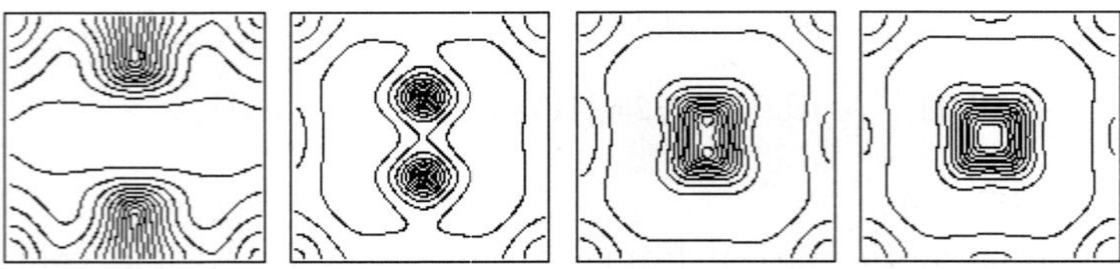

Fig. 22. Dynamics of the merger of two d wave vortices.

Numerical studies of other extensions of the G-L models have also been performed, for instance, see [7] for simulations based on the $SO(5)$ model.

6.9. *Vortex density models*

The number of vortices present in a superconducting sample of dimension, say, a millimeter, will contain a huge number vortices as the separation of individual vortices occurs on a typical length scale of 100 or so angstrom. In order to effectively model the vortex state and to allow efficient numerical simulations, the mathematical analysis and the numerical studies of the mean field models for superconducting vortices have received much attention recently [23, 30, 31, 34, 56, 64, 77, 78, 80, 81, 133, 134]. In the mean field models, the vortex structures occurring in the superconductors are homogenized and a vortex density is solved for. With the reduced complexity, the vortex simulations based on the mean field models could be a possible alternative to simulations based on other type of models such as the Ginzburg-Landau models [65, 67].

Recently, some mean field type models have been derived to describe the vortex state using a vortex density ω. These models are closely related to the G-L models as one normally first derives a discrete set of motion laws for individual vortices based on the G-L dynamics and a vortex density model is then derived when the number of vortices becomes large [31, 77]. If we let u denote the averaged magnetic field, then the simplest two-dimensional version of that type of model is given by

$$\lambda^2 u - \Delta u = \omega \quad \text{in } \Omega \times (0, T),$$
$$\omega_t - \nabla \cdot (\omega \nabla u) = 0 \quad \text{in } \Omega \times (0, T),$$

with suitable initial and boundary conditions. In [77], the effects of pinning potential and the thermal noise effect were examined based on the vortex density model.

In the limit $\lambda = 0$, the following general system has been considered:

$$\begin{cases} \partial_t \rho + \operatorname{div}(u\rho) = 0, \\ u = M(\theta)\nabla\triangle^{-1}\rho, \\ \rho|_{t=0} = \rho_0, \end{cases} \tag{6.10}$$

with ρ_0 being a bounded Radon measure and $M(\theta)$ being a constant orthogonal matrix corresponding to a rotation with angle θ:

$$M(\theta) = \begin{pmatrix} \cos\theta & -\sin\theta \\ \sin\theta & \cos\theta \end{pmatrix}.$$

The global existence of measure-valued solution and the classical weak solution to the above equations were studied in [76]. The matrix M contains information on the various time relaxation constants such as the η_i's given in the TDGL equations. In particular, $M^2 = I$ gives the two dimensional incompressible Euler equations.

Define the following cut-off function:

$$T_\epsilon(\xi) := \begin{cases} \xi, & \xi \geq -1/\epsilon, \\ -1/\epsilon, & \xi \leq -1/\epsilon. \end{cases}$$

An approximate solution sequence to the above system (6.10) can be constructed by

$$\begin{cases} \partial_t \rho_\epsilon + u_\epsilon \nabla \rho_\epsilon = -\cos\theta T_\epsilon(\rho_\epsilon)\rho_\epsilon, & (t, x) \in \mathbb{R}^+ \times \mathbb{R}^2, \\ u_\epsilon = M(\theta)\nabla\triangle^{-1}\rho_\epsilon, \\ \rho_\epsilon|_{t=0} = \rho_{0,\epsilon}, \end{cases} \tag{6.11}$$

where $\rho_{0,\epsilon} = (\rho_0\chi_\epsilon) * j_\epsilon$, $\chi_\epsilon(x) = \chi(\frac{x}{\epsilon})$, $\chi \in C_c^\infty(\mathbb{R}^2)$, and

$$\chi(x) = \begin{cases} 1, & |x| \leq 1, \\ 0, & |x| \geq 2, \end{cases}$$

and $j_\epsilon(x)$ is the standard Friedrich's mollifier with $\operatorname{supp} j_\epsilon \subset B_\epsilon(0)$.

Let $S_\epsilon(\xi) = |\xi| * j_\epsilon$. The approximate solution sequence to (6.11) may be defined by the following equation:

$$\begin{cases} \partial_t \rho_\epsilon + \mathrm{div}\,(u_\epsilon S_\epsilon(\rho_\epsilon)) = \epsilon \triangle \rho_\epsilon, & (t,x) \in \mathbb{R}^+ \times \mathbb{R}^2, \\ u_\epsilon = \nabla \triangle^{-1} \rho_\epsilon, \\ \rho_\epsilon|_{t=0} = (\rho_0 \chi_\epsilon) * j_\epsilon. \end{cases}$$

Theorem 6.2: *Let $\rho_0 \in \mathcal{M}(\mathbb{R}^2)$, the set of Radon measures and $\cos\theta > 0$. Then there exist a subsequence of $\{\rho_\epsilon, u_\epsilon\}$ constructed by (6.11) (still denoted by $\{\rho_\epsilon, u_\epsilon\}$ for convenience), functions $\rho \in L^q_{loc}(\mathbb{R}^+ \times \mathbb{R}^2) \cap L^\infty(\mathbb{R}^+, L^1(\mathbb{R}^2))$ and $u \in L^q_{loc}(\mathbb{R}^+, W^{1,q}_{loc}(\mathbb{R}^2))$ for any $q < 2$, and a positive Radon measure $\mu \in \mathcal{M}^+(\mathbb{R}^+ \times \mathbb{R}^2)$ such that the following hold:*

1. *The following convergence properties and estimates hold:*

$$\rho_\epsilon \rightharpoonup \rho \quad \text{weakly in} \quad L^q_{loc}(\mathbb{R}^+ \times \mathbb{R}^2$$

$$u_\epsilon \rightharpoonup u \quad \text{weakly in} \quad L^q_{loc}(\mathbb{R}^+, W^{1,q}_{loc}(\mathbb{R}^2)),$$

$$\left(\rho_\epsilon + \frac{1}{\epsilon}\right) \rho_\epsilon 1_{\rho_\epsilon \leq -\frac{1}{\epsilon}} \rightharpoonup \mu \quad \text{in the sense of} \quad \mathcal{M}(\mathbb{R}^+ \times \mathbb{R}^2),$$

$$\int_0^\infty \int_{\mathbb{R}^2} d\mu \leq \int_{\mathbb{R}^2} |\,d\rho_0(x)|.$$

2. *The following decay estimates hold:*

$$\rho(t,x) \leq \frac{\cos\theta}{t} \quad \text{for a.e.} \quad (t,x) \in \mathbb{R}^+ \times \mathbb{R}^2.$$

3. *For all test functions $\varphi \in C^\infty_c([0,\infty) \times \mathbb{R}^2)$, there holds*

$$\int_0^\infty \int_{\mathbb{R}^2} (\rho \partial_t \varphi + \rho u \nabla \varphi + \mu \varphi)\,dx\,dt + \int_{\mathbb{R}^2} \varphi(0,x) \rho_0\,dx = 0$$

and $u = M(\theta) \nabla \triangle^{-1} \rho$.

When pinning effect is considered, the motion of vortices due to the applied current is subject to the critical current, as discussed in earlier sections. To study such effects, models such as the Bean's critical state model have been used [18, 19]. In [136, 137], the critical current was cast as a constraint in a variational inequality formulation. In [64], the well-posedness and the finite element approximations of the following model were considered:

$$\begin{cases} \omega = H - \lambda^2 \Delta H & \text{in } \Omega \times (0,T), \\ \omega_t - \nabla \cdot (m \nabla H) = 0 & \text{in } \Omega \times (0,T), \\ m \geq 0, \quad \text{and} \quad |\nabla H| \leq J_c, \\ |\nabla H| < J_c \quad \text{whenever} \quad m = 0. \end{cases} \tag{6.12}$$

For $\lambda = 0$, it reduces to the Bean's critical state model where the term critical state indicates the fact that the current density has regions where they reach the critical value.

Let H_1 be the boundary value of H, $g = -\frac{\partial H_1}{\partial t}$, $u = H - H_1$, the approach taken in [64] is to introduce a regularized problem:

$$\frac{\partial u_\epsilon}{\partial t} - \lambda^2 \Delta \frac{\partial u_\epsilon}{\partial t} + \frac{1}{\epsilon}\beta_0(u_\epsilon) = g \quad \text{in } \Omega \times (0, T), \tag{6.13}$$

with homogeneous Dirichlet Boundary condition and initial condition u_0. Here, β_0 is defined as an operator from $L^4(0, T; W_0^{1,4}(\Omega))$ to $(L^4(0, T; W_0^{1,4}(\Omega)))'$ defined by

$$(\beta_0(u), \phi) = \int_\Omega \left(|\nabla u|^2 - 1\right)^+ \nabla u \cdot \nabla \phi \qquad \forall \phi \in L^4(0, T; W_0^{1,4}(\Omega)).$$

Here $(\psi)^+ = \psi$ whenever $\psi \geq 0$ and $(\psi)^+ = 0$ otherwise. One can easily verify that β_0 is a monotone operator. Let $\mathcal{V} = L^\infty(0, T; W_0^{1,\infty}(\Omega))$ and

$$\mathcal{K} = \{\phi \in \mathcal{V} \ : \ |\nabla \phi| \leq 1 \ a.e., \ \phi|_{\partial\Omega} = 0\}.$$

By deriving uniform bounds on the solution u_ϵ as $\epsilon \to 0$ of the regularized problem, and passing to the limit, the following theorems have been established in [64]:

Theorem 6.3: *Let $g \in L^2(0, T; H^{-1}(\Omega))$ and $u_0 \in \mathcal{K}$. Then, there exists a unique solution of the following variational inequality problem: find $u \in \mathcal{K} \cap H^1(0, T; W_0^{1,2}(\Omega))$ such that $u(0) = u_0$ and*

$$\int_0^s \left[(\frac{\partial \phi}{\partial t} - g, \phi - u) + (\lambda^2 \nabla \frac{\partial \phi}{\partial t}, \nabla \phi - \nabla u)\right] dt$$

$$\leq \frac{1}{2}\|\phi(s) - u(s)\|_{L^2(\Omega)}^2 + \frac{\lambda^2}{2}\|\nabla \phi(s) - \nabla u(s)\|_{L^2(\Omega)}^2$$

$$- \frac{1}{2}\|\phi(0) - u(0)\|_{L^2(\Omega)}^2 + \frac{\lambda^2}{2}\|\nabla \phi(0) - \nabla u(0)\|_{L^2(\Omega)}^2$$

for any $\phi \in \mathcal{K} \cap H^1(0, T; W_0^{1,2}(\Omega))$.

Moreover, let u_ϵ^h be the standard continuous piecewise linear finite element Galerkin approximation of the regularized problem, it was shown that:

Theorem 6.4: *Let $g \in L^2(0, T; H^{-1}(\Omega))$ and $u_0 \in \mathcal{K}$. Also, assume, in the two-dimensional case, that $u_0 \in W^{2,p}(\Omega)$ for some $p > 2$. Then, the finite element solution u_ϵ^h converges weakly to the solution $u = H - H_1$ where*

H is the solution of (6.12) in $L^4(0,T;W_0^{1,4}(\Omega))$ and $H^1(0,T;W_0^{1,2}(\Omega))$ as $h \to 0$ and $\epsilon \to 0$ if $h^{2-4/p}/\epsilon$ is uniformly bounded and $h/\epsilon \to 0$.

Macroscopic models like the critical state models are particularly relevant to the device design using superconductors. There have been more studies made both from analytical and computational aspects on macroscopic models, see [34, 56, 58, 76, 79, 80, 116] and the references listed there.

7. The vortex state in the Bose-Einstein condensation

The 1995 experimental confirmation of Bose-Einstein condensation (BEC) in alkali-metal gases, a Nobel prize winning work in 2001, provides another avenue to study the phenomena of quantized vortices. In recent BEC experiments, vortices have been nucleated with the help of laser stirring and rotating magnetic traps. Remarkably, many of the phenomenological properties of quantized vortices have been well captured by mathematical models such as the Gross-Pitaevskii (G-P) equations.

The breakthrough development in BEC has attracted the attention of many mathematicians and computational scientists, in particular, a large community of researchers who have worked on NLS and related problems. The recent experimental confirmation of Bose-Einstein condensation (BEC) has once again drawn spotlight to the phenomena of quantized vortices. In recent years, the studies of quantized vortices in BECs have also becoming increasingly important.

Studies of vortices in the BEC often are based on the Gross-Pitaevskii equations:

$$\eta \frac{\partial u}{\partial t} - (\nabla - i\mathbf{A}(\mathbf{x}))^2 u + \frac{1}{\epsilon^2}(|u|^2 - a_\epsilon(\mathbf{x}))u = 0 \,. \tag{7.1}$$

Here, $\eta = i$ is for the real time dynamics. ϵ is proportional to the healing length. $\mathbf{A}(\mathbf{x}) = c(y, -x, 0)/2$ with c being the angular velocity. And,

$$a_\epsilon(\mathbf{x}) = a_0 - w_1^2 x^2 - w_2^2 y^2 - w_3^2 z^2 + \epsilon^2 c^2(x^2 + y^2 + z^2)$$

is a modified harmonic trapping potential with frequencies (w_1, w_2, w_3). For problems confined by a trapping potential, no particles escape to infinity, we thus can use the boundary condition $u = 0$ on the boundary of a large enough box. Note that the mass conservation condition

$$\frac{d}{dt} \int_{\mathbb{R}^d} |u(\mathbf{x}, t)|^2 d\mathbf{x} = 0$$

is automatically satisfied by the equation (7.1).

7.1. *Vortices in BEC confined in a rotating magnetic trap*

In some recent experiments where a rotating magnetic trap is applied to the BEC cloud, vortex nucleation is observed when the angular velocity is above certain critical velocity. Mathematical and computational studies on the vortex state in such experiments have been carried out based on the Gross-Pitaevskii theory.

For the case of rotating magnetic traps, the mathematical form of the G-P equations have close resemblance with the high-kappa Ginzburg-Landau (G-L) model we have studied in the context of superconductivity modeling. Utilizing the mathematical theory and the numerical codes developed for the G-L, we were able to build up a similar mathematical framework for rigorously characterizing the critical velocities for the vortex nucleation in a BEC cloud which is subject to a rotating magnetic trap. This also allowed us to make qualitative studies of the quantized vortex state [4].

As an illustration, for a properly scaled two dimensional version of the Gross-Pitaevskii energy of the form:

$$E(u) = \int_{\mathcal{D}} \left\{ |(\nabla - i\mathbf{A})u|^2 + \frac{1}{2\epsilon^2}(a_\epsilon(\mathbf{r}) - |u|^2)^2 \right\}$$

with $\mathbf{A}(x,y) = (cy, -cx)/2$ and $a_\epsilon(x,y) = a_0 - w_1^2 x^2 - w_2^2 y^2 + \epsilon^2 c^2(x^2 + y^2)$, some rigorous analytical results have been derived in [4] based on the asymptotic behavior as $\epsilon \to 0$ since in many of the experiments, ϵ ranges in $10^{-3} \sim 10^{-2}$ and may be viewed as a small constant. In the limit $\epsilon \to 0$, for fixed c, we have

$$|u|^2(x,y) = \begin{cases} a_0 - w_1^2 x^2 - w_2^2 y^2\,, & \text{for } w_1^2 x^2 + w_2^2 y^2 \leq a_0\,, \\ 0\,, & \text{for } w_1^2 x^2 + w_2^2 y^2 > a_0 \end{cases}$$

which gives the density profile in the Thomas-Fermi limit. The main ingredient of the analysis lies in the decoupling of the energy into three sources: a part coming from the state without vortices, another part from contribution of individual vortices and an additional part produced due to the rotation.

Based on the energy expansion, we get estimates on the critical angular velocity for the nucleation of n-vortices:

$$c_n \sim (1 + \lambda^2)\sqrt{\frac{\pi}{2\lambda}} \left\{ |\ln(\epsilon)| + \frac{n-1}{2}\ln\left((1 + \lambda^2)\sqrt{\frac{\pi}{2\lambda}}|\ln(\epsilon)|\right) \right\}.$$

This is in agreement with the experimental observation that for larger values of λ, that is, more intense anisotropy, the critical angular velocities have wider bands. In addition, it was shown that near the critical angular

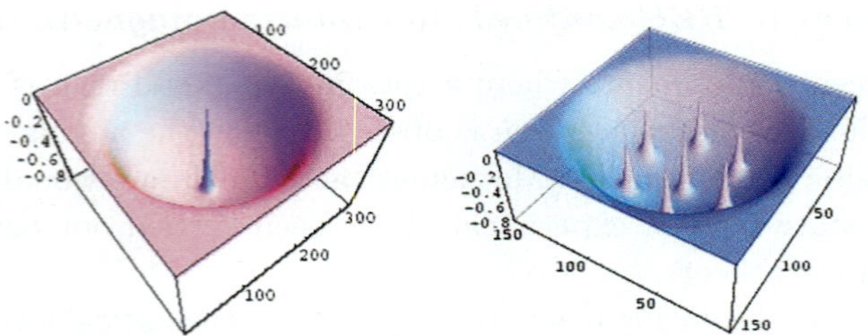

Fig. 23. 2d single vortex and multiple vortices solutions in a BEC with a rotating magnetic trap (shown $-|\psi|^2$).

Fig. 24. Vortex tubes in a BEC with a rotating magnetic trap.

velocity, the energetically favorable locations of the vortices $\{(x_i, y_i)\}$ of a n-vortex minimizer of the energy is related to the minimizer of

$$\sum_{i \neq j} \ln\left(|x_i - x_j|^2 + \frac{|y_i - y_j|^2}{\lambda^2}\right) - \alpha \sum_i (x_i^2 + y_i^2) \,.$$

where α is a given positive constant and λ is the aspect ratio of two dimensional harmonic magnetic trap. We refer to [4] for more detailed analysis and additional references.

To compute the ground state solution, one may proceed with the time dependent equation in the imaginary time, that is by taking $\eta = 1$:

$$\frac{\partial u}{\partial t} - (\nabla - i\mathbf{A})^2 u + \frac{1}{\epsilon^2}|u|^2 u - \frac{a_\epsilon(x)}{\epsilon^2} u = \mu_\epsilon(u)u \,, \tag{7.2}$$

where $\mu_\epsilon(u)$ denotes the Lagrange multiplier corresponding to the $\|u\| = 1$. Simple calculation shows that

$$\mu_\epsilon(u) = \int_{\mathcal{D}} \left\{ |(\nabla - i\mathbf{A})\,u|^2 + \frac{1}{\epsilon^2}|u|^4 - \frac{a_\epsilon(x)}{\epsilon^2}|u|^2 \right\} d\mathcal{D} \,.$$

For the time-dependent Gross-Pitaevskii equations, we have developed both finite element and finite difference spatial discretizations as well as explicit and implicit time integration schemes [4].

Another popular approach, as outlined in [57] is the splitting methods: given u_{n-1}, the numerical approximation of the 3D G-P equation at the time step $n-1$, one may proceed by alternatingly solve the following subproblems:

1 . For each position ρ, solve the ODE:

$$\eta u_t = (a_\epsilon(\rho) - |u|^2)u \quad \text{in} \quad [t, t + \delta_1] .$$

2 . For each (x, y), solve the linear equation in time and z :

$$\eta u_t = \partial_{zz} u \quad \text{in} \quad [t, t + \delta_2] .$$

3 . For each z, solve the linear equation in time and x, y :

$$\eta u_t = (\nabla - i\mathbf{A})^2 u , \quad \text{in} \quad [t, t + \delta_3] .$$

Here, δ_i's may be viewed as fractional time-steps.

Recently, we have made progress on the analysis of a class of splitting schemes for computing the ground state solutions of the BEC condensate based on the normalized gradient flow [11]. With the time-splitting scheme, it has become a popular approach to use spectral Galerkin or spectral collocation methods for the spatial discretization. For example, with a Fourier type spectral approximation, the nonlinear ODEs can be solved in real space while the linear PDEs can be solved in the spectral/frequency space. One can employ FFT to perform the transformation between the real space and frequency space representations. In [12], such methodology has been developed to solve the Gross-Pitaevskii equation with a rotation term, the same idea can also be applied to solve the time-dependent Ginzburg-Landau equations.

For real time integrations, the mass conservation constraint is automatically satisfied and the problem of simulating the Gross-Pitaevskii dynamics is similar to that of simulating the nonlinear Schrodinger equations. To preserve the long-time stability, we have also discussed symplectic and multi-symplectic schemes in [57]. For example, the three space dimensional time-dependent G-P equation possesses a multi-symplectic structure, namely, for $u = p + iq$, let $Z = (p, q, \mathbf{v}, \mathbf{v}_0)$ with $\mathbf{v} = \nabla p - \mathbf{A}q$, $\mathbf{v}_0 = \nabla q + \mathbf{A}p$), and

$$S(Z) = \frac{1}{2}(|\mathbf{v}|^2 + |\mathbf{v}_0|^2) + (\mathbf{A} \cdot \mathbf{v})q - (\mathbf{A} \cdot \mathbf{v}_0)p + \frac{1}{4}(a(\rho) - |p|^2 - |q|^2)^2 .$$

Then, the G-P equation can be rewritten as a multi-symplectic Hamiltonian system:

$$M\frac{\partial}{\partial t}Z + K_1\frac{\partial}{\partial x}Z + K_2\frac{\partial}{\partial y}Z + K_3\frac{\partial}{\partial z}Z = \nabla_Z S(Z)\,, \qquad (7.3)$$

where $\nabla_Z S(Z)$ denotes the gradient of the function $S = S(Z)$ with respect to the variable Z and

$$M = \begin{pmatrix} 0 & -1 & 0 & 0 & 0 & 0 & 0 & 0 \\ 1 & 0 & 0 & 0 & 0 & 0 & 0 & 0 \\ 0 & 0 & 0 & 0 & 0 & 0 & 0 & 0 \\ 0 & 0 & 0 & 0 & 0 & 0 & 0 & 0 \\ 0 & 0 & 0 & 0 & 0 & 0 & 0 & 0 \\ 0 & 0 & 0 & 0 & 0 & 0 & 0 & 0 \\ 0 & 0 & 0 & 0 & 0 & 0 & 0 & 0 \\ 0 & 0 & 0 & 0 & 0 & 0 & 0 & 0 \end{pmatrix}, \quad K_1 = \begin{pmatrix} 0 & 0 & 1 & 0 & 0 & 0 & 0 & 0 \\ 0 & 0 & 0 & 0 & 0 & 1 & 0 & 0 \\ -1 & 0 & 0 & 0 & 0 & 0 & 0 & 0 \\ 0 & 0 & 0 & 0 & 0 & 0 & 0 & 0 \\ 0 & 0 & 0 & 0 & 0 & 0 & 0 & 0 \\ 0 & -1 & 0 & 0 & 0 & 0 & 0 & 0 \\ 0 & 0 & 0 & 0 & 0 & 0 & 0 & 0 \\ 0 & 0 & 0 & 0 & 0 & 0 & 0 & 0 \end{pmatrix},$$

$$K_2 = \begin{pmatrix} 0 & 0 & 0 & 1 & 0 & 0 & 0 & 0 \\ 0 & 0 & 0 & 0 & 0 & 0 & 1 & 0 \\ 0 & 0 & 0 & 0 & 0 & 0 & 0 & 0 \\ -1 & 0 & 0 & 0 & 0 & 0 & 0 & 0 \\ 0 & 0 & 0 & 0 & 0 & 0 & 0 & 0 \\ 0 & 0 & 0 & 0 & 0 & 0 & 0 & 0 \\ 0 & -1 & 0 & 0 & 0 & 0 & 0 & 0 \\ 0 & 0 & 0 & 0 & 0 & 0 & 0 & 0 \end{pmatrix}, \quad K_3 = \begin{pmatrix} 0 & 0 & 0 & 0 & 1 & 0 & 0 & 0 \\ 0 & 0 & 0 & 0 & 0 & 0 & 0 & 1 \\ 0 & 0 & 0 & 0 & 0 & 0 & 0 & 0 \\ 0 & 0 & 0 & 0 & 0 & 0 & 0 & 0 \\ -1 & 0 & 0 & 0 & 0 & 0 & 0 & 0 \\ 0 & 0 & 0 & 0 & 0 & 0 & 0 & 0 \\ 0 & 0 & 0 & 0 & 0 & 0 & 0 & 0 \\ 0 & -1 & 0 & 0 & 0 & 0 & 0 & 0 \end{pmatrix}.$$

The system (7.3) has a multi-symplectic conservation law:

$$\frac{\partial}{\partial t}(-dp \wedge dq) + \nabla \cdot (dp \wedge d\mathbf{v} + dq \wedge d\mathbf{v_0}) = 0\,,$$

as well as the local energy conservation law:

$$\frac{\partial}{\partial t}\left[S - \frac{1}{2}(ZK_1Z_x + ZK_2Z_y + ZK_3Z_z)\right] + \nabla \cdot Z\left(\sum_{j=1}^{3} K_j\right)Z_t = 0\,.$$

Giving a uniform Cartesian grid, let Z_c denote the center average of the Z on the eight vertices, Z_{fx+} and Z_{fx-} be the averages of Z on the four vertices in each face in the x-direction. $Z_{fy\pm}$ and $Z_{fz\pm}$ follow similar convention. Let $Z^{n+1/2}$ be the averages of Z^n and Z^{n+1} at two consecutive time steps. Then the following difference scheme preserves the multi-symplectic property in the discrete sense:

$$M\frac{Z_c^{n+1} - Z_c^n}{\Delta t} + K_1\frac{Z_{fx+}^{n+1/2} - Z_{fx-}^{n+1/2}}{\Delta x} + K_2\frac{Z_{fy+}^{n+1/2} - Z_{fy-}^{n+1/2}}{\Delta y}$$

$$+K_3\frac{Z_{fz+}^{n+1/2} - Z_{fz-}^{n+1/2}}{\Delta y} = \nabla_Z S(Z_c^{n+1/2}) \, .$$

The development of efficient iterative schemes for the solution of the above difference approximation remains to be carried out in order to make the multi-symplectic integrator effective for the integration of multi-dimensional PDEs [57].

In a more recent, a class of accurate and efficient splitting schemes for computing the real time dynamics of the BEC condensate in a rotating trap has been studied in [12]. Numerical experiments were conducted to illustrate both the efficiency of the numerical schemes and the various interesting dynamics of the time dependent solutions.

For the imaginary time integration, by coupling existing simulation codes for the G-L models and new algorithmic improvement, we constructed a battery of computational algorithms and codes [58] to numerically simulate the experimental results of the ENS and MIT groups, see figure 25 for a particular comparison of experimental pictures and the result of numerical simulations with parameters taken from their experimental values.

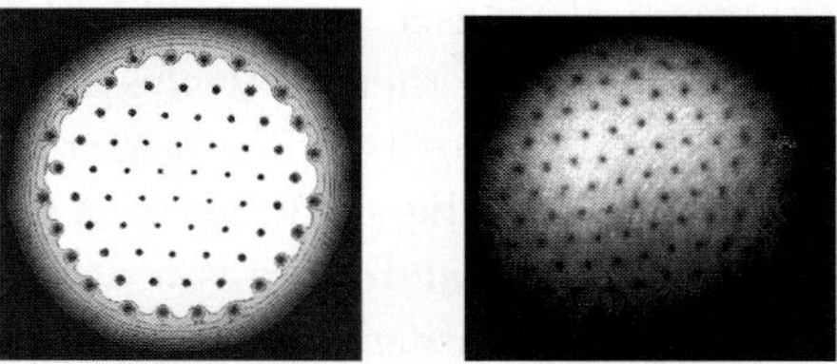

Fig. 25. Vortices in BEC: our simulation (left) vs the MIT experiment (right).

7.2. *Vortex shedding behind a stirring laser beam*

In BEC, the story of vortices, as put by Y. Pomeau, goes beyond the rotating magnetic trap. With a blue tune laser beam stirring the BEC cloud, energy dissipation has also been observed in recent experiments [131]. One may again attempt to apply the time-dependent G-P theory to provide theoretical interpretation of the experimental findings.

A time-dependent Gross-Pitaesvkii model in a moving frame and the associated Painlevé reduction were used [5]:

$$i\frac{\partial u}{\partial t} - \Delta u + iv\frac{\partial u}{\partial x_1} + \frac{1}{\epsilon^2}(|u|^2 - a(x))u = 0 \,. \qquad (7.4)$$

The equation is imposed outside an obstacle Ω, and on its boundary, we set $u = 0$. Here, v represents the velocity of the laser which takes up the interior of Ω.

The above equations provided good models for the study of the onset of dissipation, and the vortex-sound interaction. Similar analysis and simulations have been carried out for the liquid Helium where $a(x) \equiv 1$, see [88, 102]. A typical two-dimensional simulation is shown in Fig. 26.

Fig. 26.   Vortex shedding in a superflow around an obstacle.

In three dimensional BECs, due to the spatial inhomogeneities of the vortex density distribution, the onset of vortex shedding becomes much more complex. In [5], we found that there is always a drag around the laser beam for whatever values of the velocity of the stirrer and we analyzed the mechanism of vortex nucleation. The shedding of vortices (see figure 27) only starts at a threshold velocity and there is a drastic increase in drag. The critical velocity computed through our 3D simulations of the NLS dynamics is lower than the critical velocity obtained for the corresponding 2D problem at the center of the cloud and agrees well with experimental results.

8. Future challenges

No doubt, the biggest mystery in the field of superconductivity remains to be the understanding the microscopic nature of high-Tc superconductivity. This is often regarded as one of the most fundamental problems in modern physics, and progress along this direction certainly will generate new issues to be studied.

Fig. 27.   Vortex handles and shedding of vortex tubes near the condensate edge.

The further study of vortex state also bears significance in the understanding of some fundamental physical phenomena, for example, the study of vortices in superfluid is directly related to issues like superfluid turbulence, which, according to Feynman, is a state consisting of a disordered set of quantized vortices. Since the vortex structure is a topologically stable object in superfluids but only an idealization in classical fluids, superfluid turbulence may help to gain new understanding on the turbulence in general[8, 84, 106, 129, 149] such as the Kolmogorov spectrum. In addition, it also helps to study the interplay between superconductivity and ferromagnetism and antiferro-magnetism. The interests of the vortex state not only lie in the the study of the vortex structures for a few isolated vortices, but also in the study of the collective effect of a large number of vortices and vortex lattices and their interactions with material structure and defects as well as the impact of sample topology and geometry. Simulations of G-L models have been performed recently on structures like bucky balls [103]. Such studies are also related to the understanding of the vortex state in junction arrays and networks. In addition, the interactions of vortices with other matters, such as BEC vortices in optical traps, and those interacting with lasers, are also very interesting subjects to study.

In 2001, a team of Japanese physicists discovered superconductivity in an abundant Magnesium-Boron compound MgB_2, with a transition temperature about 40K, which is believed to be the limit of the conventional BCS superconductors. According to a recent account by Canfeld and Bodko: the discovery brings out a sensation *"of incredulous excitement swept over the solid-state physics community"*, *"Although 40K (or -233° C) may sound rather low, it was nearly double the record for compounds made of metals (about 23K for Niobium-based alloys, which are widely used in research and industry). A transition temperature that high can be achieved by technologies that cost much less than those needed to bring about superconductivity in the Niobium alloys"*. Numerical simulations of the G-L type models and for MgB2 thin films have also been carried out in the last few years. Due to the multi-band feature, this naturally brings out new analytical and numerical issues to be examined.

 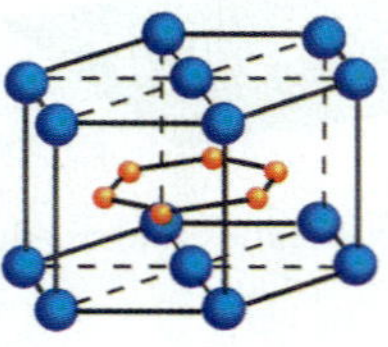

Fig. 28.   A buckyball and the atomic structure of MgB2.

There are many other mathematical issues remain to be studied further in relation to the topics presented here, for instance:

Vortex creep : thermal fluctuations cause motion, e.g., vibrations, in the vortices; these cause resistivity and could dislodge vortices from pinning sites. Though numerical simulations have been conducted, a rigorous mathematical theory is still limited. Such problems are also relevant to the study of the vortex glass and liquid states, as well as the critical fields and temperatures for transition between glass (vortices are stationary or pinned) and liquid states (movable vortices with potentially some resistivity).

Critical currents : too large applied currents or magnetic fields overcome the pinning forces and cause the vortices to move, thus inducing resistivity. Preliminary studies on the vortex states in the presence of an applied current has been made, numerical simulations of periodic vortex motion have also been carried out. A complete mathematical characterization is still lacking.

Vortices in 3D : the motion, tangling, orientation, pinning, etc. of vortex filaments in three-dimensions, as well as collective properties of a large number of filaments, their elastic and plastic flow properties.

As for problems in the technological applications of superconductivity, one may consider the optimal design of composite materials. for instance, to determine the best distribution of pinning sites so that, e.g., large currents can be supported, and to control the path and speed of vortices in motion which may be very important in device design.

On the computational side, there are still number of questions related to the rigorous analysis of the numerical algorithms, for instance the time-splitting schemes for the time-dependent nonlinear G-P equations. For the vortex density models, most of the numerical analysis have been limited to the two dimensional cases. Generalizations to the three dimensional models have not been explored so far.

From a practical point of view, numerical simulations of exceedingly large numbers of vortices based on the G-L models are of great interests but they remain computationally challenging, partly due to the intriguing properties associated with the vortex quantization effect. More specifically, for a square sample with n vortices in the interior, the phase angle of the complex order parameter will endure a change of $2n\pi$ around the boundary of the sample. If m points are needed to resolve a single period of phase winding, then $mn/4$ points will roughly be needed on each edge of the square. Thus, a uniform mesh will require up to $m^2n^2/16$ grid points. For large n and even moderate values of m, this can become very demanding computationally even for two dimensional problems, not to mention the more challenging three dimensional cases.

Naturally, adaptive schemes may save the computational cost significant, but with densely packed vortices, refinement may be required almost everywhere, and their efficiency are thus reduced. In particular, we note that it is not sufficient to only refine near the vortex cores as the correct resolution of the phase change is also very important, see Fig.29 for an interesting illustration on how the real and imaginary parts of the order parameter behave in comparison with the magnitude for a solution with 22 pairs of vortices in a spherical shell simulation [69]. It is easy to see that the variation in the phase variable starts to become increasingly dramatic when getting closer to the boundary (equator). Resolving the high oscillatory phase in order to reflect the correct topological quantization effect is a computational chanllenge that is similar to resolving the high frequency solution of Helmholtz equation [59]. Moreover, we observe an interesting phemenonon in terms of the dimension mismatch, namely, for a lattice like distribution of the vortices, as the diameter of the domain doubles, the number of vortices quadraples. Hence, even though the boundary size gets only doubled, the resolution needs to be improved four-folds. How to effectively tackling such solution behavior remains to be investigated.

On the other hand, for high-temperature superconductors, codes for mezoscale G-L models cannot hope to be of direct use in the design of devices due to the presence of large number of vortices. The number of vortices present in a superconducting sample of dimension, say, a millimeter,

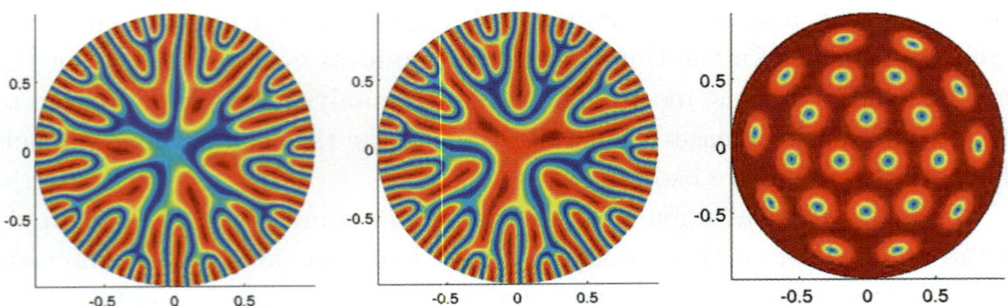

Fig. 29.   The absolute value of the real (left) and imaginary (center) parts of ψ and the magnitude $|\psi|$.

will contain a huge number vortices as the separation of individual vortices occurs on a typical length scale of 100 or so angstrom. Whenever computational or analytical results are used in such an environment, they are based on simple homogenized or macroscopic models, for example, mean field type models which describe the vortex state using a vortex density.

Since superconductivity is a phenomenon where the quantum mechanical effects show up in macroscopic behavior, it is perceivable that more efficient simulation schemes on the vortex state can be developed with a multi-scale approach that combines G-L model or vortex dynamic laws and the mean field models together.

9. Conclusion

Superconductivity is one of the grand challenges identified as being crucial to future economic prosperity and scientific leadership. In this paper, various methods for the numerical approximations of the Ginzburg-Landau models of superconductivity are discussed, with an emphasis on the application to the study of vortex dynamics.

From a practical point of view, large-scale numerical simulations of the magnetic vortices complement physical experiments due to the complex three-dimensional, time-dependent, stochastic and multi-scale nature of the phenomena. Thus, the development and analysis of efficient and reliable numerical algorithms remain important tasks. These algorithms and codes may ultimately prove to be useful by physicists and engineers in their study of superconducting phenomena and other related problems such as the BEC superfluidity.

Historically, the story of vortex dynamics in superfluids was first told almost a century ago with the study of liquid helium. For the last 50 years, the center of attention n quantized vortices has been devoted to the issues related to superconductivity. The recent experiments on BECs provided exciting breakthroughs in new techniques to probe the properties of superfluidity, there is no doubt more stories are to be told on the subject of quantized vortex dynamics in years to come.

Acknowledgment

The author wishes to thank all of his collaborators for joint works in the area covered by this lecture notes over the years, most of their names are provided in the references. He is grateful for their teaching on the subject and for sharing their insight and hard work. The author also would like to thank the Institute of Mathematical Sciences and the organizers of the special program at the National University of Singapore, in particular, Dr.Weizhu Bao, for the invitation and their warm hospitality. The research of the author was supported in part by several NSF grants (the most recent ones being the NSF DMS-0409297 and NSF-ITR DMR-0205232), grants from HKRGC and a state major basic research project in China.

References

1. A. ABRIKOSOV, *On the magnetic properties of superconductors of the second group*, Soviet Physics JETP, 5, pp.1174-1182, 1957.
2. S. ADLER AND T. PIRAN, *Relaxation methods for gauge field equilibrium equations*, Rev. Mod. Phys., 56, pp.1-40, 1984.
3. A. AFTALION AND Q. DU, *The bifurcation diagram for the Ginzburg-Landau system for superconductivity*, Physica D, 163, pp.94-105, 2001.
4. A. AFTALION AND Q. DU, *Vortices in the Bose-Einstein condensate: the critical velocities and energy diagrams in the Thomas-Fermi regime*, Physical Review A, 64, 063603(1-11), 2001
5. A. AFTALION, Q. DU AND Y. POMEAU, *Dissipative flow and vortex shedding in the Painlevé boundary layer of a Bose Einstein condensate*, Phy. Rev. Lett., 91, 090407, 2003
6. A. AFTALION, E. SANDIER AND S. SERFATY, *Pinning phenomena in the Ginzburg-Landau model of superconductivity*, J. Math. Pures Appl., 80, pp.339–372, 2001.
7. S. ALAMA, A. J. BERLINSKY, L. BRONSARD, AND T. GIORGI, *Vortices with antiferromagnetic cores in the SO(5) model of high-temperature superconductivity*, 60, pp.6901-6906, 1999.
8. T. ARAKI, M. TSUBOTA AND S. K. NEMIROVSKII, *Energy spectrum of superfluid turbulence with no normal-fluid component*, **Phys. Rev. Lett.**, 89, 145301(1-4), 2002.
9. B. BAELUS AND F. PEETERS, *Dependence of the vortex configuration on the geometry of mesoscopic flat samples*, Phys. Rev. B, 65, 104515, 2002.

10. B. BAELUS, F. PEETERS AND V. SCHWEIGERT, *Saddle-point states and energy barriers for vortex entrance and exit in superconducting disks and rings*, Phys. Rev. B, 63, 144517, 2001.

11. W. BAO AND Q. DU, *Computing the ground state of the BEC via normalized gradient flow*, SIAM J. Scientific Comp., 25, pp.1674-1697, 2004.

12. W. BAO, Q. DU AND Y. ZHANG, *Dynamics of rotating Bose-Einstein condensates and their efficient and accurate numerical computation*, SIAM J. Appl. Math., 66, pp. 758-786, 2006.

13. W. BAO, D. JAKSCH AND P. MARKOWICH, *Numerical solution of the Gross-Pitaevskii equation for Bose-Einstein condensation*, J. Comput. Phys., 187, pp. 318-342, 2003.

14. W. BAO AND J. SHEN, *A fourth-order time-splitting Laguerre-Hermite pseudo-spectral method for Bose-Einstein condensates*, SIAM J. Sci. Comput., 26, pp. 2010-2028, 2005.

15. W. BAO AND W. TANG, *Ground state solution of trapped interacting Bose-Einstein condensate by directly minimizing the energy functional*, J. Comput. Phys., 187, pp. 230-254, 2003.

16. P. BAUMAN, C. CHEN, D. PHILLIPS AND P. STERNBERG, *Vortex annihilation in nonlinear heat flow for Ginzburg-Landau systems*, Eur. J. Appl. Math. 6, pp.115-126, 1995.

17. P. BAUMAN, D. PHILLIPS, AND Q. TANG, *Stable nucleation for the Ginzburg-Landau system with an applied magnetic field*, Arch. Rational Mech. Anal., 142, pp.1–43, 1998.

18. C. BEAN, *Magnetization of hard superconductors*, **Phys. Rev. Lett.**, 8, pp. 250-253, 1962.

19. C. BEAN, *Magnetization of high-field superconductors*, *Rev. Mod. Phys.*, 36, pp. 31-39, 1964.

20. A. BERNOFF AND P. STERNBERG, *Onset of superconductivity in decreasing fields for general domains.* J. Math. Phys. 39, pp.1272–1284, 1998.

21. F. BETHUEL, H. BREZIS AND F. HELEIN, Ginzburg-Landau vortices, Birkhäuser, Boston, 1994.

22. M. BOCKO, A. HERR AND M. FELDMAN, *Prospects for quantum coherent computation using superconducting electronics*, IEEE Trans. Applied Superconductivity, 7, pp.3638-3641, 1997.

23. S. CHAPMAN, *A hierarchy of models for type-II superconductors*, SIAM Rev., 42, pp.555-598, 2000.

24. S. CHAPMAN, Q. DU AND M. GUNZBURGER, *A variable thickness thin film model for superconductivity*, ZAMP, 47, pp.410-431 1995.

25. S. CHAPMAN, Q. DU, M. GUNZBURGER, *A Ginzburg-Landau type model of superconducting/normal junctions including Josephson junctions*, Euro J Appl Math, 6, pp.97-114, 1995.

26. S. CHAPMAN, Q. DU, M. GUNZBURGER, *On the Lawrence–Doniach and anisotropic Ginzburg–Landau models for layered superconductors*, SIAM J. Appl. Math., 55, pp156-174, 1995.

27. S. CHAPMAN, Q. DU, M. GUNZBURGER AND J. PETERSON, *Simplified Ginzburg-Landau models for superconductivity valid for high kappa and high fields*, Adv. Math. Sci. Appl., 5, pp.193–218, 1995.

28. S. CHAPMAN AND D. HERON, *The motion of superconducting vortices in thin films of varying thickness*, SIAM J. Applied Mathematics, 58, pp.1808-1825, 1998

29. CHAPMAN, S., HOWISON, S. AND OCKENDON, J., *Macroscopic models of superconductivity*, SIAM Review, 34, pp.529-560, 1992.

30. S. CHAPMAN AND G. RICHARDSON: *Motion of vortices in type-II supercon-ductors*, SIAM J. Appl. Math., 55, pp. 1275-1296, 1995.

31. S. CHAPMAN, J. RUBINSTEIN, AND M. SCHATZMAN, *A mean-field model of superconducting vortices*, Euro. J. Appl. Math., 7, pp.97-111, 1996

32. Z. CHEN, *Mixed finite element methods for a dynamical Ginzburg-Landau model in superconductivity*, Numer Math, 76, pp.323-353, 1997

33. Z. CHEN AND S. DAI, *Adaptive Galerkin method with error control for a dynamical Ginzburg-Landau model in superconductivity*, SIAM J. Numer. Anal., 38, pp.1961-1985, 2001.

34. CHEN Z. AND Q. DU, *A non-conforming finite element methods for a mean field model of superconducting vortices*, Math. Modelling and Numer. Anal., 34, pp.687-706, 2000

35. Z. CHEN AND K. HOFFMANN, *Numerical solutions of an optimal control problem governed by a Ginzburg-Landau model in superconductivity*, Numer. Func Anal and Appl., 19, pp.737-757, 1998

36. sc J. Clem, *Two-dimensional vortices in a stack of thin superconducting films: A model for high-temperature superconducting multilayers*, Phys. Rev. B, 43, pp.7837-7846, 1991.

37. E. COSKUN AND M. KWONG, *Simulating vortex motion in superconducting films with the time-dependent Ginzburg-Landau equations*, Nonlinearity, 10, pp.579-593, 1997.

38. G. CRABTREE, G. LEAF, H. KAPER, V. VINOKUR, A. KOSHELEV, D. BRAUN, D. LEVINE, W. KKWOK AND J. FENDRICH, *Time-dependent Ginzburg-Landau simulations of vortex guidance by twin boundaries*, Physica C., 263, pp. 401-408, 1996

39. J. DEANG, Q. DU AND M. GUNZBURGER, *Stochastic dynamics of the Ginzburg-Landau vortices*, Physical Review B, 64, pp.52506-52510, 2001

40. J. DEANG, Q. DU AND M. GUNZBURGER, *Modeling and computation of random thermal fluctuations and material defects in the Ginzburg-Landau model for superconductivity*, J. Computational Physics, 181, pp.45-67, 2002

41. J. DEANG, Q. DU, M. GUNZBURGER AND J. PETERSON, *Vortices in su-perconductors: modeling and computer simulations*, Philos. Trans. Roy. Soc. London A, 355, pp.1957-1968, 1997;

42. K. DEMACHI, Y. YOSHIDA, H. ASAKURA AND K. MIYA, *Numerical analysis of magnetization processes in type II superconductors based on Ginzburg-Landau theory*, IEEE Transactions on Magnetics 32, pp.1156-1159, 1996

43. DING S. AND DU Q., *Critical magnetic field and asymptotic behavior for superconducting thin films*, SIAM Math. Anal., 34, pp.239-256, 2002

44. S. DING AND Q. DU, *The Global Minimizers and vortex solutions to a Ginzburg-Landau model of superconducting films*, Comm on Pure and Ap-plied Analysis, 1, pp327-340, 2002.

45. S. DING AND Z. LIU, *Asymptotic behavior of a class of Ginzburg-Landau functionals*, Chinese J. Contemp. Math., **18**, pp.239–248, 1997.

46. S. DING, Z. LIU, AND YU, W., *A variational problem related to the Ginzburg-Landau model of superconductivity with normal impurity inclusions*, SIAM J. Math. Anal., 29, pp.48-68, 1998

47. M. DORIA, J. GUBERNATIS AND D.RAINER, *Solving the Ginzburg Landau equations by simulated annealing*, Phys. Rev. B, 41 pp.6335-6340, 1990.

48. A. DORSEY, *Vortex motion and the Hall effect in type-II superconductors: A time- dependent Ginzburg-Landau theory approach*, Phys. Rev. B, 46, pp.8376–8392, 1992.

49. A. DORSEY, M. HUANG, AND M. FISHER, *Dynamics of the normal to superconducting vortex-glass transition: Mean-field theory and fluctuations*, Phys. Rev. B 45, 523, 1992.

50. Q. DU, *Time dependent Ginzburg-Landau models for superconductivity*, in Proceedings of the world congress of nonlinear analysts 1992 (Lakshmikantham ed.), de Gruyter, Berlin, pp. 3789-3801, 1996

51. Q. DU, *Finite element methods for the time dependent Ginzburg-Landau model of superconductivity*, Comp. Math. Appl., 27, pp.119-133, 1994.

52. Q. DU, *Global existence and uniqueness of solutions of the time-dependent Ginzburg-Landau equations in superconductivity*, Applicable Anal., 52, pp.1-17, 1994.

53. Q. DU, *Computational methods for the time dependent Ginzburg-Landau model for superconductivity*, in *Numerical methods for applied sciences*, edited by W. Cai, et. al., Science Press, New York, pp.51-65, 1996.

54. Q. DU, *Discrete gauge invariant approximations of a time-dependent Ginzburg-Landau model of superconductivity*, Mathematics of Computation, 67, pp.965-986, 1998.

55. Q. DU, *Studies of a Ginzburg-Landau model for d-wave superconductors*, SIAM J. Applied Math., 59, pp.1225-1250, 1999.

56. Q. DU, *Convergence analysis of a numerical method for a mean field model of superconducting vortices*, SIAM J. Numer. Anal., 37, pp. 911–926, 2000

57. Q. DU, *Numerical computations of quantized vortices in Bose-Einstein condensate*, in Recent Progress in computational and applied PDEs, Proceedings for the international conference held in ZhangJiaJie 2001, edited by T.Chan, et. al., Kluwer Academic Publisher, pp.155-168, 2002.

58. Q. DU, *Diverse vortex dynamics in superfluids*, Contemp Math. 329, pp.105-117, AMS, 2003

59. Q. DU, *Numerical approximations of Ginzburg-Landau models of superconductivity*, Journal of Mathematical Physics, to appear

60. Q. DU, V. FABER, AND M. GUNZBURGER, *Centroidal Voronoi tessellations: Applications and algorithms*, SIAM Review, 41, pp.637-676, 1999.

61. Q. DU AND P. GRAY, *High-kappa limit of the time dependent Ginzburg-Landau model for superconductivity*, SIAM J. Appl. Math., 56, pp.1060-1093, 1996.

62. Q. DU AND P. GRAY, *Numerical algorithmss of the of Lawrence-Doniach models and its parallel implementation*, SIAM J. Sci. Comp., 20, pp.2122-2139, 1999.

63. Q. Du, M. Gunzburger, and L. Ju, *Constrained centroidal Voronoi tessellations on general surfaces*, SIAM J. Sci. Comput., 24, pp.1488-1506, 2003.

64. Q. Du, M. Gunzburger, and H. Lee, *Analysis and computation of a mean field model for superconductivity*, Numer. Math., 81, pp539-560, 1999.

65. Q. Du, M. Gunzburger M. and J. Peterson, *Analysis and approximation of the Ginzburg-Landau model of superconductivity*, SIAM Review, 34, pp. 54-81, 1992.

66. Q. Du, M. Gunzburger and J. Peterson, *Solving the Ginzburg-Landau equations by finite element methods*, Physical Rev. B., 46, pp.9027-9034, 1992.

67. Q. Du, M. Gunzburger and J. Peterson, *Computational simulations of type-II superconductivity including pinning mechanisms*, Phys. Rev. B, 51, pp. 16194-16203, 1995.

68. Q. Du and L. Ju, *Numerical simulation of the quantized vortices on a thin superconducting hollow sphere*, J. Computational Phys., 201, pp.511-530, 2004.

69. Q. Du and L. Ju, *Approximations of a Ginzburg-Landau model for superconducting hollow spheres based on spherical centroidal Voronoi tessellations*, Mathematics of Computation, 74, pp.1257-1280, 2005.

70. Q. Du, and C. Liu, *Stability of Ginzburg-Landau vortices under the influence of a small current*, in preparation.

71. Q. Du, R. Nicolaides and X. Wu, *Analysis and convergence of a covolume approximation of the Ginzburg-Landau models of superconductivity*, SIAM J. Numer. Anal., 35, pp. 1049-1072, 1998.

72. Q. Du and J. Remski, *Simplified models of superconducting normal superconducting junctions and their numerical approximations*, European J. Appl. Math., 10 , pp.1-25, 1999.

73. Q. Du and J. Remski, *Limiting models for Josephson junctions and superconducting weak links*, J. Math Anal Appl, 266, pp.357-382, 2002.

74. Q. Du and X. Wu, *Numerical approximation of the three dimensional Ginzburg-Landau equations using artificial boundary conditions*, SIAM J. Numer. Anal., 36, pp.1482-1506, 1999

75. Q. Du and Y. Yang, *The critical temperature and gap solution of the Bardeen-Cooper-Schrieffer theory of superconductivity*, Lett. Math. Phys., 29, pp133-150, 1993

76. Q. Du and P. Zhang, *Existence of weak solutions to some vortex density models*, SIAM Journal on Math. Anal., 34, pp.1278-1298, 2003.

77. W. E., *Dynamics of vortices in Ginzburg-Landau theories with applications to superconductivity*, Physic. D, 77, pp. 383-404, 1994.

78. W. E: *Dynamics of vortex liquid in Ginzburg-Landau theories, with applications to superconductivity*, Phys. Rev. B, 50, pp.1126–1135, 1994

79. C. Elliott, D. Kay and V. Styles, *A finite element approximation of a variational inequality formulation of Bean's model for superconductivity*, SIAM J. Numerical Analysis, 42, pp.1324-1341, 2004

80. C. Elliott and V. Styles, *Numerical analysis of a mean field model of superconductivity*, IMA J. Numer. Anal., 21, pp. 1–51, 2001.

81. C. Elliott, R. Schätzle and B. Stoth, *Viscosity solutions of a degenerate parabolic-elliptic system arising in the mean-field theory of superconductivity*, Arch. Rational Mech. Anal., 145, pp.99-127, 1998.

82. Y. Enomoto and R. Kato, *Computer simulation of a two-dimensional type-II super-conductor in a magnetic field*, J. Phys.: Condens. Matter 3, pp.375-380, 1991.

83. R. Feynman, *Application of quantum mechanics to liquid helium*, in Progress in Low Temperature Physics, edited by C. Groter, North-Holland, Amsterdam, Vol 1, chapter 2.

84. A. Finne, T. Araki, B. Blaauwgeers, V. Eltsov, N. Kopnin, M. Krusius, L.Skrbek, M. Tsubota and G. Volovik, *An intrinsic velocity-independent criterion for superfluid turbulence*, **Nature**, 424, 1022-1025, 2003.

85. J. Fleckinger, H. Kaper, P.Takac, *Dynamics of the Ginzburg-Landau Equation of Superconductivity*, Nonlinear anal. TMA, 32, pp.647-665, 1998.

86. A. Filippov, A. Radievsky, and A. Zelster, *Nucleation at the fluctuation induced first order phase transition to superconductivity*, Phy. Letters A, 192, 131, 1994.

87. H. Frahm, S. Ullah and A. Dorsey, *Flux dynamics and the growth of the superconducting phase*, Phys. Rev. Lett., 66, pp.3067–3070, 1991.

88. T. Frisch, Y. Pomeau, and S. Rica, *Transition to dissipation in a model of superflow*, Phys. Rev. Lett. 69, 1644, 1992.

89. V. Ginzburg and L. Landau, Theory of superconductivity, Zh. Eksp. Teor. Fiz. , 20, pp.1064-1082, 1950

90. D. Givoli and J. Keller, *A finite element method for large domains*, Comp. Meth. Appl. Mech. Engng., 76 pp.41-66, 1989.

91. W. Gropp, , H. Kaper, G. Leaf, D. Levine, M. Plumbo, and V. Vinokur, *Numerical simulation of vortex dynamics in type-II superconductors*, J. Comp. Phys., 123, pp. 254-266, 1996.

92. M. Gunzburger, L. Hou and S. Ravindran, *Analysis and approximation of optimal control problems for a simplified Ginzburg-Landau model of superconductivity*, Numer Math, 77, pp.243-268, 1997

93. D. Gunter, H. Kaper, G. Leaf, *Implicit integration of the time-dependent Ginzburg-Landau equations of superconductivity*, SIAM Journal Scientific Computing 23, pp.1943-1958, 2002.

94. H. Han and W. Bao, *Error estimates for the finite element approximation of problems in unbounded domains*, SIAM J. Numer. Anal., Vol. 37, pp.1101-1119, 2000.

95. P. Hohenberg and B. Halperin, *Theory of dynamic critical phenomena*, Rev. of Mod. Phy., 49, pp.453, 1977.

96. K. Hoffmann and J. Zou, *Finite element analysis on the Lawrence-Doniach model for layered superconductors*, Num Func Anal and Optimization, 18, pp.567-589, 1997

97. Y. Huang and W. Xue, *Convergence of finite element approximations and multilevel linearization for Ginzburg-Landau model of d-wave superconductors*, Adv in Compu Math, 17, pp.309-330, 2002

98. J. IVARSSON, *A posteriori error analysis of a finite element method for the time-dependent Ginzburg-Landau equations*, preprint 1998-24, Department of Mathematics, Chalmers University of Technology.

99. H. JADALLAH, J. RUBINSTEIN, AND P. STERNBERG, *Phase Transition Curves for Mesoscopic Superconducting Samples*, Phys. Rev. Lett, 82, pp. 2935-2938, 1999.

100. R. JERRARD AND M. SONER, *Dynamics of Ginzburg-Landau vortices*, Arch. Rational Mech. Anal. 142 (1998), no. 2, 99–125.

101. H. JIAN AND B. SONG, *Vortex dynamics of Ginzburg-Landau equations in inho mogeneous superconductivity*, J. Diff. Eqns., 170(1), (2001), 123-141.

102. C. JOSSERAND, Y. POMEAU AND S. RICA, *Vortex shedding in a model of superflow*, Physica D. 134, 111 (1999).

103. M. KATO AND O. SATO, *Numerical solution of Ginzburg-Landau equation for superconducting networks*, Physica C, 392, pp.396-400, 2003

104. R. KATO, Y. ENOMOTO AND S. MAEKAWA, *Computer simulations of dynamics of flux lines in type-II superconductors*, Phys. Rev. B, 44, pp.6916-6920, 1991.

105. S. KIM, C. HU AND M. ANDREWS, *Steady-state and equilibrium vortex configurations, transitions, and evolution in a mesoscopic superconducting cylinder*, Physical Review B 69, 094521, 2004.

106. M. KOBAYASHI AND M. TSUBOTA, *Kolmogorov spectrum of superfluid turbulence: numerical analysis of the Gross-Pitaevskii equation with the small scale dissipation*, cond-mat/0411750

107. A. LARKIN, *Effect of inhomogeneities on the structure of the mixed state of superconductors*, Soviet Physics JETP (English translation), 31, 784, 1970.

108. W. LAWRENCE AND S. DONIACH, *Theory of layer structure superconductors*, Proc. 12th Inter. Conf. on Low Temperature Physics, Academic Press of Japan, Kyoto, pp.361-362, 1971.

109. M. MACHIDA AND H. KABURAKI, *Direct numerical experiment on two-dimensional pinning dynamics of a three-dimensional vortex line in layered superconductors*, Physical Review Letters, 75, 3178-3181, 1995.

110. H. LIAO, S. ZHOU, X. SHI, *Simulating the time-dependent Ginzburg-Landau equations for type-II superconductors by finite-difference method*, Chinese Phy., 13, pp.737-745, 2004.

111. K. LIKHAREV, *Superconducting weak links*, Rev. Mod. Phys., 51, pp101-161, 1979.

112. F. LIN, *Complex Ginzburg-Landau equations and dynamics of vortices, fila ments, and codimension-2 submanifolds*, Comm. Pure Appl. Math., 51, pp.385-441, 1998

113. F. LIN AND Q. DU, *Ginzburg-Landau vortices, dynamics, pinning and hysteresis*, SIAM J. Mathematical Analysis, 28, pp. 1265-1293, 1999.

114. F. LIN AND C. LIN, *Vortex state of d-wave superconductors in the Ginzburg-Landau energy*, SIAM J. Mathematical Analysis, 32, pp. 493-503, 2001.

115. F. LIN, AND J. XIN, *A unified approach to vortex motion laws of complex scalar field equations*, Math. Res. Lett. 5, pp.455–460, 1998

116. F. LIN AND P. ZHANG, *On the hydrodynamic limit of Ginzburg-Landau vortices*, Discrete Contin. Dynam. Systems, 6, pp. 121–142, 2000

117. F. LIU, M. MONDELLO AND N. GOLDENFELD, *Kinetics of the superconducting transition*, Phys. Rev. Lett. 66, pp.3071-3074, 1991.

118. K. LU AND X. PAN, *Estimates of the upper critical field for the Ginzburg-Landau equations of superconductivity*, Phys. D, 127 pp.73–104, 1999.

119. M. MACHIDA AND H. KABURAKI, *Direct simulation of the time-dependent Ginzburg-Landau equation for type-II superconducting thin film*, Phys. Rev. Lett. 71, pp.3206-3209, 1993.

120. S. MAURER, N. YEH, T. TOMBRELLO, *Vortex pinning by cylindrical defects in type-II superconductors: Numerical solutions to the Ginzburg-Landau equations*, Physical Review B 54, pp.15372-15379, 1996.

121. V. MOSHCHALKOV, L. GIELEN, C. STRUNK, R. JONCKHEERE, X. QIU, C. VAN HAESENDONCK AND Y. BRUYNSERAEDE, *Effect of sample topology on the critical fields of mesoscopic superconductors*, Nature, 373, pp.319–321, 1995.

122. M. MU *A Linearized Crank-Nicolson-Galerkin Method for the Ginzburg-Landau Model*, SIAM J. Sci Comp, 18, pp.1028-1039, 1997.

123. M. MU AND Y. HUANG, *An alternating Crank-Nicolson method for decoupling the Ginzburg-Landau equations*, SIAM J. Numer Anal, 35, pp1740-1761, 1998

124. J. NEU, *Vortices in complex scalar fields*, Physica D., 43, pp.385–406, 1990.

125. J. NEUBERGER AND R. RENKA, *Sobolev Gradients and the Ginzburg-Landau Functional*, SIAM J. Sci. Comput. 20, pp.582-590,1998.

126. J. NEUBERGER AND R. RENKA, *Numerical approximation of critical points of the Ginzburg-Landau functional*, Nonlinear Analysis- theory methods and appl, 47, pp.3259-3270, 2001

127. J. NEUBERGER AND R. RENKA, *Computational simulation of vortex phenomena in superconductors*, Electronic Journal of Differential Equations, 10, pp 241-250, 2003.

128. R. NICOLAIDES, *Direct discretization of planar div-curl problems*, SIAM J. Numer. Anal., 29, pp. 32-56, 1992

129. C. NORE, M. ABID AND M. BRACHET, *Decaying Kolmogorov turbulence in a model of superflow*, **Phys. Fluids**, 9, 2644-2669, 1997.

130. C. OLSON, G.ZIMANYI, A.KOLTONAND N. GRONBECH-JENSEN, *Static and Dynamic Coupling Transitions of Vortex Lattices in Disordered Anisotropic Superconductors*, Phys. Rev. Lett. 85, pp5416-5420, 2000

131. R. ONOFRIO, C. RAMAN, J. M. VOGELS, J. R. ABO-SHAEER, A. P. CHIKKATUR, AND W. KETTERLE, *Observation of Superfluid Flow in a Bose-Einstein Condensed Gas*, Phys. Rev. Lett. 85, pp.2228-2231, 2000.

132. R. OUBOTER, *Heike Kamerlingh Onnes's Discovery of Superconductivity*, Scientific American, pp.98-103, 1997.

133. L. PERES, AND J. RUBINSTEIN, *Vortex Dynamics for U(1)- Ginzburg-Landau Models*, Phys. D. , 64, pp.299–309, 1993

134. L. PISMEN, AND J. D. RODRIGUEZ, *Mobilities of singularities in dissipative Ginzburg-Landau equations*, Phys. Rev. A , 42, pp.2471–2474, 1990

135. X. PAN, *Surface superconductivity in applied magnetic fields above Hc2*, Comm. Math. Phys. 228, pp.327-370, 2002

136. L. PRIGOZHIN, *The Bean model in superconductivity: Variational formulation and numerical solution,* **J. Comput. Phys.**, 129, pp. 190-200, 1996.

137. L. PRIGOZHIN, *Analysis of critical-state problems in type-II superconductivity,* **IEEE Trans. Appl. Super.**, 7, pp. 3866-3873, 1997.

138. Y. REN, J. XU AND C. TING, *Ginzburg-Landau Equations and Vortex Structure of a $d_{x^2-y^2}$ Superconductor,* Phys. Rev. Lett., 74, pp.3680-3683, 1995

139. J. RUBINSTEIN, M. SCHATZMAN AND P. STERNBERG, *Ginzburg-Landau model in thin loops with narrow constrictions,* to appear in SIAM, J. Appl Math, 2005

140. J. RUBINSTEIN AND P. STERNBERG, *On the Slow Motion of Vortices in the Ginzburg–Landau Heat Flow,* SIAM Journal on Mathematical Analysis, 26, pp.1452-1466, 1995

141. J. RUBINSTEIN AND P. STERNBERG, *Limits of a Ginzburg-Landau model with codimension-one defects,* J. Math. Phys., 44, pp.1240-1251, 2003.

142. D. SAINT-JAMES AND P. DE GENNES, *Onset of superconductivity in decreasing fields,* Phys. Lett., 7, pp.306-307, 1963

143. E. SANDIER AND S. SERFATY, *On the energy of type-II superconductors in the mixed phase,* Revs. Math. Phys., 12, pp.1219-1257, 2000

144. V. SCHWEIGERT, F. PEETERS AND P. DEO, *Vortex Phase Diagram for Mesoscopic Superconducting Disks,* Phys. Rev. Lett., 81, pp. 2783-2786, 1998.

145. S. SERFATY, *Local minimizer for the Ginzburg-Landau energy near critical magnetic field: Part I and II,* Comm. Contemp. Math., 1, (1999), 213-254. and 295-333.

146. S. SERFATY, *Stable configurations in superconductivity: uniqueness, multiplicity, and vortex-nucleation,* Arch. Rat. Mech. Anal., 149, (1999), 329-365.

147. Q. TANG AND S. WANG, *Time dependent Ginzburg-Landau superconductivity equations,* Physica D, 88, pp.139-166, 1995.

148. M. TINKHAM, *Introduction to Superconductivity,* 2nd ed., McGraw-Hill, New York, 1994.

149. M. TSUBOTA, T. ARAKI AND C. BARENGHI, *Rotating superfluid turbulence,* **Phys. Rev. Lett.**, 90, 205301(1-4), 2003.

150. Z. WANG AND C. HU, *A numerical relaxation approach for solving the general Ginzburg-Landau equations for type-II superconductors,* Phys. Rev. B 44, pp.11918-11923, 1991

151. Z. WANG AND Q. WANG, *Vortex state and dynamics of a d-wave superconductor: Finite-element analysis,* Phy. Rev B 55, pp.11756-11765, 1997

152. T. WINIECKI AND C. ADAMS, *A fast semi-implicit finite difference method for the TDGL equations,* J. Comp. Phys, 179, pp127-139, 2002.

153. D. YU, Natural Boundary Integral Method and Its Applications, Kluwer Publisher, V 539, 2002.

154. M. ZHAN, *Finite element analysis and approximations of phase-lock equations of superconductivity,* Discrete Continuous Dynamical Systems B, 2, pp.95-108, 2002.

The Nonlinear Schrödinger Equation and Applications in Bose-Einstein Condensation and Plasma Physics

Weizhu Bao

Department of Mathematics
Center for Computational Science and Engineering
National University of Singapore
117543 Singapore
E-mail: bao@math.nus.edu.sg,
URL: http://www.math.nus.edu.sg/~bao/

Contents

1. Introduction

The Schrödinger equation was proposed to model a system when the quantum effect was considered. For a system with N particles, the Schrödinger equation is defined in $3N + 1$ dimensions. With such high dimensions, even use today's supercomputer, it is impossible to solve the Schrödinger equation for dynamics of N particles with $N > 10$. After assumed Hatree or Hatree-Fork ansatz, the $3N + 1$ dimensions linear Schrödinger equation was approximated by a $3 + 1$ dimensions nonlinear Schrödinger equation (NLSE) or Schrödinger-Poisson (S-P) system. Although nonlinearity in NLSE brought some new difficulties, but the dimensions were reduced significantly compared with the original problem. This opened a light to study dynamics of N particles when N is large. Later, it was found that NLSE had applications in different subjects, e.g. quantum mechanics, solid state physics, condensed matter physics, quantum chemistry, nonlinear optics, wave propagation, optical communication, protein folding and bending, semiconductor industry, laser propagation, nano technology and industry, biology etc. Currently, the study of NLSE including analysis, numerics and applications becomes a very important subject in applied and computational mathematics. This study has very important impact to the progress of other science and technology subjects.

A typical application of NLSE is for wave motion and interaction in plasma physics where the Zakharov system (ZS) was derived by Zakharov [121] in 1972 for governing the coupled dynamics of the electric-field amplitude and the low-frequency density fluctuations of ions. Then it has become commonly accepted that ZS is a general model to govern interaction of dispersive wave and nondispersive (acoustic) wave. It has important applications in plasma physics (interaction between Langmuir and ion acoustic waves [121, 101]), in the theory of molecular chains (interaction of the intramolecular vibrations forming Davydov solitons with the acoustic disturbances in the chain [39]), in hydrodynamics (interaction between short-wave and long-wave gravitational disturbances in the atmosphere [110, 40]), and so on. In three spatial dimensions, ZS was also derived to model the collapse of caverns [121]. Later, the standard ZS was extended to generalized Zakharov system (GZS) [72, 73], vector Zakharov system (VZS) [113] and vector Zakharov system for multi-component plasma (VZSM) [72, 73].

In this chapter, we first review derivation of NLSE from wave propagation and Bose-Einstein condensation (BEC). Then we present variational formulation of NLSE including conservation laws, Lagrangian structure,

Hamiltonian structure and variance identity. Plane and soliton wave solutions, existence/blowup results of NLSE are then presented. Ground, excited and central vortex states of NLSE with an external potential are studied. We also study formally semiclassical limits of NLSE by WKB expansion and Wigner transform when the (scaled) Planck constant $\varepsilon \to 0$. In addition, numerical methods for computing ground states and dynamics of NLSE are presented and numerical results are also reported. Then we review derivation of VZS from the two-fluid model [113] for ion-electron dynamics in plasma physics and generalize VZS to VZSM, reduce VZSM to generalized vector Zakharov system (GVZS), GVZS to GZS or vector nonlinear Schrödinger (VNLS) equations, and GZS to NLSE, as well as generalize GZS and GVZS with a linear damping term to arrest blowup. Conservation laws of the systems and well-posedness of GZS are presented, and plane wave, soliton wave and periodic wave solutions of GZS are reviewed. Furthermore, we present a time-splitting spectral (TSSP) method to discretize GZS and compare it with the standard Crank-Nicolson finite difference (CNFD) method.

Throughout this notes, we use f^*, Re(f) and Im(f) denote the conjugate, real part and imaginary part of a complex function f respectively. We also adopt the standard Sobolev norms.

2. Derivation of NLSE from wave propagation

In this section, we review briefly derivation of NLSE from wave propagation, i.e. parabolic or paraxial approximation for forward propagation time harmonic waves, to analyze high frequency asymptotics.

The wave equation

$$\frac{1}{c^2} \frac{\partial^2 u(\mathbf{x}, t)}{\partial t^2} - \triangle u(\mathbf{x}, t) = 0, \qquad \mathbf{x} \in \mathbb{R}^3, \tag{2.1}$$

where $\mathbf{x} = (x, y, z)$ is the Cartesian coordinate, t is time and $c = c(\mathbf{x}, |u|)$ is the propagation speed, has time harmonic solutions of the form $e^{i\omega t} u(\mathbf{x})$ with the complex wave function u satisfying the Helmholtz or reduced wave equation

$$\triangle u(\mathbf{x}) + \frac{\omega^2}{c^2}\, u = 0, \qquad \mathbf{x} \in \mathbb{R}^3. \tag{2.2}$$

Let c_0 be a uniform reference speed, $k_0 = \omega/c_0$ be the wave number and $n(\mathbf{x}, |u|) = c_0/c(\mathbf{x}, |u|)$ be the index of refraction. The reduced wave equation has then the form

$$\triangle u(\mathbf{x}) + k_0^2 n^2(\mathbf{x}, |u|)u = 0. \tag{2.3}$$

When wave propagates in a uniform medium, $n(\mathbf{x}, |u|) = 1$; in a linear medium, $n(\mathbf{x}, |u|) = n(\mathbf{x})$; and in a Kerr medium, $n(\mathbf{x}, |u|) = \sqrt{1 + 4n_2|u|^2/n_0}$ with n_0 linear index of refraction and n_2 Kerr coefficient.

When waves are approximately plane and move in one direction primarily, say the z direction, e.g. propagation of laser beams, we look for solutions of the form

$$u(x, y, z) = e^{ik_0 z}\psi(\mathbf{x}, z) \tag{2.4}$$

where $\mathbf{x} = (x, y)$ denotes the transverse variables. We insert (2.4) into the reduced wave equation (2.3) and get

$$2ik_0\psi_z + \triangle_\perp \psi + k_0^2\mu(\mathbf{x}, z, |\psi|)\psi + \psi_{zz} = 0, \tag{2.5}$$

where $\triangle_\perp$ is the Laplacian in the transverse variables and $\mu(\mathbf{x}, z, |\psi|) = n^2(\mathbf{x}, z, |\psi|) - 1$ is the fluctuation in the refractive index. Note that the direction of propagation z ploys the role of time and $-k_0^2\mu(\mathbf{x}, z, |\psi|)$ is the (time dependent) potential.

Introduce nondimensional variables:

$$\tilde{x} = \frac{x}{r_0}, \qquad \tilde{y} = \frac{y}{r_0}, \qquad \tilde{t} = \frac{z}{k_0 r_0^2}, \qquad \tilde{\psi}(\tilde{x}, \tilde{y}, \tilde{t}) = \frac{\psi(x, y, z)}{\psi_s}, \tag{2.6}$$

where r_0 is the dimensionless length unit, e.g. width of the input laser beam, and ψ_s is dimensionless unit for ψ to be determined. Plugging (2.6) into (2.5), multiplying by $r_0^2/2$, and then removing all $\tilde{}$, we get the following dimensionless equation:

$$i\psi_t = -\frac{1}{2}\triangle_\perp \psi + f(\mathbf{x}, t, |\psi|)\psi - \frac{\delta}{2}\psi_{tt}, \tag{2.7}$$

where $\delta = 1/r_0^2 k_0^2$ and the real-valued function f depends on μ. Due to the input beam width $r_0 \gg \lambda = 2\pi/k_0$, we get

$$\delta/2 = \lambda^2/8\pi^2 r_0^2 \ll 1.$$

Thus we drop the nonparaxial term ψ_{tt} in (2.7) and obtain the NLSE:

$$i\psi_t = -\frac{1}{2}\triangle_\perp \psi + f(\mathbf{x}, t, |\psi|)\psi, \tag{2.8}$$

Of course (2.7) is only an approximation to the full reduced wave equation and it is valid when the variations in the index of refraction are smooth and the bulk of the wave energy is away from boundaries. This important and very useful approximation for wave propagation is well suited for numerical approximation since we now have an initial value problem for ψ, assuming that $\psi(\mathbf{x}, 0)$ is known, rather than a boundary value problem for u.

When $n(\mathbf{x}, z, |u|) = 1$ in (2.3), then $\mu(\mathbf{x}, z, |\psi|) = 0$ in (2.5) and $f(\mathbf{x}, t, |\psi|) = 0$ in (2.7), thus (2.7) collapses to the free Schrödinger equation:

$$i\psi_t = -\frac{1}{2} \triangle_\perp \psi. \tag{2.9}$$

When $n(\mathbf{x}, z, |u|) = n(\mathbf{x}, z)$ in (2.3), then $\mu(\mathbf{x}, z, |\psi|) = \mu(\mathbf{x}, z)$ in (2.5) and $f(\mathbf{x}, t, |\psi|) = V(\mathbf{x}, t)$ in (2.7), thus (2.7) collapses to a linear Schrödinger equation with potential $V(\mathbf{x}, t)$:

$$i\psi_t = -\frac{1}{2} \triangle_\perp \psi + V(\mathbf{x}, t)\psi. \tag{2.10}$$

When $n(\mathbf{x}, z, |u|) = \sqrt{1 + 4n_2|u|^2/n_0}$, i.e. laser beam in Kerr medium, then $\mu(\mathbf{x}, z, |\psi|) = 2n_2 r_0^2 k_0^2 |\psi|^2/n_0$ in (2.5) and $f(\mathbf{x}, t, |\psi|) = -|\psi|^2$ in (2.7) by choosing $\psi_s = \sqrt{n_0}/r_0 k_0 \sqrt{2n_2}$, thus (2.7) collapses to NLSE with a cubic focusing nonlinearity:

$$i\psi_t = -\frac{1}{2} \triangle_\perp \psi - |\psi|^2\psi. \tag{2.11}$$

The wave energy or power of the beam is conserved:

$$N(\psi) = \int_{\mathbb{R}^2} |\psi(\mathbf{x}, t)|^2 \, d\mathbf{x} \equiv \int_{\mathbb{R}^2} |\psi(\mathbf{x}, 0)|^2 \, d\mathbf{x}, \qquad t \geq 0. \tag{2.12}$$

Remark 2.1: When we consider high frequency asymptotics which concerns approximate solutions of (2.10) that are good approximations to oscillatory solutions. For such solutions the propagation distance is long compared to the wavelength, the propagation time is large compared to the period and the potential $V(\mathbf{x})$ varies slowly. To make this precise, we introduce slow time and space variables $t \to t/\varepsilon$, $\mathbf{x} \to \mathbf{x}/\varepsilon$ with $0 < \varepsilon \ll 1$ the (scaled) Planck constant and the scaled wave function $\psi^\varepsilon(\mathbf{x}, t) = \psi(\mathbf{x}/\varepsilon, t/\varepsilon)$ which satisfies the NLSE in the semiclassical regime

$$i\varepsilon\psi_t^\varepsilon = -\frac{\varepsilon^2}{2} \triangle_\perp \psi^\varepsilon + V^\varepsilon(\mathbf{x}, t)\psi^\varepsilon, \qquad \mathbf{x} \in \mathbb{R}^2, \ t > 0, \tag{2.13}$$

where $V^\varepsilon(\mathbf{x}, t) = V(\mathbf{x}/\varepsilon, t/\varepsilon)$.

3. Derivation of NLSE from BEC

Since its realization in dilute bosonic atomic gases [3, 26], BEC of alkali atoms and hydrogen has been produced and studied extensively in the laboratory [71], and has spurred great excitement in the atomic physics community and renewed the interest in studying the collective dynamics of macroscopic ensembles of atoms occupying the same one-particle quantum state [99, 34, 68]. The condensate typically consists of a few thousands to millions of atoms which are confined by a trap potential. In fact, beside the effects of the internal interactions between the atoms, the macroscopic behavior of BEC matter is highly sensitive to the shape of this external trapping potential. Theoretical predictions of the properties of a BEC like the density profile [19], collective excitations [43] and the formation of vortices [105] can now be compared with experimental data [3]. Needless to say that this dramatic progress on the experimental front has stimulated a wave of activity on both the theoretical and the numerical front.

At temperatures T much smaller than the critical temperature T_c [84], a BEC is well described by the macroscopic wave function $\psi = \psi(\mathbf{x}, t)$ whose evolution is governed by a self-consistent, mean field NLSE known as the Gross-Pitaevskii equation (GPE) [69, 103]

$$i\hbar \frac{\partial \psi(\mathbf{x}, t)}{\partial t} = \frac{\delta H(\psi)}{\delta \psi^*}$$

$$= -\frac{\hbar^2}{2m} \nabla^2 \psi(\mathbf{x}, t) + V(\mathbf{x})\psi(\mathbf{x}, t) + NU_0|\psi(\mathbf{x}, t)|^2 \psi(\mathbf{x}, t), \quad (3.1)$$

where $\mathbf{x} = (x, y, z)$, m is the atomic mass, $\hbar$ is the Planck constant, N is the number of atoms in the condensate, $V(\mathbf{x})$ is an external trapping potential. When a harmonic trap potential is considered, $V(\mathbf{x}) = \frac{m}{2}\left(\omega_x^2 x^2 + \omega_y^2 y^2 + \omega_z^2 z^2\right)$ with ω_x, ω_y and ω_z being the trap frequencies in x, y and z-direction, respectively. The Hamiltonian (or energy) of the system $H(\psi)$ per particle is defined as

$$H(\psi) = \int_{\mathbb{R}^3} \psi^*(\mathbf{x}, t) \left[-\frac{\hbar^2}{2m} \nabla^2 + V(\mathbf{x}) \right] \psi(\mathbf{x}, t)d\mathbf{x}$$

$$+ \frac{1}{2} \int_{\mathbb{R}^3 \times \mathbb{R}^3} \psi^*(\mathbf{x}, t)\, \psi^*(\mathbf{x}', t)\Phi(\mathbf{x} - \mathbf{x}')\psi(\mathbf{x}', t)\psi(\mathbf{x}, t)d\mathbf{x}d\mathbf{x}', \quad (3.2)$$

where the interaction potential is taken as the Fermi form

$$\Phi(\mathbf{x}) = (N - 1)U_0\delta(\mathbf{x}). \tag{3.3}$$

$U_0 = 4\pi\hbar^2 a_s/m$ describes the interaction between atoms in the condensate with the s-wave scattering length a_s (positive for repulsive interaction and negative for attractive interaction). It is convenient to normalize the wave function by requiring

$$\int_{\mathbb{R}^3} |\psi(\mathbf{x}, t)|^2 \, d\mathbf{x} = 1. \tag{3.4}$$

3.1. *Dimensionless GPE*

In order to scale the Eq. (3.1) under the normalization (3.4), we introduce

$$\tilde{t} = \omega_m t, \quad \tilde{\mathbf{x}} = \frac{\mathbf{x}}{a_0}, \quad \tilde{\psi}(\tilde{\mathbf{x}}, \tilde{t}) = a_0^{3/2}\psi(\mathbf{x}, t), \quad \text{with } a_0 = \sqrt{\hbar/m\omega_m}, \qquad (3.5)$$

where $\omega_m = \min\{\omega_x, \omega_y, \omega_z\}$, a_0 is the length of the harmonic oscillator ground state. In fact, we choose $1/\omega_m$ and a_0 as the dimensionless time and length units, respectively. Plugging (3.5) into (3.1), multiplying by $1/m\omega_m^2 a_0^{1/2}$, and then removing all $\tilde{\ }$, we get the following dimensionless GPE under the normalization (3.4) in three dimension

$$i\,\frac{\partial\psi(\mathbf{x}, t)}{\partial t} = -\frac{1}{2}\nabla^2\psi(\mathbf{x}, t) + V(\mathbf{x})\psi(\mathbf{x}, t) + \beta\,|\psi(\mathbf{x}, t)|^2\psi(\mathbf{x}, t), \qquad (3.6)$$

where $\beta = \frac{U_0 N}{a_0^3 \hbar \omega_m} = \frac{4\pi a_s N}{a_0}$ and

$$V(\mathbf{x}) = \frac{1}{2}\left(\gamma_x^2 x^2 + \gamma_y^2 y^2 + \gamma_z^2 z^2\right), \quad \text{with } \gamma_\alpha = \frac{\omega_\alpha}{\omega_m} \ (\alpha = x, y, z).$$

There are two extreme regimes of the interaction parameter β: (1) $\beta = o(1)$, the Eq. (3.6) describes a weakly interacting condensation; (2) $\beta \gg 1$, it corresponds to a strongly interacting condensation or to the semiclassical regime.

There are two typical extreme regimes between the trap frequencies: (1) $\gamma_x = 1$, $\gamma_y \approx 1$ and $\gamma_z \gg 1$, it is a disk-shaped condensation; (2) $\gamma_x = 1$, $\gamma_y \gg 1$ and $\gamma_z \gg 1$, it is a cigar-shaped condensation. In these two cases, the three-dimensional (3D) GPE (3.6) can be approximately reduced to a 2D and 1D equation respectively [85, 8, 5] as explained below.

3.2. *Reduction to lower dimension*

Case I (disk-shaped condensation):

$$\omega_x \approx \omega_y, \quad \omega_z \gg \omega_x, \qquad \Longleftrightarrow \qquad \gamma_x = 1, \ \gamma_y \approx 1, \quad \gamma_z \gg 1.$$

Here, the 3D GPE (3.6) can be reduced to a 2D GPE with $\mathbf{x} = (x, y)$ by assuming that the time evolution does not cause excitations along the z-axis, since the excitations along the z-axis have large energy (of order $\hbar\omega_z$) compared to that along the x and y-axis with energies of order $\hbar\omega_x$. Thus, we may assume that the condensation wave function along the z-axis is always well described by the harmonic oscillator ground state wave function, and set

$$\psi = \psi_2(x, y, t)\phi_{\text{ho}}(z) \quad \text{with} \quad \phi_{\text{ho}}(z) = (\gamma_z/\pi)^{1/4}\, e^{-\gamma_z z^2/2}. \qquad (3.7)$$

Plugging (3.7) into (3.6), multiplying by $\phi_{\text{ho}}^*(z)$, integrating with respect to z over $(-\infty, \infty)$, we get

$$i\,\frac{\partial \psi_2(\mathbf{x}, t)}{\partial t} = -\frac{1}{2}\nabla^2 \psi_2 + \frac{1}{2}\left(\gamma_x^2 x^2 + \gamma_y^2 y^2 + C\right)\psi_2 + \beta_2 |\psi_2|^2 \psi_2, \qquad (3.8)$$

where

$$\beta_2 = \beta \int_{-\infty}^{\infty} \phi_{\text{ho}}^4(z)\,dz = \beta\sqrt{\frac{\gamma_z}{2\pi}}, \quad C = \int_{-\infty}^{\infty}\left(\gamma_z^2 z^2 |\phi_{\text{ho}}(z)|^2 + \left|\frac{d\phi_{\text{ho}}}{dz}\right|^2\right)dz.$$

Since this GPE is time-transverse invariant, we can replace $\psi_2 \to \psi\,e^{-i\frac{Ct}{2}}$ so that the constant C in the trap potential disappears, and we obtain the 2D effective GPE:

$$i\,\frac{\partial \psi(\mathbf{x}, t)}{\partial t} = -\frac{1}{2}\nabla^2 \psi + \frac{1}{2}\left(\gamma_x^2 x^2 + \gamma_y^2 y^2\right)\psi + \beta_2 |\psi|^2 \psi. \qquad (3.9)$$

Note that the observables, e.g. the position density $|\psi|^2$, are not affected by dropping the constant C in (3.8).

Case II (cigar-shaped condensation):

$$\omega_y \gg \omega_x, \quad \omega_z \gg \omega_x \quad \Longleftrightarrow \quad \gamma_x = 1, \quad \gamma_y \gg 1, \quad \gamma_z \gg 1.$$

Here, the 3D GPE (3.6) can be reduced to a 1D GPE with $\mathbf{x} = x$. Similarly as in the 2D case, we can derive the following 1D GPE [85, 8, 5]:

$$i\,\frac{\partial \psi(x, t)}{\partial t} = -\frac{1}{2}\psi_{xx}(x, t) + \frac{\gamma_x^2 x^2}{2}\psi(x, t) + \beta_1 |\psi(x, t)|^2 \psi(x, t), \qquad (3.10)$$

where $\beta_1 = \beta\sqrt{\gamma_y \gamma_z}/2\pi$.

The 3D GPE (3.6), 2D GPE (3.9) and 1D GPE (3.10) can be written in a unified form:

$$i\,\frac{\partial \psi(\mathbf{x}, t)}{\partial t} = -\frac{1}{2}\nabla^2 \psi + V_d(\mathbf{x})\psi + \beta_d\,|\psi|^2\psi, \quad \mathbf{x} \in \mathbb{R}^d, \qquad (3.11)$$

$$\psi(\mathbf{x}, 0) = \psi_0(\mathbf{x}), \qquad \mathbf{x} \in \mathbb{R}^d, \qquad (3.12)$$

with

$$\beta_d = \beta \begin{cases} \sqrt{\gamma_y \gamma_z}/2\pi, \\ \sqrt{\gamma_z/2\pi}, \\ 1, \end{cases} \quad V_d(\mathbf{x}) = \begin{cases} \gamma_x^2 x^2/2, & d = 1, \\ \left(\gamma_x^2 x^2 + \gamma_y^2 y^2\right)/2, & d = 2, \\ \left(\gamma_x^2 x^2 + \gamma_y^2 y^2 + \gamma_z^2 z^2\right)/2, & d = 3, \end{cases}$$

$$(3.13)$$

where $\gamma_x > 0$, $\gamma_y > 0$ and $\gamma_z > 0$ are constants. The normalization condition for (3.11) is

$$N(\psi) = \|\psi(\cdot, t)\|^2 = \int_{\mathbb{R}^d} |\psi(\mathbf{x}, t)|^2\,d\mathbf{x} \equiv \int_{\mathbb{R}^d} |\psi_0(\mathbf{x})|^2\,d\mathbf{x} = 1. \qquad (3.14)$$

Remark 3.1: When $\beta_d \gg 1$, i.e. in a strongly repulsive interacting condensation or in semiclassical regime, another scaling of the GPE (3.11) is also very useful. In fact, after a rescaling in (3.11) under the normalization (3.14): $\mathbf{x} \to \varepsilon^{-1/2}\mathbf{x}$ and $\psi \to \varepsilon^{d/4}\psi$ with $\varepsilon = \beta_d^{-2/(d+2)}$, then the GPE (3.11) can be rewritten as

$$i\varepsilon \frac{\partial \psi(\mathbf{x}, t)}{\partial t} = -\frac{\varepsilon^2}{2}\nabla^2 \psi + V_d(\mathbf{x})\psi + |\psi|^2\psi, \quad \mathbf{x} \in \mathbb{R}^d. \tag{3.15}$$

4. The NLSE and variational formulation

Consider the following NLSE:

$$i\psi_t = -\frac{1}{2}\triangle \psi + V(\mathbf{x})\psi + \beta|\psi|^{2\sigma}\psi, \quad \mathbf{x} \in \mathbb{R}^d, \quad t \geq 0, \tag{4.1}$$

$$\psi(\mathbf{x}, 0) = \psi_0(\mathbf{x}), \quad \mathbf{x} \in \mathbb{R}^d, \tag{4.2}$$

where $\sigma > 0$ is a positive constant ($\sigma = 1$ corresponds to a cubic nonlinearity and $\sigma = 2$ corresponds to a quintic nonlinearity), $V(\mathbf{x})$ is a real-valued potential whose shape is determined by the type of system under investigation, β positive/negative corresponds to defocusing/focusing NLSE.

4.1. *Conservation laws*

Two important invariants of (4.1) are the *normalization of the wave function*

$$N(\psi(\cdot, t)) = \int_{\mathbb{R}^d} |\psi(\mathbf{x}, t)|^2 \, d\mathbf{x} \equiv N = N(\psi_0), \quad t \geq 0 \tag{4.3}$$

and the *energy*

$$E(\psi(\cdot, t)) = \int_{\mathbb{R}^d} \left[\frac{1}{2}|\nabla\psi(\mathbf{x}, t)|^2 + V(\mathbf{x})|\psi(\mathbf{x}, t)|^2 + \frac{\beta}{\sigma + 1}|\psi(\mathbf{x}, t)|^{2\sigma+2} \right] d\mathbf{x}$$
$$= E(\psi_0), \quad t \geq 0. \tag{4.4}$$

When $V(\mathbf{x}) \equiv 0$, another important invariant of (4.1) is the *momentum*

$$\mathbf{P}(\psi(\cdot, t)) = \frac{i}{2} \int_{\mathbb{R}^d} (\psi\nabla\psi^* - \psi^*\nabla\psi) \, d\mathbf{x} \equiv \mathbf{P}(\psi_0), \quad t \geq 0. \tag{4.5}$$

Define the *mass center*

$$\bar{\mathbf{x}}(t) = \frac{1}{N} \int_{\mathbb{R}^d} \mathbf{x}|\psi(\mathbf{x}, t)|^2 \, d\mathbf{x}. \tag{4.6}$$

Note that the *mass center* obeys

$$N\frac{d\bar{\mathbf{x}}}{dt} = \int \mathbf{x}\partial_t |\psi|^2 \, d\mathbf{x} = -\frac{i}{2}\int \mathbf{x}\nabla \cdot [\psi\nabla\psi^* - \psi^* \triangle \psi] \, d\mathbf{x}$$
$$= \frac{i}{2}\int [\psi\nabla\psi^* - \psi^* \triangle \psi] \, d\mathbf{x} = \mathbf{P}(\psi_0) \tag{4.7}$$

and thus moves at a constant speed.

To get more conservation laws, one can use the Noether theorem [113].

4.2. *Lagrangian structure*

Define the Lagrangian density $\mathcal{L}$ associated to (4.1) in terms of the real and imaginary parts u and v of ψ, or equivalently in terms of ψ and ψ^* viewed as independent variables in the form

$$\mathcal{L} = \frac{i}{2}(\psi^*\psi_t - \psi\psi_t^*) - \frac{1}{2}\nabla\psi \cdot \nabla\psi^* - V(\mathbf{x})\psi\psi^* - \frac{\beta}{\sigma+1}\psi^{\sigma+1}(\psi^*)^{\sigma+1}. \tag{4.8}$$

Consider the action

$$\mathcal{S}\{\psi,\psi^*\} = \int_{t_0}^{t_1} \int_{\mathbb{R}^d} \mathcal{L} \, d\mathbf{x}dt \tag{4.9}$$

as a functional on all admissible regular function satisfying the prescribed conditions $\psi(\mathbf{x},t_0) = \psi_0(\mathbf{x})$ and $\psi(\mathbf{x},t_1) = \psi_1(\mathbf{x})$. Its variation

$$\delta\mathcal{S} = \mathcal{S}\{\psi + \delta\psi, \psi^* + \delta\psi^*\} - S\{\psi,\psi^*\} \tag{4.10}$$

for infinitesimal $\delta\psi$ and $\delta\psi^*$ reads

$$\delta\mathcal{S} = \int_{t_0}^{t_1} \int_{\mathbb{R}^d} \left[\frac{\partial\mathcal{L}}{\partial\psi}\delta\psi + \frac{\partial\mathcal{L}}{\partial\nabla\psi} \cdot \nabla\delta\psi + \frac{\partial\mathcal{L}}{\partial\psi_t}\delta\psi_t\right] d\mathbf{x}dt + \text{c.c.}$$
$$= \int_{t_0}^{t_1} \int_{\mathbb{R}^d} \left[\frac{\partial\mathcal{L}}{\partial\psi} - \nabla \cdot \left(\frac{\partial\mathcal{L}}{\partial\nabla\psi}\right) - \partial_t\left(\frac{\partial\mathcal{L}}{\partial\psi_t}\right)\right] \delta\psi \, d\mathbf{x}dt$$
$$+ \left[\frac{\partial\mathcal{L}}{\partial\psi_t}\delta\psi\right]_{t_0}^{t_1} + \text{c.c.} \tag{4.11}$$

A necessary and sufficient condition for a function $\psi(\mathbf{x},t)$ to lead to an extremum for the action $\mathcal{S}$ among the functions with prescribed values $\psi(\cdot,t_0)$ and $\psi(\cdot,t_1)$, thus reduces to the Euler-Lagrange equations

$$\frac{\partial\mathcal{L}}{\partial\psi} = \nabla \cdot \left(\frac{\partial\mathcal{L}}{\partial\nabla\psi}\right) + \partial_t\left(\frac{\partial\mathcal{L}}{\partial\psi_t}\right), \tag{4.12}$$

which, when the Lagrangian (4.8) is used, reduces to the NLSE (4.1). This system is easily rewritten in terms of the real fields $u = (\psi + \psi^*)/2$ and $v = (\psi - \psi^*)/2i$ as

$$\frac{\partial \mathcal{L}}{\partial u} = \nabla \cdot \left(\frac{\partial \mathcal{L}}{\partial \nabla u} \right) + \partial_t \left(\frac{\partial \mathcal{L}}{\partial u_t} \right), \tag{4.13}$$

$$\frac{\partial \mathcal{L}}{\partial v} = \nabla \cdot \left(\frac{\partial \mathcal{L}}{\partial \nabla v} \right) + \partial_t \left(\frac{\partial \mathcal{L}}{\partial v_t} \right). \tag{4.14}$$

4.3. *Hamiltonian structure*

As usual, a Hamiltonian structure is easily derived from the existence of a Lagrangian. Writing $\psi = u + iv$ in order to deal with real fields, the Hamiltonian density $\mathcal{H} = \frac{i}{2}(\psi^* \partial_t \psi - \psi \partial_t \psi^*) - \mathcal{L}$ becomes

$$\mathcal{H} = v \partial_t u - u \partial_t v - \mathcal{L}. \tag{4.15}$$

Introducing the canonical variables

$$q_1 \equiv u, \qquad\qquad p_1 \equiv \frac{\partial \mathcal{L}}{\partial(\partial_t q_1)}, \tag{4.16}$$

$$q_2 \equiv v, \qquad\qquad p_2 \equiv \frac{\partial \mathcal{L}}{\partial(\partial_t q_2)}, \tag{4.17}$$

it takes the form

$$\mathcal{H} = \sum_j p_j \partial_t q_j - \mathcal{L}. \tag{4.18}$$

Define

$$\rho_j \equiv \frac{\partial \mathcal{L}}{\partial(\nabla q_j)}, \tag{4.19}$$

and rewrite the Euler-Lagrange equations as

$$\frac{\partial \mathcal{L}}{\partial q} = \nabla \cdot \rho_j + \partial_t p_j. \tag{4.20}$$

Using that

$$\partial_t \mathcal{L} = \sum_j \frac{\partial \mathcal{L}}{\partial q_j} \partial_t q_j + \frac{\partial \mathcal{L}}{\partial \nabla q_j} \partial_t \nabla q_j + \frac{\partial \mathcal{L}}{\partial(\partial_t q_j)} \partial_{tt} q_j, \tag{4.21}$$

and the Euler-Lagrange equations, we get

$$\partial_t \mathcal{H} = -\nabla \cdot \sum_j \rho_j \partial_t q_j, \tag{4.22}$$

which ensures that the conservation fo the Hamiltonian or energy $H = \int_{\mathbb{R}^d} \mathcal{H} \, d\mathbf{x}$.

Similarly, from the variation of the Lagrangian density

$$\delta\mathcal{L} = \sum_j \frac{\delta\mathcal{L}}{\delta q_j}\delta q_j + \frac{\delta\mathcal{L}}{\delta\nabla q_j}\nabla\delta q_j + \frac{\delta\mathcal{L}}{\delta(\partial_t q_j)}\partial_t\delta q_j, \qquad (4.23)$$

the Euler-Lagrange equations, and the definition of p_j we obtain the variation of the Hamiltonian H, in the form

$$\delta H = \sum_j \int (\partial_t q_j \delta p_j - \partial_t p_j \delta q_j) d\mathbf{x}, \qquad (4.24)$$

which leads to the Hamilton equaitons

$$\frac{\partial q_j}{\partial t} = \frac{\delta H}{\delta p_j}, \qquad \frac{\partial p_j}{\partial t} = -\frac{\delta H}{\delta q_j}, \qquad (4.25)$$

or in complex form,

$$i\partial_t\psi = \frac{\delta H}{\delta\psi^*}. \qquad (4.26)$$

4.4. *Variance identity*

Define the 'variance' (or 'momentum of inertia' in a context where N is referred to as the mass of the wave packet) as

$$\delta_V = \int_{\mathbb{R}^d} |\mathbf{x}|^2 |\psi|^2 \, d\mathbf{x} = \sum_{j=1}^{d} \delta_j, \ \delta_j = \int_{\mathbb{R}^d} x_j^2 |\psi|^2 \, d\mathbf{x}, \ j = 1, \cdots, d \quad (4.27)$$

and the square width of the wave packet

$$\delta_{\mathbf{x}} = \frac{1}{N}\int_{\mathbb{R}^d} |\mathbf{x}-\bar{\mathbf{x}}|^2 |\psi|^2 \, d\mathbf{x} = \frac{1}{N}\int_{\mathbb{R}^d} (|\mathbf{x}|^2 - |\bar{\mathbf{x}}|^2)|\psi|^2 \, d\mathbf{x} = \frac{\delta_V}{N} - |\bar{\mathbf{x}}|^2. \quad (4.28)$$

Here we use $\mathbf{x} = (x_1, \cdots, x_d) \in \mathbb{R}^d$.

When $V(\mathbf{x}) \equiv 0$ in (4.1), due to the conservation of the wave energy N and of the linear momentum $\mathbf{P}$, we have

$$\frac{d^2\delta_{\mathbf{x}}}{dt^2} = \frac{1}{N}\frac{d^2\delta_V}{dt^2} - 2\frac{|\mathbf{P}|^2}{N^2}. \qquad (4.29)$$

Lemma 4.1: *Suppose $\psi(\mathbf{x}, t)$ be the solution of the problem (4.1), (4.2), then we have*

$$\frac{d^2 \delta_j(t)}{dt^2} = \int_{\mathbb{R}^d} \left[2|\partial_{x_j}\psi|^2 + \frac{2\sigma\beta}{\sigma+1}|\psi|^{2\sigma+2} - 2x_j|\psi|^2 \partial_{x_j}(V(\mathbf{x})) \right] d\mathbf{x}, \quad (4.30)$$

$$\delta_j(0) = \delta_j^{(0)} = \int_{\mathbb{R}^d} x_j^2 |\psi_0(\mathbf{x})|^2 \, d\mathbf{x}, \qquad j = 1, \cdots, d, \quad (4.31)$$

$$\delta_j'(0) = \delta_j^{(1)} = 2 \operatorname{Im} \left[\int_{\mathbb{R}^d} x_j \psi_0^* \, \partial_{x_j} \psi_0 d\mathbf{x} \right]. \quad (4.32)$$

Proof: Differentiate (4.27) with respect to t, notice (4.1), integrate by parts, we have

$$\begin{aligned}
\frac{d\delta_j(t)}{dt} &= \frac{d}{dt} \int_{\mathbb{R}^d} x_j^2 |\psi(\mathbf{x}, t)|^2 \, d\mathbf{x} = \int_{\mathbb{R}^d} x_j^2 \left(\psi \, \partial_t \psi^* + \psi^* \, \partial_t \psi \right) d\mathbf{x} \\
&= \frac{i}{2} \int_{\mathbb{R}^d} x_j^2 \left(\psi^* \triangle \psi - \psi \triangle \psi^* \right) d\mathbf{x} \\
&= i \int_{\mathbb{R}^d} x_j \left(\psi \, \partial_{x_j} \psi^* - \psi^* \, \partial_{x_j} \psi \right) d\mathbf{x}.
\end{aligned} \quad (4.33)$$

Similarly, differentiate (4.33) with respect to t, notice (4.1), integrate by parts, we have

$$\begin{aligned}
\frac{d^2 \delta_j(t)}{dt^2} &= i \int_{\mathbb{R}^d} x_j \left[\partial_t\psi \, \partial_{x_j}\psi^* + \psi \partial_{x_j\,t}\psi^* - \partial_t\psi^* \, \partial_{x_j}\psi - \psi^* \partial_{x_j\,t}\psi \right] d\mathbf{x} \\
&= \int_{\mathbb{R}^d} \left[2ix_j \left(\partial_t\psi \, \partial_{x_j}\psi^* - \partial_t\psi^* \, \partial_{x_j}\psi \right) + i \left(\psi^* \, \partial_t\psi - \psi \, \partial_t\psi^* \right) \right] d\mathbf{x} \\
&= \int_{\mathbb{R}^d} \left[-x_j \left(\partial_{x_j}\psi^* \triangle \psi + \partial_{x_j}\psi \triangle \psi^* \right) - \frac{1}{2} \left(\psi^* \triangle \psi + \psi \triangle \psi^* \right) \right. \\
&\qquad\quad + 2x_j V(\mathbf{x}) \left(\psi \, \partial_{x_j}\psi^* + \psi^* \, \partial_{x_j}\psi \right) + 2V(\mathbf{x})|\psi|^2 + 2\beta|\psi|^{2\sigma+2} \\
&\qquad\quad \left. + 2\beta x_j |\psi|^{2\sigma} \left(\psi \, \partial_{x_j}\psi^* + \psi^* \, \partial_{x_j}\psi \right) \right] d\mathbf{x} \\
&= \int_{\mathbb{R}^d} \left[2|\partial_{x_j}\psi|^2 - |\nabla\psi|^2 - |\psi|^2 \partial_{x_j}(2x_j V(\mathbf{x})) + |\nabla\psi|^2 \right. \\
&\qquad\quad \left. - \frac{2\beta}{\sigma+1}|\psi|^{2\sigma+2} + 2V(\mathbf{x})|\psi|^2 + 2\beta|\psi|^{2\sigma+2} \right] d\mathbf{x} \\
&= \int_{\mathbb{R}^d} \left[2|\partial_{x_j}\psi|^2 + \frac{2\sigma\beta}{\sigma+1}|\psi|^{2\sigma+2} - 2x_j|\psi|^2 \, \partial_{x_j}(V(\mathbf{x})) \right] d\mathbf{x}. \quad (4.34)
\end{aligned}$$

Thus we obtain the desired equality (4.30). Setting $t = 0$ in (4.27) and (4.33), we get (4.31) and (4.32) respectively. $\qquad\square$

Lemma 4.2: *When $V(\mathbf{x}) \equiv 0$ in the NLSE (4.1), we have*

$$\frac{d^2\delta_V}{dt^2} = 4E(\psi_0) + \frac{2\beta(d\sigma - 2)}{\sigma + 1} \int_{\mathbb{R}^d} |\psi|^{2\sigma+2} \, d\mathbf{x}. \qquad (4.35)$$

Proof: Sum (4.30) from $j = 1$ to d, we get

$$\begin{aligned}
\frac{d^2\delta_V(t)}{dt^2} &= \sum_{j=1}^{d} \frac{d^2\delta_j(t)}{dt^2} = \sum_{j=1}^{d} \int_{\mathbb{R}^d} \left(2|\partial_{x_j}\psi|^2 + \frac{2\sigma\beta}{\sigma+1}|\psi|^{2\sigma+2} \right) d\mathbf{x} \\
&= \int_{\mathbb{R}^d} \left[2|\nabla\psi|^2 + \frac{2d\sigma\beta}{\sigma+1}|\psi|^{2\sigma+2} \right] d\mathbf{x} \\
&= 4E + \frac{2\beta(\sigma d - 2)}{\sigma + 1} \int_{\mathbb{R}^d} |\psi|^{2\sigma+2} d\mathbf{x}. \qquad (4.36)
\end{aligned}$$

Here we use conservation of energy of the NLSE. $\qquad\square$

From this lemma, when $V(\mathbf{x}) \equiv 0$ and at critical dimension, i.e. $d\sigma - 2 = 0$, (4.35) reduces to

$$\frac{d^2\delta_V}{dt^2} = 4E, \qquad (4.37)$$

leading to

$$\delta_V(t) = 2Et^2 + \delta_V'(0)t + \delta_V(0). \qquad (4.38)$$

When the external potential $V(\mathbf{x})$ is chosen as harmonic oscillator (3.13) and $\sigma = 1$ in (4.1), we have

Lemma 4.3: *(i) In 1D without interaction, i.e. $d = 1$ and $\beta = 0$ in (4.1), we have*

$$\delta_x(t) = \frac{E(\psi_0)}{\gamma_x^2} + \left(\delta_x^{(0)} - \frac{E(\psi_0)}{\gamma_x^2} \right) \cos(2\gamma_x t) + \frac{\delta_x^{(1)}}{2\gamma_x} \sin(2\gamma_x t), \qquad t \geq 0.$$
$$(4.39)$$

(ii) In 2D with radial symmetry, i.e. $d = 2$ and $\gamma_x = \gamma_y := \gamma_r$ in (4.1), for any initial data $\psi_0(x, y)$ in (4.2), we have

$$\delta_r(t) = \frac{E(\psi_0)}{\gamma_r^2} + \left(\delta_r^{(0)} - \frac{E(\psi_0)}{\gamma_r^2} \right) \cos(2\gamma_r t) + \frac{\delta_r^{(1)}}{2\gamma_r} \sin(2\gamma_r t), \qquad t \geq 0,$$
$$(4.40)$$

where

$$\begin{aligned}
\delta_r(t) &= \delta_x(t) + \delta_y(t), \\
\delta_r^{(0)} &:= \delta_r(0) = \delta_x(0) + \delta_y(0), \\
\delta_r^{(1)} &:= \delta_r'(0) = \delta_x'(0) + \delta_y'(0).
\end{aligned}$$

Furthermore, when $d = 2$ and $\gamma_x = \gamma_y$ in (4.1) and the initial data $\psi_0(\mathbf{x})$ in (4.2) satisfying

$$\psi_0(x, y) = f(r)e^{im\theta} \qquad with \quad m \in \mathbb{Z} \quad and \quad f(0) = 0 \text{ when } m \neq 0, \quad (4.41)$$

we have for $t \geq 0$

$$\delta_x(t) = \delta_y(t) = \frac{1}{2}\delta_r(t)$$

$$= \frac{E(\psi_0)}{2\gamma_x^2} + \left(\delta_x^{(0)} - \frac{E(\psi_0)}{2\gamma_x^2}\right)\cos(2\gamma_x t) + \frac{\delta_x^{(1)}}{2\gamma_x}\sin(2\gamma_x t). \quad (4.42)$$

(iiii) For all other cases, we have for $t \geq 0$

$$\delta_j(t) = \frac{E(\psi_0)}{\gamma_{x_j}^2} + \left(\delta_j^{(0)} - \frac{E(\psi_0)}{\gamma_{x_j}^2}\right)\cos(2\gamma_{x_j} t) + \frac{\delta_j^{(1)}}{2\gamma_{x_j}}\sin(2\gamma_{x_j} t) + g_j(t),$$
$$(4.43)$$

where $g_j(t)$ is a solution of the following problem

$$\frac{d^2 g_j(t)}{dt^2} + 4\gamma_{x_j}^2 g_j(t) = f_j(t), \qquad g_j(0) = \frac{dg_j(0)}{dt} = 0, \quad (4.44)$$

with

$$f_j(t) = \int_{\mathbb{R}^d} \left[2|\partial_{x_j}\psi|^2 - 2|\nabla\psi|^2 - \beta|\psi|^4 + (2\gamma_{x_j}^2 x_j^2 - 4V(\mathbf{x}))|\psi|^2\right] d\mathbf{x}$$

satisfying

$$|f_\alpha(t)| < 4E_\beta(\psi_0), \qquad t \geq 0.$$

Proof: (i) From (4.30) with $d = 1$ and $\beta_1 = 0$, we have

$$\frac{d^2\delta_x(t)}{dt^2} = 4E(\psi_0) - 4\gamma_x^2\delta_x(t), \qquad t > 0, \quad (4.45)$$

$$\delta_x(0) = \delta_x^{(0)}, \qquad \delta_x'(0) = \delta_x^{(1)}. \quad (4.46)$$

Thus (4.39) is the unique solution of this ordinary differential equation (ODE).

(ii). From (4.30) with $d = 2$, we have

$$\frac{d^2\delta_x(t)}{dt^2} = -2\gamma_x^2\delta_x(t) + \int_{\mathbb{R}^d} \left(2|\partial_x\psi|^2 + \beta|\psi|^4\right) d\mathbf{x}, \quad (4.47)$$

$$\frac{d^2\delta_y(t)}{dt^2} = -2\gamma_y^2\delta_y(t) + \int_{\mathbb{R}^d} \left(2|\partial_y\psi|^2 + \beta|\psi|^4\right) d\mathbf{x}. \quad (4.48)$$

Sum (4.47) and (4.48), notice (4.4) and $\gamma_x = \gamma_y$, we have the ODE for $\delta_r(t)$:

$$\frac{d^2\delta_r(t)}{dt^2} = 4E(\psi_0) - 4\gamma_x^2\delta_r(t), \qquad t > 0, \tag{4.49}$$

$$\delta_r(0) = 2\delta_x^{(0)}, \qquad \delta_r'(0) = 2\delta_x^{(1)}. \tag{4.50}$$

Thus (4.40) is the unique solution of the second order ODE (4.49) with the initial data (4.50). Furthermore, when the initial data $\psi_0(\mathbf{x})$ in (4.2) satisfies (4.41), due to the radial symmetry, the solution $\psi(\mathbf{x}, t)$ of (4.1)-(4.2) satisfies

$$\psi(x, y, t) = g(r, t)e^{im\theta} \qquad \text{with} \quad g(r, 0) = f(r). \tag{4.51}$$

This implies

$$\delta_x(t) = \int_{\mathbb{R}^2} x^2 \, |\psi(x, y, t)|^2 \, d\mathbf{x} = \int_0^\infty \int_0^{2\pi} r^2 \cos^2\theta \, |g(r, t)|^2 r \, d\theta dr$$

$$= \pi \int_0^\infty r^2 |g(r, t)|^2 r \, dr = \int_0^\infty \int_0^{2\pi} r^2 \sin^2\theta \, |g(r, t)|^2 r \, d\theta dr$$

$$= \int_{\mathbb{R}^2} y^2 |\psi(x, y, t)|^2 \, d\mathbf{x} = \delta_y(t), \qquad t \geq 0. \tag{4.52}$$

Thus the equality (4.42) is a combination of (4.52) and (4.40).

(iii). From (4.30), notice the energy conservation (4.4) of the GPE (4.1), we have

$$\frac{d^2\delta_j(t)}{dt^2} = 4E(\psi_0) - 4\gamma_{x_j}^2\delta_j(t) + f_j(t), \qquad t \geq 0. \tag{4.53}$$

Thus (4.43) is the unique solution of this ODE (4.53). $\qquad\qquad\square$

5. Plane and solitary wave solutions of NLSE

For simplicity, we assume $V(\mathbf{x}) \equiv 0$, $d = 1$ and $\sigma = 1$ in this section. In this case, the NLSE (4.1) collapses to

$$i\psi_t = -\frac{1}{2}\psi_{xx} + \beta|\psi|^2\psi. \tag{5.1}$$

To find the plane wave solution of (5.1), we take the ansatz

$$\psi = Ae^{i(kx - \omega t)}, \tag{5.2}$$

where A, ω and k are amplitude, angular frequency and wavenumber respectively. Plugging (5.2) into (5.1), we get the dispersive relation

$$\omega = \frac{1}{2}k^2 + \beta|A|^2 \tag{5.3}$$

This implies that the dispersive relation depends on wavenumber and amplitude. Define the group velocity

$$c_g \equiv \frac{d\omega}{dk} = k. \tag{5.4}$$

So the NLSE has the plane wave solution (5.2) provided the dispersive relation is satisfied. In fact, (5.3) can be viewed as zeroth-order approximation of the NLSE (5.1), and (5.2) can be viewed as zeroth-order solution of the NLSE (5.1).

To find the solitary wave solution, we take the ansatz

$$\psi = \phi(\xi)e^{i(kx-\omega t)}, \qquad \xi = x - c_g t, \tag{5.5}$$

where ϕ is a real-valued function. Plugging (5.5) into (5.1), we get

$$\frac{1}{2}\frac{d^2\phi}{d\xi^2} + (\omega - k^2/2)\phi - \beta\phi^3 + i(k - c_g)\frac{d\phi}{d\xi} = 0. \tag{5.6}$$

This implies

$$-\frac{d^2\phi}{d\xi^2} + \gamma\phi + 2\beta\phi^3 = 0, \qquad \gamma = k^2 - 2\omega > 0; \qquad c_g = k. \tag{5.7}$$

When $\beta < 0$, we have a solution for (5.7)

$$\phi(\xi) = \pm\sqrt{\frac{\gamma}{-\beta(2 - k^2)}}\,\mathrm{dn}\left(\sqrt{\frac{\gamma}{2 - k^2}}(\xi - \xi_0), k\right), \tag{5.8}$$

where dn is the Jacobian elliptic function. Letting $k \to 1$, we have

$$\phi(\xi) = \pm\sqrt{\frac{\gamma}{-\beta}}\,\mathrm{sech}\sqrt{\gamma}(\xi - \xi_0). \tag{5.9}$$

Thus we get a solitary wave solution for the NLSE (5.1) with $\beta < 0$:

$$\psi(x,t) = \sqrt{\frac{\gamma}{-\beta}}\,\mathrm{sech}\sqrt{\gamma}(x - t - x_0)e^{i[x-(1-\gamma)t/2]}, \tag{5.10}$$

where $\gamma > 0$ is a constant.

For $\beta > 0$, one can get a traveling wave in a similar manner.

6. Existence/blowup results of NLSE

For simplicity, in this section, we assume $V(\mathbf{x}) \equiv 0$ in (4.1).

6.1. Integral form

When $\beta = 0$ in (4.1), the free Schrödinger equation is solved as

$$\psi(\mathbf{x}, t) = U(t)\psi_0(\mathbf{x}), \qquad \mathbf{x} \in \mathbb{R}^d, \quad t \geq 0, \tag{6.1}$$

where free Schrödinger operator $U(t) = e^{it\triangle/2}$ given by

$$U(t)\psi_0(\mathbf{x}) = \left(\frac{1}{4\pi i t}\right)^{d/2} \int_{\mathbb{R}^d} e^{i\frac{|\mathbf{x}-\mathbf{x}'|^2}{4t}} \psi_0(\mathbf{x}') \, d\mathbf{x}' \tag{6.2}$$

defines a unitary transformation group in L^2.

Theorem 6.1: *(Decay estimates) For conjugate p and p' $(\frac{1}{p} + \frac{1}{p'} = 1)$, with $2 \leq p \leq \infty$, and $t \neq 0$, the transformation $U(t)$ maps continuously $L^{p'}(\mathbb{R}^d)$ into $L^p(\mathbb{R}^d)$ and*

$$\|U(t)\psi_0\|_{L^p} \leq (4\pi|t|)^{-d(\frac{1}{2} - \frac{1}{p})} \|\psi_0\|_{L^{p'}}. \tag{6.3}$$

Proof (scratch): Use the conservation of L^2-norm $\|\psi(t)\|_{L^2} = \|\psi_0\|_{L^2}$, the estimate $|\psi(\mathbf{x}, t)| \leq (4\pi|t|)^{-d/2}\|\psi_0\|_{L^1}$ and the Riesz-Thorin interpolation theorem. $\qquad\square$

When $\beta \neq 0$ in (4.1), the problem is conveniently rewritten in the integral form

$$\psi(t) = U(t)\psi_0 - i\beta \int_0^t U(t - t')|\psi(t')|^{2\sigma}\psi(t') \, dt'. \tag{6.4}$$

6.2. Existence results

Based on a fixed point theorem to (6.4), the following existence results for NLSE is proved [113]:

Theorem 6.2: *(Solution in H^1) For $0 \leq \sigma < \frac{2}{d-2}$ (no condition on σ when $d = 1$ or 2) and an initial condition $\psi_0 \in H^1(\mathbb{R}^d)$, there exists, locally in time, a unique maximal solution ψ in $C((-T^*, T^*), H^1(\mathbb{R}^d))$, where maximal means that if $T^* < \infty$, then $\|\psi\|_{H^1} \to \infty$ as t approaches T^*. In addition, ψ satisfies the normalization and energy (or Hamiltonian) conservation laws*

$$N(\psi) \equiv \int_{\mathbb{R}^d} |\psi|^2 \, d\mathbf{x} = N(\psi_0), \tag{6.5}$$

$$E(\psi) \equiv \int_{\mathbb{R}^d} \left[\frac{1}{2}|\nabla\psi|^2 + \frac{\beta}{\sigma + 1}|\psi|^{2\sigma+2}\right] d\mathbf{x} = E(\psi_0), \tag{6.6}$$

and depends continuously on the initial condition ψ in H^1.

If in addition, the initial condition ψ_0 belongs to the space $\sum = \{f, f \in H^1(\mathbb{R}^d), |\mathbf{x}f(\mathbf{x})| \in L^2(\mathbb{R}^d)\}$ of the functions in H^1 with finite variance, the above maximal solution belongs to $C((-T^*, T^*), \sum)$. The variance $\delta_V(t) = \int_{\mathbb{R}^d} |\mathbf{x}|^2 |\psi|^2 \, d\mathbf{x}$ belongs to $C^2(-T^*, T^*)$ and satisfies the identity

$$\frac{d^2 \delta_V}{dt^2} = 4E(\psi_0) + \frac{2\beta(d\sigma - 2)}{\sigma + 1} \int_{\mathbb{R}^d} |\psi|^{2\sigma+2} \, d\mathbf{x}. \tag{6.7}$$

Theorem 6.3: *(Solution in L^2) For $0 \le \sigma < \frac{2}{d}$ and an initial condition $\psi_0 \in L^2(\mathbb{R}^d)$, there exist a unique solution ψ in $C((-T^*, T^*), L^2(\mathbb{R}^d)) \cap L^q((-T^*, T^*), L^{2\sigma+2}(\mathbb{R}^d))$ with $q = \frac{4(\sigma+1)}{d\sigma}$, satisfying the L^2-norm conservation law (6.5).*

Theorem 6.4: *(Global existence in H^1) Assume $0 \le \sigma < 2/(d-2)$ if $\beta < 0$ (attractive nonlinearity), or $0 \le \sigma < 2/d$ if $\beta > 0$ (repulsive nonlinearity). For any $\psi \in H^1(\mathbb{R}^d)$, there exists a unique solution ψ in $C(\mathbb{R}, H^1(\mathbb{R}^d))$. It satisfies the conservation laws (6.5) and (6.6) and depends continuously on initial conditions in $H^1(\mathbb{R}^d)$.*

Theorem 6.5: *(Global existence in L^2) For $0 \le \sigma < 2/d$ and $\psi_0 \in L^2(\mathbb{R}^d)$, there exists a unique solution ψ in $C(\mathbb{R}, L^2(\mathbb{R}^d)) \cap L^q_{\text{loc}}(\mathbb{R}, L^{2\sigma+2}(\mathbb{R}^d))$ with $q = 4(\sigma + 1)/d\sigma$ that satisfies the L^2-norm conservation (6.5) and depends continuously on initial conditions in L^2.*

6.3. *Finite time blowup results*

Classical blowup results are based on the "variance identity", also known as the "viral theorem", and "uncertainty principle". Define the variance $\delta_V(t) = \int_{\mathbb{R}^d} |\mathbf{x}|^2 |\psi|^2 \, d\mathbf{x}$, we have the identity

$$\frac{d^2}{dt^2} \delta_V(t) = 4E + \frac{2\beta(d\sigma - 2)}{\sigma + 1} \int_{\mathbb{R}^d} |\psi|^{2\sigma+2} \, d\mathbf{x}. \tag{6.8}$$

Theorem 6.6: *Suppose that $\beta < 0$ and $d\sigma \ge 2$. Consider an initial condition $\psi_0 \in H^1$ with $\delta_V(0)$ finite that satisfies one of the conditions below:*
    (i) $E(\psi_0) < 0$,
    (ii) $E(\psi_0) = 0$ and $\delta'_V(0) = 2 \operatorname{Re} \int_{\mathbb{R}^d} \psi_0^*(\mathbf{x} \cdot \nabla \psi_0) d\mathbf{x} < 0$,
    (iii) $E(\psi_0) > 0$ and $\left|\delta'_V(0)\right| \ge 2\sqrt{2E(\psi_0)\delta_V(0)} = 2\sqrt{2E(\psi_0)}\|\mathbf{x}\psi_0\|_{L^2}$.
Then, there exists a time $t_* < \infty$ such that

$$\lim_{t \to t_*} \|\nabla \psi\|_{L^2} = \infty \qquad \text{and} \qquad \lim_{t \to t_*} \|\psi\|_{L^\infty} = \infty. \tag{6.9}$$

Proof: If $\beta < 0$ and $d\sigma \geq 2$,

$$\frac{d^2}{dt^2}\delta_V(t) \leq 4E, \tag{6.10}$$

and by time integration,

$$\delta_V(t) \leq 2Et^2 + \delta_V'(0)t + \delta_V(0). \tag{6.11}$$

Under any of the hypotheses (i)-(iii) of the above theorem, there exists a time t_0 such that the right-hand side of (6.11) vanishes, and thus also $t_1 \leq t_0$ such that

$$\lim_{t \to t_1} \delta_V(t) = 0. \tag{6.12}$$

Furthermore, from the equality

$$\int_{\mathbb{R}^d} |f|^2 \, d\mathbf{x} = \frac{1}{d}\int_{\mathbb{R}^d} (\nabla \cdot \mathbf{x})|f|^2 \, d\mathbf{x} = -\frac{1}{d}\int_{\mathbb{R}^d} \mathbf{x} \cdot \nabla(|f|^2) \, d\mathbf{x}, \tag{6.13}$$

one gets the "uncertainty principle"

$$\|f\|_{L^2}^2 \leq \frac{2}{d}\|\nabla f\|_{L^2} \, \|\mathbf{x}f\|_{L^2}. \tag{6.14}$$

When this inequality is applied to a solution ψ, one gets from (6.14) and from the conservation of $\|\psi\|_{L^2}^2$, that there exists a time $t_* \leq t_1$ such that $\lim_{t \to t_*} \|\nabla \psi\|_{L^2} = \infty$. The conservation of E then ensures that $\lim_{t \to t_*} \|\psi\|_{L^{2\sigma+2}}^{2\sigma+2} = \infty$, and since $\|\psi\|_{L^2}^2$ is conserved, this implies that $\lim_{t \to t_*} \|\psi\|_{L^\infty} = \infty$. $\qquad\square$

7. WKB expansion and quantum hydrodynamics

In this section, we consider the NLSE in semiclassical regime

$$i\varepsilon\psi_t^\varepsilon = -\frac{\varepsilon^2}{2}\triangle\psi^\varepsilon + V(\mathbf{x})\psi^\varepsilon + f(|\psi^\varepsilon|^2)\psi^\varepsilon, \qquad \mathbf{x} \in \mathbb{R}^d, \quad t \geq 0, \tag{7.1}$$

$$\psi^\varepsilon(\mathbf{x}, 0) = \psi_0^\varepsilon(\mathbf{x}), \qquad \mathbf{x} \in \mathbb{R}^d, \tag{7.2}$$

where $0 < \varepsilon \ll 1$ is the (scaled) Planck constant, $f(\rho)$ is a given real-valued function; and find its semiclassical limit by using WKB expansion.

Suppose that the initial datum ψ_0^ε in (7.2) is rapidly oscillating on the scale ε, given in WKB form:

$$\psi_0^\varepsilon(\mathbf{x}) = A_0(\mathbf{x})\exp\left(\frac{i}{\varepsilon}S_0(\mathbf{x})\right), \qquad \mathbf{x} \in \mathbb{R}^d, \tag{7.3}$$

where the amplitude A_0 and the phase S_0 are smooth real-valued functions. Plugging the radial-representation of the wave-function

$$\psi^\varepsilon(\mathbf{x}, t) = A^\varepsilon(\mathbf{x}, t) \exp\left(\frac{i}{\varepsilon} S^\varepsilon(\mathbf{x}, t)\right) = \sqrt{\rho^\varepsilon(\mathbf{x}, t)} \exp\left(\frac{i}{\varepsilon} S^\varepsilon(\mathbf{x}, t)\right) \quad (7.4)$$

into (7.1), one obtains the following quantum hydrodynamic (QHD) form of the NLSE for $\rho^\varepsilon = |A^\varepsilon|^2$, $\mathbf{J}^\varepsilon = \rho^\varepsilon \nabla S^\varepsilon$ [58, 41, 79]

$$\rho_t^\varepsilon + \operatorname{div} \mathbf{J}^\varepsilon = 0, \quad (7.5)$$

$$\mathbf{J}_t^\varepsilon + \operatorname{div}\left(\frac{\mathbf{J}^\varepsilon \otimes \mathbf{J}^\varepsilon}{\rho^\varepsilon}\right) + \nabla P(\rho^\varepsilon) + \rho^\varepsilon \nabla V = \frac{\varepsilon^2}{4} \operatorname{div}(\rho^\varepsilon \nabla^2 \log \rho^\varepsilon); \quad (7.6)$$

with initial data

$$\rho^\varepsilon(\mathbf{x}, 0) = \rho_0^\varepsilon(\mathbf{x}) = |A_0(\mathbf{x})|^2, \quad \mathbf{J}^\varepsilon(\mathbf{x}, 0) = \rho_0^\varepsilon(\mathbf{x}) \nabla S_0(\mathbf{x}) = |A_0(\mathbf{x})|^2 \nabla S_0(\mathbf{x}), \quad (7.7)$$

(see Grenier [66], Jüngel [79, 80], for mathematical analyses of this system). Here the hydrodynamic pressure $P(\rho)$ is related to the nonlinear potential $f(\rho)$ by

$$P(\rho) = \rho f(\rho) - \int_0^\rho f(s)\, ds, \quad (7.8)$$

i.e. f' is the enthalpy. Letting $\varepsilon \to 0+$, one obtains formally the following Euler system

$$\rho_t + \operatorname{div} \mathbf{J} = 0, \quad (7.9)$$

$$\mathbf{J}_t + \operatorname{div}\left(\frac{\mathbf{J} \otimes \mathbf{J}}{\rho}\right) + \nabla P(\rho) + \rho \nabla V = 0. \quad (7.10)$$

which can be viewed formally as the dispersive (semiclassical) limit of the NLSE (7.1). In the case $f' > 0$ we expect (7.9), (7.10) to be the 'rigorous' semiclassical limit of (7.1) as long as caustics do not occur, i.e. in the pre-breaking regime. After caustics the dispersive behavior of the NLSE takes over and (7.9), (7.10) is not correct any more.

8. Wigner transform and semiclassical limit

In this section, we consider the linear Schrödinger equation in semiclassical regime

$$i\varepsilon \psi_t^\varepsilon = -\frac{\varepsilon^2}{2} \triangle \psi^\varepsilon + V(\mathbf{x})\psi^\varepsilon, \qquad \mathbf{x} \in \mathbb{R}^d, \quad t \geq 0, \quad (8.1)$$

$$\psi^\varepsilon(\mathbf{x}, 0) = \psi_0^\varepsilon(\mathbf{x}), \qquad \mathbf{x} \in \mathbb{R}^d, \quad (8.2)$$

and find its semiclassical limit by using Wigner transformation.

Let $f, g \in L^2(\mathbb{R}^d)$. Then the Wigner-transform of (f, g) on the scale $\varepsilon > 0$ is defined as the phase-space function:

$$w^\varepsilon(f, g)(\mathbf{x}, \xi) = \frac{1}{(2\pi)^d} \int_{\mathbb{R}^d} f^*\left(\mathbf{x} + \frac{\varepsilon}{2}\mathbf{y}\right) g\left(\mathbf{x} - \frac{\varepsilon}{2}\mathbf{y}\right) e^{i\mathbf{y}\cdot\xi}\, d\mathbf{y} \qquad (8.3)$$

(cf. [58], [91] for a detailed analysis of the Wigner-transform). It is well-known that the estimate

$$\|w^\varepsilon(f, g)\|_{\mathcal{E}^*} \leq \|f\|_{L^2(\mathbb{R}^d)} \|g\|_{L^2(\mathbb{R}^d)} \qquad (8.4)$$

holds, where $\mathcal{E}$ is the Banach space

$$\mathcal{E} := \{\phi \in C_0(\mathbb{R}^d_{\mathbf{x}} \times \mathbb{R}^d_\xi) \ : \ (\mathcal{F}_{\xi \to \mathbf{v}}\phi)(\mathbf{x}, \mathbf{v}) \in L^1(\mathbb{R}^d_{\mathbf{v}}; C_0(\mathbb{R}^d_{\mathbf{x}}))\},$$

$$\|\phi\|_{\mathcal{E}} := \int_{\mathbb{R}^d_{\mathbf{v}}} \sup_{\mathbf{x} \in \mathbb{R}^d_{\mathbf{x}}} |(\mathcal{F}_{\xi \to \mathbf{v}}\phi)(\mathbf{x}, \mathbf{v})| \ d\mathbf{v},$$

(cf. [91]). $\mathcal{E}^*$ denotes the dual space of $\mathcal{E}$ and $(\mathcal{F}_{\xi \to \mathbf{v}}\sigma)(\mathbf{v}) := \int_{\mathbb{R}^d_\xi} \sigma(\xi) \, e^{-i\mathbf{v}\cdot\xi} \, d\xi$ the Fourier transform.

Now let $\psi^\varepsilon(t)$ be the solution of the linear Schrödinger equation (8.1), (8.2) and denote

$$w^\varepsilon(t) := w^\varepsilon(\psi^\varepsilon(t), \psi^\varepsilon(t)). \qquad (8.5)$$

Then w^ε satisfies the Wigner equation

$$w^\varepsilon_t + \xi \cdot \nabla_{\mathbf{x}} w^\varepsilon + \Theta^\varepsilon[V]w^\varepsilon = 0, \qquad (\mathbf{x}, \xi) \in \mathbb{R}^d_{\mathbf{x}} \times \mathbb{R}^d_\xi, \quad t \in \mathbb{R}, \quad (8.6)$$

$$w^\varepsilon(t = 0) = w^\varepsilon(\psi^\varepsilon_0, \psi^\varepsilon_0), \qquad (8.7)$$

where $\Theta^\varepsilon[V]$ is the pseudo-differential operator:

$$\Theta^\varepsilon[V]w^\varepsilon(\mathbf{x}, \xi, t) := \frac{i}{(2\pi)^d} \int_{\mathbb{R}^d_\alpha} \frac{V(\mathbf{x} + \frac{\varepsilon}{2}\alpha) - V(\mathbf{x} - \frac{\varepsilon}{2}\alpha)}{\varepsilon} \, \hat{w}^\varepsilon(\mathbf{x}, \alpha, t)e^{i\alpha\cdot\xi}d\alpha,$$

$$(8.8)$$

here $\hat{w}^\varepsilon$ stands for the Fourier-transform

$$\mathcal{F}_{\xi \to \alpha} \, w^\varepsilon(\mathbf{x}, \cdot, t) := \int_{\mathbb{R}^d_\xi} w^\varepsilon(\mathbf{x}, \xi, t)e^{-i\alpha\cdot\xi} \, d\xi.$$

The main advantage of the formulation (8.6), (8.7) is that the semiclassical limit $\varepsilon \to 0$ can easily be carried out. Taking ε to 0 gives the Vlasov-equation (or Liouville equation):

$$w^0_t + \xi \cdot \nabla_{\mathbf{x}} w^0 - \nabla_{\mathbf{x}} V(\mathbf{x}) \cdot \nabla_\xi w^0 = 0, \quad (\mathbf{x}, \xi) \in \mathbb{R}^d_{\mathbf{x}} \times \mathbb{R}^d_\xi, \ t \in \mathbb{R}, \quad (8.9)$$

$$w^0(t = 0) = w_0 := \lim_{\varepsilon \to 0} w^\varepsilon(\psi^\varepsilon_0, \psi^\varepsilon_0), \qquad (8.10)$$

(cf. [58], [91]), where

$$w^0 := \lim_{\varepsilon \to 0} w^\varepsilon.$$

Here, the limits hold in an appropriate weak sense (i.e. in $\mathcal{E}^* - \omega^*$) and have to be understood for subsequences $(\varepsilon_{n_k}) \to 0$ of sequence ε_n. We recall that w_0, $w^0(t)$ are positive bounded measures on the phase-space $\mathbb{R}^d_{\mathbf{x}} \times \mathbb{R}^d_\xi$.

When the initial Wigner distribution has the high frequency form

$$w_0 = |A_0(\mathbf{x})|^2 \delta(\xi - \nabla S_0(\mathbf{x})), \tag{8.11}$$

then it is easy to see that the solution of (8.9) is given that

$$w^0(\mathbf{x}, \xi, t) = |A(\mathbf{x}, t)|^2 \delta(\xi - \nabla S(\mathbf{x}, t)), \tag{8.12}$$

where $A(\mathbf{x}, t)$ is the solution of the transport equation

$$(|A|^2)_t + \nabla \cdot (|A|^2 \nabla S) = 0, \qquad |A(\mathbf{x}, 0)|^2 = |A_0(\mathbf{x})|^2 \tag{8.13}$$

and $S(\mathbf{x}, t)$ is the solution of the Eiconal equation

$$S_t + \frac{1}{2}|\nabla S|^2 + V(\mathbf{x}) = 0, \qquad S(\mathbf{x}, 0) = S_0(\mathbf{x}). \tag{8.14}$$

Define the moments

$$\rho(\mathbf{x}, t) = \int_{\mathbb{R}^d_\xi} w^0(\mathbf{x}, \xi, t)\, d\xi, \tag{8.15}$$

$$\mathbf{J}(\mathbf{x}, t) = \int_{\mathbb{R}^d_\xi} \xi w^0(\mathbf{x}, \xi, t)\, d\xi. \tag{8.16}$$

Then ρ and $\mathbf{J}$ satisfy the pressureless Euler equation:

$$\rho_t + \operatorname{div} \mathbf{J} = 0, \tag{8.17}$$

$$\mathbf{J}_t + \operatorname{div}\left(\frac{\mathbf{J} \otimes \mathbf{J}}{\rho}\right) + \rho \nabla V = 0; \tag{8.18}$$

with initial data

$$\rho(\mathbf{x}, 0) = \rho_0(\mathbf{x}) = |A_0(\mathbf{x})|^2, \quad \mathbf{J}(\mathbf{x}, 0) = \rho_0(\mathbf{x}) \nabla S_0(\mathbf{x}) = |A_0(\mathbf{x})|^2 \nabla S_0(\mathbf{x}). \tag{8.19}$$

9. Ground, excited and central vortex states of GPE

For simplicity, in this section, we take $\sigma = 1$ and the potential $V(\mathbf{x})$ as a harmonic oscillator (4.1), i.e. NLSE is considered in terms of BEC setup.

9.1. *Stationary states*

To find a stationary solution of (4.1), we write

$$\psi(\mathbf{x}, t) = e^{-i\mu t}\phi(\mathbf{x}), \tag{9.1}$$

where μ is the chemical potential and ϕ is a function independent of time. Inserting (9.1) into (4.1) gives the following equation for $\phi(\mathbf{x})$

$$\mu\,\phi(\mathbf{x}) = -\frac{1}{2}\triangle\phi(\mathbf{x}) + V(\mathbf{x})\,\phi(\mathbf{x}) + \beta|\phi(\mathbf{x})|^2\phi(\mathbf{x}), \qquad \mathbf{x} \in \mathbb{R}^d, \tag{9.2}$$

under the normalization condition

$$\|\phi\|^2 = \int_{\mathbb{R}^d} |\phi(\mathbf{x})|^2\,d\mathbf{x} = 1. \tag{9.3}$$

This is a nonlinear eigenvalue problem under a constraint and any eigenvalue μ can be computed from its corresponding eigenfunction ϕ by

$$\mu = \mu(\phi) = \int_{\mathbb{R}^d} \left[\frac{1}{2}|\nabla\phi(\mathbf{x})|^2 + V(\mathbf{x})\,|\phi(\mathbf{x})|^2 + \beta\,|\phi(\mathbf{x})|^4\right]d\mathbf{x}$$

$$= E(\phi) + \int_{\mathbb{R}^d} \frac{\beta}{2}|\phi(\mathbf{x})|^4\,d\mathbf{x}. \tag{9.4}$$

In fact, the eigenfunctions of (9.2) under the constraint (9.3) are equivalent to the critical points of the energy functional over the unit sphere $S = \{\phi \mid \|\phi\| = 1, E(\phi) < \infty\}$. Furthermore, as noted in [6], they are equivalent to the steady state solutions of the following continuous normalized gradient flow (CNGF):

$$\partial_t\phi = \frac{1}{2}\triangle\phi - V(\mathbf{x})\phi - \beta\,|\phi|^2\phi + \frac{\mu(\phi)}{\|\phi(\cdot, t)\|^2}\,\phi, \; \mathbf{x} \in \mathbb{R}^d,\; t \geq 0, \tag{9.5}$$

$$\phi(\mathbf{x}, 0) = \phi_0(\mathbf{x}), \quad \mathbf{x} \in \mathbb{R}^d \qquad \text{with} \qquad \|\phi_0\| = 1. \tag{9.6}$$

9.2. *Ground state*

The BEC ground state wave function $\phi_g(\mathbf{x})$ is found by minimizing the energy functional $E(\phi)$ over the unit sphere $S = \{\phi \mid \|\phi\| = 1, E(\phi) < \infty\}$:

(V) Find $(\mu_\beta^g, \phi_\beta^g \in S)$ such that

$$E_\beta^g = E(\phi_\beta^g) = \min_{\phi \in S} E(\phi), \qquad \mu_\beta^g = \mu(\phi_\beta^g) = E(\phi_\beta^g) + \int_{\mathbb{R}^d} \frac{\beta}{2}|\phi_\beta^g|^2\,d\mathbf{x}. \tag{9.7}$$

In the case of a defocusing condensate, i.e. $\beta \geq 0$, the energy functional $E(\phi)$ is positive, coercive and weakly lower semicontinuous on S, thus the existence of a minimum follows from the standard theory. For understanding the uniqueness question note that $E(\alpha\phi_\beta^g) = E(\phi_\beta^g)$ for all $\alpha \in \mathbb{C}$ with $|\alpha| = 1$. Thus an additional constraint has to be introduced to show uniqueness. For non-rotating BECs, the minimization problem (9.7) has a unique real valued nonnegative ground state solution $\phi_\beta^g(\mathbf{x}) > 0$ for $\mathbf{x} \in \mathbb{R}^d$ [87].

When $\beta = 0$, the ground state solution is given explicitly [5]

$$\mu_0^g = \frac{1}{2} \begin{cases} \gamma_x, & d = 1, \\ \gamma_x + \gamma_y, & d = 2, \\ \gamma_x + \gamma_y + \gamma_z, & d = 3, \end{cases} \tag{9.8}$$

$$\phi_0^g(\mathbf{x}) = \frac{1}{\pi^{d/4}} \begin{cases} \gamma_x^{1/4} e^{-\gamma_x x^2/2}, & d = 1, \\ (\gamma_x \gamma_y)^{1/4} e^{-(\gamma_x x^2 + \gamma_y y^2)/2}, & d = 2, \\ (\gamma_x \gamma_y \gamma_z)^{1/4} e^{-(\gamma_x x^2 + \gamma_y y^2 + \gamma_z z^2)/2}, & d = 3. \end{cases} \tag{9.9}$$

In fact, this solution can be viewed as an approximation of the ground state for weakly interacting condensate, i.e. $|\beta_d| \ll 1$. For a condensate with strong repulsive interaction, i.e. $\beta \gg 1$ and $\gamma_\alpha = O(1)$ ($\alpha = x, y, z$), the ground state can be approximated by the Thomas-Fermi approximation in this regime [5]:

$$\phi_\beta^{\text{TF}}(\mathbf{x}) = \begin{cases} \sqrt{(\mu_\beta^{\text{TF}} - V(\mathbf{x}))/\beta}, & V(\mathbf{x}) < \mu_\beta^{\text{TF}}, \\ 0, & \text{otherwise}, \end{cases} \tag{9.10}$$

$$\mu_\beta^{\text{TF}} = \frac{1}{2} \begin{cases} (3\beta\gamma_x/2)^{2/3}, & d = 1, \\ (4\beta\gamma_x\gamma_y/\pi)^{1/2}, & d = 2, \\ (15\beta\gamma_x\gamma_y\gamma_z/4\pi)^{2/5}, & d = 3. \end{cases} \tag{9.11}$$

Due to ϕ_β^{TF} is not differentiable at $V(\mathbf{x}) = \mu_\beta^{\text{TF}}$, as noticed in [5, 8], $E(\phi_\beta^{\text{TF}}) = \infty$ and $\mu(\phi_\beta^{\text{TF}}) = \infty$. This shows that we **can't** use (4.4) to define the energy of the Thomas-Fermi approximation (9.10). How to define the energy of the Thomas-Fermi approximation is not clear in the literatures. Using (9.4), (9.11) and (9.10), here we present a way to define the energy of the Thomas-Fermi approximation (9.10):

$$E_\beta^{\text{TF}} = \mu_\beta^{\text{TF}} - \int_{\mathbb{R}^d} \frac{\beta}{2} |\phi_\beta^{\text{TF}}(\mathbf{x})|^4 \, d\mathbf{x} = \int_{\mathbb{R}^d} \left[V(\mathbf{x}) |\phi_\beta^{\text{TF}}(\mathbf{x})|^2 + \frac{\beta}{2} |\phi_\beta^{\text{TF}}(\mathbf{x})|^4 \right] d\mathbf{x}$$

$$= \frac{d+2}{d+4} \mu_\beta^{\text{TF}}, \qquad d = 1, 2, 3. \tag{9.12}$$

From the numerical results in [6, 5], when $\gamma_x = O(1)$, $\gamma_y = O(1)$ and $\gamma_z = O(1)$, we can get

$$E_\beta^g - E_\beta^{\text{TF}} = E(\phi_\beta^g) - E_\beta^{\text{TF}} \to 0, \qquad \text{as} \quad \beta_d \to \infty.$$

Any eigenfunction $\phi(\mathbf{x})$ of (9.2) under constraint (9.3) whose energy $E(\phi) > E(\phi_\beta^g)$ is usually called as excited states in physical literatures.

9.3. *Central vortex states*

To find central vortex states in 2D with radial symmetry, i.e. $d = 2$ and $\gamma_x = \gamma_y = 1$ in (4.1), we write

$$\psi(\mathbf{x}, t) = e^{-i\mu_m t}\phi_m(x, y) = e^{-i\mu_m t}\phi_m(r)e^{im\theta}, \tag{9.13}$$

where (r, θ) is the polar coordinate, $m \neq 0$ is an integer and called as index or winding number, μ_m is the chemical potential, and $\phi_m(r)$ is a real function independent of time. Inserting (9.13) into (4.1) gives the following equation for $\phi_m(r)$ with $0 < r < \infty$

$$\mu_m\,\phi_m(r) = \left[-\frac{1}{2r}\frac{d}{dr}\left(r\frac{d}{dr}\right) + \frac{1}{2}\left(r^2 + \frac{m^2}{r^2}\right) + \beta_2|\phi_m|^2\right]\phi_m, \tag{9.14}$$

$$\phi_m(0) = 0, \qquad \lim_{r\to\infty}\phi_m(r) = 0. \tag{9.15}$$

under the normalization condition

$$2\pi\int_0^\infty |\phi_m(r)|^2\, r\, dr = 1. \tag{9.16}$$

In order to find the central vortex state $\phi_\beta^m(x, y) = \phi_\beta^m(r)e^{im\theta}$ with index m, we find a real nonnegative function $\phi_\beta^m(r)$ which minimizes the energy functional

$$
\begin{aligned}
E^m(\phi(r)) &= E(\phi(r)e^{im\theta}) \\
&= \pi\int_0^\infty \left[|\phi'(r)|^2 + \left(r^2 + \frac{m^2}{r^2}\right)|\phi(r)|^2 + \beta_2|\phi(r)|^4\right] r\,dr,
\end{aligned} \tag{9.17}
$$

over the set $S_0 = \{\phi \mid 2\pi\int_0^\infty |\phi(r)|^2 r\, dr = 1,\ \phi(0) = 0,\ E^m(\phi) < \infty\}$. The existence and uniqueness of nonnegative minimizer for this minimization problem can be obtained similarly as for the ground state [87]. Note that the set $S_m = \{\phi(r)e^{im\theta} \mid \phi \in S_0\} \subset S$ is a subset of the unit sphere, so $\phi_\beta^m(r)e^{im\theta}$ is a minimizer of the energy functional E_β over the set $S_m \subset S$. When $\beta_2 = 0$ in (4.1), $\phi_0^m(r) = \frac{1}{\sqrt{\pi|m|!}}r^{|m|}e^{-r^2/2}$ [6].

Similarly, in order to find central vortex line states in 3D with cylindrical symmetry, i.e. $d = 3$ and $\gamma_x = \gamma_y = 1$ in (4.1), we write

$$\psi(\mathbf{x}, t) = e^{-i\mu_m t}\phi_m(x, y, z) = e^{-i\mu_m t}\phi_m(r, z)e^{im\theta}, \tag{9.18}$$

where $m \neq 0$ is an integer and called as index, μ_m is the chemical potential, and $\phi_m(r, z)$ is a real function independent of time. Inserting (9.18) into (4.1) with $d = 3$ gives the following equation for $\phi_m(r, z)$

$$\mu_m \, \phi_m = \left[-\frac{1}{2r} \frac{\partial}{\partial r} \left(r \frac{\partial}{\partial r} \right) - \frac{\partial^2}{2 \, \partial z^2} + \frac{1}{2} \left(r^2 + \frac{m^2}{r^2} + \gamma_z^2 z^2 \right) + \beta_3 |\phi_m|^2 \right] \phi_m,$$

(9.19)

$$\phi_m(0, z) = 0, \qquad \lim_{r \to \infty} \phi_m(r, z) = 0, \qquad -\infty < z < \infty, \tag{9.20}$$

$$\lim_{|z| \to \infty} \phi_m(r, z) = 0, \qquad 0 \leq r < \infty, \tag{9.21}$$

under the normalization condition

$$2\pi \int_0^\infty \int_{-\infty}^\infty |\phi_m(r, z)|^2 \, r \, dr dz = 1. \tag{9.22}$$

In order to find the central vortex line state $\phi_\beta^m(x, y, z) = \phi_\beta^m(r, z) e^{im\theta}$ with index m, we find a real nonnegative function $\phi_\beta^m(r, z)$ which minimizes the energy functional

$$E^m(\phi(r, z)) = E(\phi(r, z) e^{im\theta}) \tag{9.23}$$

$$= \pi \int_0^\infty \int_{-\infty}^\infty \left[|\partial_r \phi|^2 + |\partial_z \phi|^2 + \left(r^2 + \gamma_z^2 z^2 + \frac{m^2}{r^2} \right) |\phi|^2 + \beta_3 |\phi|^4 \right] r \, dr dz,$$

over the set $S_0 = \{ \phi \mid 2\pi \int_0^\infty \int_{-\infty}^\infty |\phi(r, z)|^2 r \, dr dz = 1, \ \phi(0, z) = 0, \ -\infty < z < \infty, \ E_\beta^m(\phi) < \infty \}$. The existence and uniqueness of nonnegative minimizer for this minimization problem can be obtained similarly as for the ground state [87]. Note that the set $S_m = \{ \phi(r, z) e^{im\theta} \mid \phi \in S_0 \} \subset S$ is a subset of the unit sphere, so $\phi_\beta^m(r, z) e^{im\theta}$ is a minimizer of the energy functional E_β over the set S_m. When $\beta_3 = 0$ in (3.11), $\phi_0^m(r, z) = \frac{\gamma_z^{1/4}}{\pi^{3/4} \sqrt{|m|!}} r^{|m|} e^{-(r^2 + \gamma_z z^2)/2}$ [6].

9.4. *Variation of stationary states over the unit sphere*

For the stationary states of (9.2), we have the following lemma:

Lemma 9.1: *Suppose $\beta = 0$ and $V(\mathbf{x}) \geq 0$ for $\mathbf{x} \in \mathbb{R}^d$, we have*
    (i) The ground state ϕ_g is a global minimizer of $E(\phi)$ over S.
    (ii) Any excited state ϕ_j is a saddle point of $E(\phi)$ over S.

Proof: Let ϕ_e be an eigenfunction of the eigenvalue problem (9.2) and (9.3). The corresponding eigenvalue is μ_e. For any function ϕ such that $E(\phi) < \infty$

and $\|\phi_e + \phi\| = 1$, notice (9.3), we have that

$$\|\phi\|^2 = \|\phi + \phi_e - \phi_e\|^2 = \|\phi + \phi_e\|^2 - \|\phi_e\|^2 - \int_{\mathbb{R}^d} (\phi^*\phi_e + \phi\phi_e^*) \, d\mathbf{x}$$

$$= -\int_{\mathbb{R}^d} (\phi^*\phi_e + \phi\phi_e^*) \, d\mathbf{x}. \tag{9.24}$$

From (4.4) with $\psi = \phi_e + \phi$ and $\beta = 0$, notice (9.3) and (9.24), integration by parts, we get

$$E(\phi_e + \phi) = \int_{\mathbb{R}^d} \left[\frac{1}{2}|\nabla\phi_e + \nabla\phi|^2 + V(\mathbf{x})|\phi_e + \phi|^2 \right] \, d\mathbf{x}$$

$$= \int_{\mathbb{R}^d} \left[\frac{1}{2}|\nabla\phi_e|^2 + V(\mathbf{x})|\phi_e|^2 \right] + \int_{\mathbb{R}^d} \left[\frac{1}{2}|\nabla\phi|^2 + V(\mathbf{x})|\phi|^2 \right] \, d\mathbf{x}$$

$$+ \int_{\mathbb{R}^d} \left[\left(-\frac{1}{2}\triangle\phi_e + V(\mathbf{x})\phi_e \right)^* \phi + \left(-\frac{1}{2}\triangle\phi_e + V(\mathbf{x})\phi_e \right)\phi^* \right] \, d\mathbf{x}$$

$$= E(\phi_e) + E(\phi) + \int_{\mathbb{R}^d} (\mu_e\phi_e^* + \mu_e\phi_e\phi^*) \, d\mathbf{x}$$

$$= E(\phi_e) + E(\phi) - \mu_e\|\phi\|^2$$

$$= E(\phi_e) + [E(\phi/\|\phi\|) - \mu_e] \, \|\phi\|^2. \tag{9.25}$$

(i) Taking $\phi_e = \phi_g$ and $\mu_e = \mu_g$ in (9.25) and noticing $E(\phi/\|\phi\|) \geq E(\phi_g) = \mu_g$ for any $\phi \neq 0$, we get immediately that ϕ_g is a global minimizer of $E(\phi)$ over S.

(ii). Taking $\phi_e = \phi_j$ and $\mu_e = \mu_j$ in (9.25), since $E(\phi_g) < E(\phi_j)$ and it is easy to find an eigenfunction ϕ of (9.2) such that $E(\phi) > E(\phi_j)$, we get immediately that ϕ_j is a saddle point of the functional $E(\phi)$ over S. $\qquad\square$

9.5. *Conservation of angular momentum expectation*

Another important quantity for studying dynamics of BEC in 2&3d, especially for measuring the appearance of vortex, is the angular momentum expectation value defined as

$$\langle L_z \rangle(t) := \int_{\mathbb{R}^d} \psi^*(\mathbf{x},t) L_z \psi(\mathbf{x},t) \, d\mathbf{x}, \qquad t \geq 0, \qquad d = 2,3, \tag{9.26}$$

where $L_z = i\,(y\partial_x - x\partial_y)$ is the z-component angular momentum.

Lemma 9.2: *Suppose $\psi(\mathbf{x},t)$ is the solution of the problem (4.1), (4.2) with $d = 2$ or 3, then we have*

$$\frac{d\langle L_z \rangle(t)}{dt} = (\gamma_x^2 - \gamma_y^2)\,\delta_{xy}(t), \quad \delta_{xy}(t) = \int_{\mathbb{R}^d} xy|\psi(\mathbf{x},t)|^2 \, d\mathbf{x}, \quad t \geq 0. \tag{9.27}$$

This implies that, at least in the following two cases, the angular momentum expectation is conserved:

i) For any given initial data $\psi_0(\mathbf{x})$ in (4.2), if the trap is radial symmetric in 2d, and resp., cylindrical symmetric in 3d, i.e. $\gamma_x = \gamma_y$;

ii) For any given $\gamma_x > 0$ and $\gamma_y > 0$ in (3.13), if the initial data $\psi_0(\mathbf{x})$ in (4.2) is either odd or even in the first variable x or second variable y.

Proof: Differentiate (9.26) with respect to t, notice (4.1), integrate by parts, we have

$$
\begin{aligned}
\frac{d\langle L_z\rangle(t)}{dt} &= \int_{\mathbb{R}^d} \left[(i\psi_t^*)\,(y\partial_x - x\partial_y)\psi + \psi^*\,(y\partial_x - x\partial_y)(i\psi_t) \right]\, d\mathbf{x} \\
&= \int_{\mathbb{R}^d} \left[\left(\frac{1}{2}\nabla^2\psi^* - V(\mathbf{x})\psi^* - \beta|\psi|^2\psi^* \right)(y\partial_x - x\partial_y)\psi \right. \\
&\qquad\qquad \left. +\psi^*\,(y\partial_x - x\partial_y)\left(-\frac{1}{2}\nabla^2\psi + V(\mathbf{x})\psi + \beta|\psi|^2\psi \right) \right]\, d\mathbf{x} \\
&= \int_{\mathbb{R}^d} \frac{1}{2}\left[\nabla^2\psi^*\,(y\partial_x - x\partial_y)\psi - \psi^*\,(y\partial_x - x\partial_y)\nabla^2\psi \right]\, d\mathbf{x} \\
&\quad + \int_{\mathbb{R}^d} \left[\psi^*\,(y\partial_x - x\partial_y)\left(V(\mathbf{x})\psi + \beta|\psi|^2\psi \right) \right. \\
&\qquad\qquad \left. - \left(V(\mathbf{x})\psi^* + \beta|\psi|^2\psi^* \right)(y\partial_x - x\partial_y)\psi \right]\, d\mathbf{x} \\
&= \int_{\mathbb{R}^d} |\psi|^2 (y\partial_x - x\partial_y)\left(V(\mathbf{x}) + \beta|\psi|^2 \right)\, d\mathbf{x} \\
&= \int_{\mathbb{R}^d} |\psi|^2 (y\partial_x - x\partial_y)V_d(\mathbf{x})\, d\mathbf{x} = \int_{\mathbb{R}^d} |\psi|^2 (\gamma_x^2 - \gamma_y^2)xy\, d\mathbf{x} \\
&= (\gamma_x^2 - \gamma_y^2) \int_{\mathbb{R}^d} xy|\psi|^2\, d\mathbf{x}, \qquad t \geq 0. \tag{9.28}
\end{aligned}
$$

For case i), since $\gamma_x = \gamma_y$, we get the conservation of $\langle L_z\rangle$ immediately from the first order ODE:

$$
\frac{d\langle L_z\rangle(t)}{dt} = 0, \qquad t \geq 0. \tag{9.29}
$$

For case ii), we know the solution $\psi(\mathbf{x}, t)$ is either odd or even in the first variable x or second variable y due to the assumption of the initial data and symmetry of $V(\mathbf{x})$. Thus $|\psi(\mathbf{x}, t)|$ is even in either x or y, which immediately implies that $\langle L_z\rangle$ satisfies the first order ODE (9.29). $\square$

10. Numerical methods for computing ground states of GPE

In this section, we present the continuous normalized gradient flow (CNGF), prove its energy diminishing and propose its semi-discretization for computing ground states in BEC. For simplicity, we take $\sigma = 1$ in (4.1).

10.1. *Gradient flow with discrete normalization (GFDN)*

Various algorithms for computing the minimizer of the energy functional $E(\phi)$ under the constraint (9.3) have been studied in the literature. For instance, second order in time discretization scheme that preserves the normalization and energy diminishing properties were presented in [2, 6]. Perhaps one of the more popular technique for dealing with the normalization constraint (9.3) is through the following construction: choose a time sequence $0 = t_0 < t_1 < t_2 < \cdots < t_n < \cdots$ with $\Delta t_n = t_{n+1} - t_n > 0$ and $k = \max_{n \geq 0} \Delta t_n$. To adapt an algorithm for the solution of the usual gradient flow to the minimization problem under a constraint, it is natural to consider the following splitting (or projection) scheme which was widely used in physical literatures [6] for computing the ground state solution of BEC:

$$\phi_t = -\frac{1}{2} \frac{\delta E(\phi)}{\delta \phi} = \frac{1}{2} \triangle \phi - V(\mathbf{x})\phi - \beta \, |\phi|^2 \phi,$$

$$\mathbf{x} \in \Omega, \ t_n < t < t_{n+1}, \ n \geq 0, \tag{10.1}$$

$$\phi(x, t_{n+1}) \stackrel{\triangle}{=} \phi(\mathbf{x}, t_{n+1}^+) = \frac{\phi(\mathbf{x}, t_{n+1}^-)}{\|\phi(\cdot, t_{n+1}^-)\|}, \qquad \mathbf{x} \in \Omega, \quad n \geq 0, \tag{10.2}$$

$$\phi(\mathbf{x}, t) = 0, \quad \mathbf{x} \in \Gamma = \partial\Omega, \qquad \phi(\mathbf{x}, 0) = \phi_0(\mathbf{x}), \quad \mathbf{x} \in \Omega; \tag{10.3}$$

where $\phi(\mathbf{x}, t_n^{\pm}) = \lim_{t \to t_n^{\pm}} \phi(\mathbf{x}, t)$, $\|\phi_0\| = 1$ and $\Omega \subset \mathbb{R}^d$. In fact, the gradient flow (10.1) can be viewed as applying the steepest decent method to the energy functional $E(\phi)$ without constraint and (10.2) then projects the solution back to the unit sphere in order to satisfying the constraint (9.3). From the numerical point of view, the gradient flow (10.1) can be solved via traditional techniques and the normalization of the gradient flow is simply achieved by a projection at the end of each time step.

10.2. *Energy diminishing of GFDN*

Let

$$\tilde{\phi}(\cdot, t) = \frac{\phi(\cdot, t)}{\|\phi(\cdot, t)\|}, \qquad t_n \leq t \leq t_{n+1}, \qquad n \geq 0. \tag{10.4}$$

For the gradient flow (10.1), it is easy to establish the following basic facts:

Lemma 10.1: *Suppose $V(\mathbf{x}) \geq 0$ for all $\mathbf{x} \in \Omega$, $\beta \geq 0$ and $\|\phi_0\| = 1$, then*
(i). $\|\phi(\cdot, t)\| \leq \|\phi(\cdot, t_n)\| = 1$ for $t_n \leq t \leq t_{n+1}$, $n \geq 0$.
(ii). For any $\beta \geq 0$,

$$E(\phi(\cdot, t)) \leq E(\phi(\cdot, t')), \qquad t_n \leq t' < t \leq t_{n+1}, \qquad n \geq 0. \tag{10.5}$$

(iii). For $\beta = 0$,

$$E(\tilde{\phi}(\cdot, t)) \leq E(\tilde{\phi}(\cdot, t_n)), \qquad t_n \leq t \leq t_{n+1}, \qquad n \geq 0. \tag{10.6}$$

Proof: (i) and (ii) follows the standard techniques used for gradient flow. As for (iii), from (4.4) with $\psi = \tilde{\phi}$ and $\beta = 0$, (10.1), (10.3) and (10.4), integration by parts and Schwartz inequality, we obtain

$$\frac{d}{dt} E(\tilde{\phi}) = \frac{d}{dt} \int_\Omega \left[\frac{|\nabla \phi|^2}{2\|\phi\|^2} + \frac{V(\mathbf{x})\phi^2}{\|\phi\|^2} \right] d\mathbf{x}$$

$$= 2 \int_\Omega \left[\frac{\nabla\phi \cdot \nabla\phi_t}{2\|\phi\|^2} + \frac{V(\mathbf{x})\phi\,\phi_t}{\|\phi\|^2} \right] d\mathbf{x} - \left(\frac{d}{dt}\|\phi\|^2 \right) \int_\Omega \left[\frac{|\nabla\phi|^2}{2\|\phi\|^4} + \frac{V(\mathbf{x})\phi^2}{\|\phi\|^4} \right] d\mathbf{x}$$

$$= 2 \int_\Omega \frac{\left[-\frac{1}{2}\triangle\phi + V(\mathbf{x})\phi \right]\phi_t}{\|\phi\|^2} d\mathbf{x} - \left(\frac{d}{dt}\|\phi\|^2 \right) \int_\Omega \frac{\frac{1}{2}|\nabla\phi|^2 + V(\mathbf{x})\phi^2}{\|\phi\|^4} d\mathbf{x}$$

$$= -2\frac{\|\phi_t\|^2}{\|\phi\|^2} + \frac{1}{2\|\phi\|^4} \left(\frac{d}{dt}\|\phi\|^2 \right)^2 = \frac{2}{\|\phi\|^4} \left[\left(\int_\Omega \phi\,\phi_t\,d\mathbf{x} \right)^2 - \|\phi\|^2\|\phi_t\|^2 \right]$$

$$\leq 0, \qquad t_n \leq t \leq t_{n+1}. \tag{10.7}$$

This implies (10.6). $\qquad\qquad\qquad\qquad\qquad\qquad\qquad\qquad\qquad\qquad\qquad\quad\square$

Remark 10.2: The property (10.5) is often referred as the energy diminishing property of the gradient flow. It is interesting to note that (10.6) implies that the energy diminishing property is preserved even with the normalization of the solution of the gradient flow for $\beta = 0$, that is, for linear evolution equations.

Remark 10.3: When $\beta > 0$, the solution of (10.1)-(10.3) may not preserve the normalized energy diminishing property

$$E(\tilde{\phi}(\cdot, t)) \leq E(\tilde{\phi}(\cdot, t')), \qquad 0 \leq t' < t \leq t_1$$

for any $t_1 > 0$ [6].

From Lemma 10.1, we get immediately

Theorem 10.4: *Suppose $V(\mathbf{x}) \geq 0$ for all $\mathbf{x} \in \Omega$ and $\|\phi_0\| = 1$. For $\beta = 0$, GFDN (10.1)-(10.3) is energy diminishing for any time step k and initial data ϕ_0, i.e.*

$$E(\phi(\cdot, t_{n+1})) \leq E(\phi(\cdot, t_n)) \leq \cdots \leq E(\phi(\cdot, 0)) = E(\phi_0), \quad n = 0, 1, 2, \cdots.$$
$$(10.8)$$

10.3. *Continuous normalized gradient flow (CNGF)*

In fact, the normalized step (10.2) is equivalent to solve the following ODE exactly

$$\phi_t(\mathbf{x}, t) = \mu_\phi(t, k)\phi(\mathbf{x}, t), \qquad \mathbf{x} \in \Omega, \quad t_n < t < t_{n+1}, \quad n \geq 0, \quad (10.9)$$
$$\phi(\mathbf{x}, t_n^+) = \phi(\mathbf{x}, t_{n+1}^-), \qquad \mathbf{x} \in \Omega; \tag{10.10}$$

where

$$\mu_\phi(t, k) \equiv \mu_\phi(t_{n+1}, \triangle t_n) = -\frac{1}{2 \triangle t_n} \ln \|\phi(\cdot, t_{n+1}^-)\|^2, \qquad t_n \leq t \leq t_{n+1}.$$
$$(10.11)$$

Thus the GFDN (10.1)-(10.3) can be viewed as a first-order splitting method for the gradient flow with discontinuous coefficients:

$$\phi_t = \frac{1}{2} \triangle \phi - V(\mathbf{x})\phi - \beta |\phi|^2 \phi + \mu_\phi(t, k)\phi, \quad \mathbf{x} \in \Omega, \ t \geq 0, \quad (10.12)$$
$$\phi(\mathbf{x}, t) = 0, \qquad \mathbf{x} \in \Gamma, \qquad \phi(\mathbf{x}, 0) = \phi_0(\mathbf{x}), \qquad \mathbf{x} \in \Omega. \quad (10.13)$$

Let $k \to 0$, we see that

$$\mu_\phi(t) := \lim_{k \to 0^+} \mu_\phi(t, k)$$

$$= \frac{1}{\|\phi(\cdot, t)\|^2} \int_\Omega \left[\frac{1}{2} |\nabla \phi(\mathbf{x}, t)|^2 + V(\mathbf{x})\phi^2(\mathbf{x}, t) + \beta\phi^4(\mathbf{x}, t) \right] d\mathbf{x}. \quad (10.14)$$

This suggests us to consider the following continuous normalized gradient flow:

$$\phi_t = \frac{1}{2} \triangle \phi - V(\mathbf{x})\phi - \beta |\phi|^2 \phi + \mu_\phi(t)\phi, \qquad \mathbf{x} \in \Omega, \quad t \geq 0, \quad (10.15)$$
$$\phi(\mathbf{x}, t) = 0, \qquad \mathbf{x} \in \Gamma, \qquad \phi(\mathbf{x}, 0) = \phi_0(\mathbf{x}), \qquad \mathbf{x} \in \Omega. \quad (10.16)$$

In fact, the right hand side of (10.15) is the same as (9.2) if we view $\mu_\phi(t)$ as a Lagrange multiplier for the constraint (9.3). Furthermore for the above CNGF, as observed in [6], the solution of (10.15) also satisfies the following theorem:

Theorem 10.5: *Suppose $V(\mathbf{x}) \geq 0$ for all $\mathbf{x} \in \Omega$, $\beta \geq 0$ and $\|\phi_0\| = 1$. Then the CNGF (10.15)-(10.16) is normalization conservation and energy diminishing, i.e.*

$$\|\phi(\cdot,t)\|^2 = \int_\Omega \phi^2(\mathbf{x},t)\,d\mathbf{x} = \|\phi_0\|^2 = 1, \qquad t \geq 0, \qquad (10.17)$$

$$\frac{d}{dt}E(\phi) = -2\,\|\phi_t(\cdot,t)\|^2 \leq 0\,, \qquad t \geq 0, \qquad (10.18)$$

which in turn implies

$$E(\phi(\cdot,t_1)) \geq E(\phi(\cdot,t_2)), \qquad 0 \leq t_1 \leq t_2 < \infty.$$

Remark 10.6: We see from the above theorem that the energy diminishing property is preserved in the continuous dynamic system (10.15).

Using argument similar to that in [88, 106], we may also get as $t \to \infty$, ϕ approaches to a steady state solution which is a critical point of the energy. In non-rotating BEC, it has a unique real valued nonnegative ground state solution $\phi_g(\mathbf{x}) \geq 0$ for all $\mathbf{x} \in \Omega$ [87]. We choose the initial data $\phi_0(\mathbf{x}) \geq 0$ for $\mathbf{x} \in \Omega$, e.g. the ground state solution of linear Schrödinger equation with a harmonic oscillator potential [5, 8]. Under this kind of initial data, the ground state solution ϕ_g and its corresponding chemical potential μ_g can be obtained from the steady state solution of the CNGF (10.15)-(10.16), i.e.

$$\phi_g(\mathbf{x}) = \lim_{t \to \infty} \phi(\mathbf{x},t), \quad \mathbf{x} \in \Omega, \ \mu_g = \mu_\beta(\phi_g) = E(\phi_g) + \frac{\beta}{2}\int_\Omega |\phi_g(\mathbf{x})|^4\,d\mathbf{x}.$$
$$(10.19)$$

10.4. *Semi-implicit time discretization*

To further discretize the equation (10.1), we here consider the following semi-implicit time discretization scheme:

$$\frac{\tilde{\phi}^{n+1} - \phi^n}{k} = \frac{1}{2}\,\triangle\,\tilde{\phi}^{n+1} - V(\mathbf{x})\tilde{\phi}^{n+1} - \beta\,|\phi^n|^2\tilde{\phi}^{n+1}\,, \quad \mathbf{x} \in \Omega, \quad (10.20)$$

$$\tilde{\phi}^{n+1}(\mathbf{x}) = 0, \quad \mathbf{x} \in \Gamma, \quad \phi^{n+1}(\mathbf{x}) = \tilde{\phi}^{n+1}(\mathbf{x})/\|\tilde{\phi}^{n+1}\|\,, \quad \mathbf{x} \in \Omega \quad (10.21)$$

Notice that since the equation (10.20) becomes linear, the solution at the new time step becomes relatively simple. In other words, in each discrete time interval, we may view (10.20) as a discretization of a linear gradient flow with a modified potential $\tilde{V}_n(\mathbf{x}) = V(\mathbf{x}) + \beta|\phi^n(\mathbf{x})|^2$.

We now first present the following lemma:

Lemma 10.7: *Suppose $\beta \geq 0$ and $V(\mathbf{x}) \geq 0$ for all $\mathbf{x} \in \Omega$ and $\|\phi^n\| = 1$. Then,*

$$\int_\Omega |\tilde{\phi}^{n+1}|^2 \, d\mathbf{x} \leq \int_\Omega \phi^n \, \tilde{\phi}^{n+1} \, d\mathbf{x}, \qquad \int_\Omega |\tilde{\phi}^{n+1}|^4 \, d\mathbf{x} \leq \int_\Omega |\phi^n|^2 \, |\tilde{\phi}^{n+1}|^2 \, d\mathbf{x}.$$
$$(10.22)$$

Proof: Multiplying both sides of (10.20) by $\tilde{\phi}^{n+1}$, integrating over Ω, and applying integration by parts, we obtain

$$\int_\Omega \left(|\tilde{\phi}^{n+1}|^2 - \phi^n \tilde{\phi}^{n+1} \right) d\mathbf{x} = -k \int_\Omega \left[\frac{1}{2} |\nabla \tilde{\phi}^{n+1}|^2 + \tilde{V}_n(\mathbf{x}) |\tilde{\phi}^{n+1}|^2 \right] d\mathbf{x} \leq 0 \,,$$

which leads to the first inequality in (10.22). Similarly,

$$\int_\Omega |\tilde{\phi}^{n+1}|^2 |\phi^n|^2 d\mathbf{x} = \int_\Omega |\tilde{\phi}^{n+1}|^2 \left| \tilde{\phi}^{n+1} - \frac{k}{2} \triangle \tilde{\phi}^{n+1} + k\tilde{V}_n(\mathbf{x})\tilde{\phi}^{n+1} \right|^2 d\mathbf{x}$$

$$= \int_\Omega |\tilde{\phi}^{n+1}|^2 \left[|\tilde{\phi}^{n+1}|^2 - 2 \frac{k}{2} \tilde{\phi}^{n+1} \triangle \tilde{\phi}^{n+1} + 2k\tilde{V}_n(\mathbf{x})|\tilde{\phi}^{n+1}|^2 \right] d\mathbf{x}$$

$$+ \int_\Omega |\tilde{\phi}^{n+1}|^2 \left| \frac{k}{2} \triangle \tilde{\phi}^{n+1} - k\tilde{V}_n(\mathbf{x})\tilde{\phi}^{n+1} \right|^2 d\mathbf{x}$$

$$= \int_\Omega |\tilde{\phi}^{n+1}|^2 \left[|\tilde{\phi}^{n+1}|^2 + 3k|\nabla \tilde{\phi}^{n+1}|^2 + 2k\tilde{V}_n(\mathbf{x})|\tilde{\phi}^{n+1}|^2 \right] d\mathbf{x}$$

$$+ \int_\Omega |\tilde{\phi}^{n+1}|^2 \left| \frac{k}{2} \triangle \tilde{\phi}^{n+1} - k\tilde{V}_n(\mathbf{x})\tilde{\phi}^{n+1} \right|^2 d\mathbf{x}$$

$$\geq \int_\Omega |\tilde{\phi}^{n+1}|^4 d\mathbf{x} \,. \tag{10.23}$$

This implies the second inequality in (10.22). $\qquad\square$

Given a linear self-adjoint operator A in a Hilbert space H with inner product $(\cdot, \cdot)$, and assume that A is positive definite in the sense that for some positive constant c, $(u, Au) \geq c(u, u)$ for any $u \in H$. We now present a simple lemma:

Lemma 10.8: *For any $k > 0$, and $(I + kA)u = v$, we have*

$$\frac{(u, Au)}{(u, u)} \leq \frac{(v, Av)}{(v, v)} \,. \tag{10.24}$$

Proof: Since A is self-adjoint and positive definite, by Hölder inequality, we have for any $p, q \geq 1$ with $p + q = pq$, that

$$(u, Au) \leq (u, u)^{1/p} (u, A^q u)^{1/q} \,,$$

which leads to

$$(u, Au) \le (u, u)^{1/2} \left(u, A^2 u\right)^{1/2} , \qquad (u, Au)\left(u, A^2 u\right) \le (u, u)\left(u, A^3 u\right) .$$

Direct calculation then gives

$$(u, Au)\left((I + kA)u, (I + kA)u\right)$$
$$= (u, Au)\,(u, u) + 2k\,(u, Au)^2 + k^2\,(u, Au)\left(u, A^2 u\right)$$
$$\le (u, Au)\,(u, u) + 2k\,(u, u)\left(u, A^2 u\right) + k^2\,(u, u)\left(u, A^3 u\right)$$
$$= (u, u)\left((I + kA)u, A(I + kA)u\right) . \tag{10.25}$$

$\square$

Let us define a modified energy $\tilde{E}_{\phi^n}$ as

$$\tilde{E}_{\phi^n}(u) = \int_\Omega \left[\frac{1}{2}|\nabla u|^2 + \tilde{V}_n(\mathbf{x})|u|^2 \right] d\mathbf{x}$$
$$= \int_\Omega \left[\frac{1}{2}|\nabla u|^2 + V(\mathbf{x})|u|^2 + \beta|\phi^n|^2|u|^2 \right] d\mathbf{x},$$

we then get from the above lemma that

Lemma 10.9: *Suppose $V(\mathbf{x}) \ge 0$ for all $\mathbf{x} \in \Omega$, $\beta \ge 0$ and $\|\phi^n\| = 1$. Then,*

$$\tilde{E}_{\phi^n}(\tilde{\phi}^{n+1}) \le \frac{\tilde{E}_{\phi^n}(\tilde{\phi}^{n+1})}{\|\tilde{\phi}^{n+1}\|} = \tilde{E}_{\phi^n}\left(\frac{\tilde{\phi}^{n+1}}{\|\tilde{\phi}^{n+1}\|} \right)$$
$$= \tilde{E}_{\phi^n}(\phi^{n+1}) \le \tilde{E}_{\phi^n}(\phi_n) . \tag{10.26}$$

Using the inequality (10.22), we in turn get:

Lemma 10.10: *Suppose $V(\mathbf{x}) \ge 0$ for all $\mathbf{x} \in \Omega$ and $\beta \ge 0$, then,*

$$\tilde{E}(\tilde{\phi}^{n+1}) \le \tilde{E}(\phi^n),$$

where

$$\tilde{E}(u) = \int_\Omega \left[\frac{1}{2}|\nabla u|^2 + V(\mathbf{x})|u|^2 + \beta|u|^4 \right] d\mathbf{x} .$$

Remark 10.11: As we noted earlier, for $\beta = 0$, the energy diminishing property is preserved in the GFDN (10.1)-(10.3) and semi-implicit time discretization (10.20)-(10.21). For $\beta > 0$, the energy diminishing property in general does **not** hold uniformly for all ϕ_0 and all step size $k > 0$, a justification on the energy diminishing is presently only possible for a modified energy within two adjacent steps.

10.5. *Discretized normalized gradient flow (DNGF)*

Consider a discretization for the GFDN (10.20)-(10.21) (or a fully discretization of (10.15)-(10.16))

$$\frac{\tilde{U}^{n+1} - U^n}{k} = -A\tilde{U}^{n+1}, \qquad U^{n+1} = \frac{\tilde{U}^{n+1}}{\|\tilde{U}^{n+1}\|}, \qquad n = 0, 1, 2, \cdots ;$$

$$(10.27)$$

where $U^n = (u_1^n, u_2^n, \cdots, u_{M-1}^n)^T$, $k > 0$ is time step and A is an $(M-1) \times (M-1)$ symmetric positive definite matrix. We adopt the inner product, norm and energy of vectors $U = (u_1, u_2, \cdots, u_{M-1})^T$ and $V = (v_1, v_2, \cdots, v_{M-1})^T$ as

$$(U, V) = U^T V = \sum_{j=1}^{M-1} u_j \, v_j, \quad \|U\|^2 = U^T U = (U, U), \quad (10.28)$$

$$E(U) = U^T A U = (U, AU), \tag{10.29}$$

respectively. Using the finite dimensional version of the lemmas given in the previous subsection, we have

Theorem 10.12: *Suppose $\|U^0\| = 1$ and A is symmetric positive definite. Then the DNGF (10.27) is energy diminishing, i.e.*

$$E\left(U^{n+1}\right) \leq E\left(U^n\right) \leq \cdots \leq E\left(U^0\right), \qquad n = 0, 1, 2, \cdots . \tag{10.30}$$

Furthermore if $I + kA$ is an M-matrix, then $(I + kA)^{-1}$ is a nonnegative matrix (i.e. with nonnegative entries). Thus the flow is monotone, i.e. if U^0 is a non-negative vector, then U^n is also a non-negative vector for all $n \geq 0$.

Remark 10.13: If a discretization for the GFDN (10.20)-(10.21) reads

$$\frac{\tilde{U}^{n+1} - U^n}{k} = -BU^n, \qquad U^{n+1} = \frac{\tilde{U}^{n+1}}{\|\tilde{U}^{n+1}\|}, \qquad n = 0, 1, 2, \cdots . \tag{10.31}$$

For symmetric, positive definite B with $\rho(kB) < 1$ ($\rho(B)$ being the spectral radius of B), (10.30) is satisfied by choosing

$$A = \frac{1}{k}\left((I - kB)^{-1} - I\right) = (I - kB)^{-1} B.$$

Remark 10.14: If a discretization for the GFDN (10.20)-(10.21) reads

$$\tilde{U}^{n+1} = BU^n, \qquad U^{n+1} = \frac{\tilde{U}^{n+1}}{\|\tilde{U}^{n+1}\|}, \qquad n = 0, 1, 2, \cdots . \tag{10.32}$$

For symmetric, positive definite B with $\rho(B) < 1$, (10.30) is satisfied by choosing

$$A = \frac{1}{k}\left(B^{-1} - I\right).$$

Remark 10.15: If a discretization for the GFDN (10.20)-(10.21) reads

$$\frac{\tilde{U}^{n+1} - U^n}{k} = -B\tilde{U}^{n+1} - CU^n, \quad U^{n+1} = \frac{\tilde{U}^{n+1}}{\|\tilde{U}^{n+1}\|}, \quad n = 0, 1, 2, \cdots .$$

$$(10.33)$$

Suppose B and C are symmetric, positive definite and $\rho(kC) < 1$. Then (10.30) is satisfied by choosing

$$A = (I - kC)^{-1}(B + C).$$

10.6. *Numerical methods*

In this section, we will present two numerical methods to discretize the GFDN (10.1)-(10.3) (or a full discretization of the CNGF (10.15)-(10.16)). For simplicity of notation we introduce the methods for the case of one spatial dimension ($d = 1$) with homogeneous periodic boundary conditions. Generalizations to higher dimension are straightforward for tensor product grids and the results remain valid without modifications. For $d = 1$, we have

$$\phi_t = \frac{1}{2}\phi_{xx} - V(x)\phi - \beta\,|\phi|^2\phi,$$

$$x \in \Omega = (a, b),\ t_n < t < t_{n+1},\ n \geq 0, \qquad (10.34)$$

$$\phi(x, t_{n+1}) \stackrel{\triangle}{=} \phi(x, t_{n+1}^+) = \frac{\phi(x, t_{n+1}^-)}{\|\phi(\cdot, t_{n+1}^-)\|}, \quad a \leq x \leq b,\ n \geq 0, \quad (10.35)$$

$$\phi(x, 0) = \phi_0(x),\ a \leq x \leq b, \quad \phi(a, t) = \phi(b, t) = 0,\ t \geq 0; \quad (10.36)$$

with

$$\|\phi_0\|^2 = \int_a^b \phi_0^2(x)\,dx = 1.$$

We choose the spatial mesh size $h = \triangle x > 0$ with $h = (b - a)/M$ and M an even positive integer, the time step is given by $k = \triangle t > 0$ and define grid points and time steps by

$$x_j := a + j\,h, \qquad t_n := n\,k, \qquad j = 0, 1, \cdots, M, \qquad n = 0, 1, 2, \cdots$$

Let ϕ_j^n be the numerical approximation of $\phi(x_j, t_n)$ and ϕ^n the solution vector at time $t = t_n = nk$ with components ϕ_j^n.

Backward Euler finite difference (BEFD) We use backward Euler for time discretization and second-order centered finite difference for spatial derivatives. The detail scheme is:

$$\frac{\phi_j^* - \phi_j^n}{k} = \frac{1}{2h^2}\left[\phi_{j+1}^* - 2\phi_j^* + \phi_{j-1}^*\right] - V(x_j)\phi_j^* - \beta\left(\phi_j^n\right)^2\phi_j^*,$$
$$j = 1, \cdots, M-1,$$
$$\phi_0^* = \phi_M^* = 0, \qquad \phi_j^0 = \phi_0(x_j), \qquad j = 0, 1, \cdots, M,$$
$$\phi_j^{n+1} = \frac{\phi_j^*}{\|\phi^*\|}, \qquad j = 0, \cdots, M, \qquad n = 0, 1, \cdots; \tag{10.37}$$

where the norm is defined as $\|\phi^*\|^2 = h\sum_{j=1}^{M-1}\left(\phi_j^*\right)^2$.

Time-splitting sine-spectral method (TSSP) From time $t = t_n$ to time $t = t_{n+1}$, the equation (10.34) is solved in two steps. First, one solves

$$\phi_t = \frac{1}{2}\phi_{xx}, \tag{10.38}$$

for one time step of length k, then followed by solving

$$\phi_t(x,t) = -V(x)\phi(x,t) - \beta|\phi|^2\phi(x,t), \qquad t_n \leq t \leq t_{n+1}, \tag{10.39}$$

again for the same time step. Equation (10.38) is discretized in space by the sine-spectral method and integrated in time *exactly*. For $t \in [t_n, t_{n+1}]$, multiplying the ODE (10.39) by $\phi(x,t)$, one obtains with $\rho(x,t) = \phi^2(x,t)$

$$\rho_t(x,t) = -2V(x)\rho(x,t) - 2\beta\rho^2(x,t), \qquad t_n \leq t \leq t_{n+1}. \tag{10.40}$$

The solution of the ODE (10.40) can be expressed as

$$\rho(x,t) = \begin{cases} \dfrac{V(x)\rho(x,t_n)}{(V(x) + \beta\rho(x,t_n))\,e^{2V(x)(t-t_n)} - \beta\rho(x,t_n)} & V(x) \neq 0, \\[2em] \dfrac{\rho(x,t_n)}{1 + 2\beta\rho(x,t_n)(t - t_n)}, & V(x) = 0. \end{cases} \tag{10.41}$$

Combining the splitting step via the standard second-order Strang splitting for solving the GFDN (10.34)-(10.36), in detail, the steps for obtaining ϕ_j^{n+1}

from ϕ_j^n are given by

$$\phi_j^* = \begin{cases} \sqrt{\dfrac{V(x_j)e^{-kV(x_j)}}{V(x_j) + \beta(1 - e^{-kV(x_j)})|\phi_j^n|^2}} \; \phi_j^n & V(x_j) \neq 0, \\[4mm] \dfrac{1}{\sqrt{1 + \beta k|\phi_j^n|^2}} \; \phi_j^n, & V(x_j) = 0, \end{cases}$$

$$\phi_j^{**} = \sum_{l=1}^{M-1} e^{-k\mu_l^2/2} \; \widehat{\phi_l^*} \; \sin(\mu_l(x_j - a)), \qquad j = 1, 2, \cdots, M-1,$$

$$\phi_j^{***} = \begin{cases} \sqrt{\dfrac{V(x_j)e^{-kV(x_j)}}{V(x_j) + \beta(1 - e^{-kV(x_j)})|\phi_j^{**}|^2}} \; \phi_j^{**} & V(x_j) \neq 0, \\[4mm] \dfrac{1}{\sqrt{1 + \beta k|\phi_j^{**}|^2}} \; \phi_j^{**}, & V(x_j) = 0, \end{cases}$$

$$\phi_j^{n+1} = \frac{\phi_j^{***}}{\|\phi^{***}\|}, \qquad j = 0, \cdots, M, \qquad n = 0, 1, \cdots; \tag{10.42}$$

where $\widehat{U}_l$ are the sine-transform coefficients of a real vector $U = (u_0, u_1, \cdots, u_M)^T$ with $u_0 = u_M = 0$ which are defined as

$$\mu_l = \frac{\pi l}{b - a}, \quad \widehat{U}_l = \frac{2}{M} \sum_{j=1}^{M-1} u_j \; \sin(\mu_l(x_j - a)), \quad l = 1, 2, \cdots, M-1 \tag{10.43}$$

and

$$\phi_j^0 = \phi(x_j, 0) = \phi_0(x_j), \qquad j = 0, 1, 2, \cdots, M.$$

Note that the only time discretization error of TSSP is the splitting error, which is second order in k.

For comparison purposes we review a few other numerical methods which are currently used for solving the GFDN (10.34)-(10.36). One is the

Crank-Nicolson finite difference (CNFD) scheme:

$$\frac{\phi_j^* - \phi_j^n}{k} = \frac{1}{4h^2} \left[\phi_{j+1}^* - 2\phi_j^* + \phi_{j-1}^* + \phi_{j+1}^n - 2\phi_j^n + \phi_{j-1}^n \right]$$

$$- \frac{V(x_j)}{2} \left[\phi_j^* + \phi_j^n \right] - \frac{\beta \left| \phi_j^n \right|^2}{2} \left[\phi_j^* + \phi_j^n \right], \qquad j = 1, \cdots, M-1,$$

$$\phi_0^* = \phi_M^* = 0, \qquad \phi_j^0 = \phi_0(x_j), \qquad j = 0, 1, \cdots, M,$$

$$\phi_j^{n+1} = \frac{\phi_j^*}{\|\phi^*\|}, \qquad j = 0, \cdots, M, \qquad n = 0, 1, \cdots. \tag{10.44}$$

Another one is the forward Euler finite difference (FEFD) method:

$$\frac{\phi_j^* - \phi_j^n}{k} = \frac{1}{2h^2} \left[\phi_{j+1}^n - 2\phi_j^n + \phi_{j-1}^n \right] - V(x_j)\phi_j^n - \beta \left| \phi_j^n \right|^2 \phi_j^n,$$

$$j = 1, \cdots, M-1,$$

$$\phi_0^* = \phi_M^* = 0, \qquad \phi_j^0 = \phi_0(x_j), \quad j = 0, 1, \cdots, M,$$

$$\phi_j^{n+1} = \frac{\phi_j^*}{\|\phi^*\|}, \qquad j = 0, \cdots, M, \qquad n = 0, 1, \cdots; \tag{10.45}$$

10.7. *Energy diminishing of DNGF*

First we analyze the energy diminishing of the different numerical methods for linear case, i.e. $\beta = 0$ in (10.34). Introducing

$$\Phi^n = \left(\phi_1^n, \ \phi_2^n, \ \cdots, \ \phi_{M-1}^n \right)^T,$$

$$D = (d_{jl})_{(M-1)\times(M-1)}, \ \text{with } d_{jl} = \frac{1}{2h^2} \begin{cases} 2, & j = l, \\ -1, & |j-l| = -1, \\ 0, & \text{otherwise,} \end{cases}$$

$$E = \text{diag}\left(V(x_1), \ V(x_2), \ \cdots, \ V(x_{M-1}) \right),$$

$$F(\Phi) = \text{diag}\left(\phi_1^2, \ \phi_2^2, \ \cdots, \ \phi_{M-1}^2 \right), \quad \text{with } \Phi = \left(\phi_1, \ \phi_2, \ \cdots, \ \phi_{M-1} \right)^T,$$

$$G = (g_{jl})_{(M-1)\times(M-1)}, \quad \text{with } g_{jl} = \frac{2}{M} \sum_{m=1}^{M-1} \sin \frac{\pi m j}{M} \sin \frac{\pi m l}{M} e^{-k\mu_m^2/2},$$

$$H = \text{diag}\left(e^{-kV(x_1)/2}, \ e^{-kV(x_2)/2}, \ \cdots, \ e^{-kV(x_{M-1})/2} \right).$$

Then the BEFD discretization (10.37) (called as BEFD normalized flow) with $\beta = 0$ can be expressed as

$$\frac{\Phi^* - \Phi^n}{k} = -(D+E)\Phi^*, \qquad \Phi^{n+1} = \frac{\Phi^*}{\|\Phi^*\|}, \qquad n = 0, 1, \cdots. \tag{10.46}$$

The TSSP discretization (10.42) (called as TSSP normalized flow) with $\beta = 0$ can be expressed as

$$\Phi^{***} = H\Phi^{**} = HG\Phi^* = HGH\Phi^n, \qquad \Phi^{n+1} = \frac{\Phi^*}{\|\Phi^*\|}, \qquad n = 0, 1, \cdots .$$
$$(10.47)$$

The CNFD discretization (10.44) (called as CNFD normalized flow) with $\beta = 0$ can be expressed as

$$\frac{\Phi^* - \Phi^n}{k} = -\frac{1}{2}(D+E)\Phi^* - \frac{1}{2}(D+E)\Phi^n, \quad \Phi^{n+1} = \frac{\Phi^*}{\|\Phi^*\|}, \qquad n = 0, 1, \cdots .$$
$$(10.48)$$

The FEFD discretization (10.45) (called as FEFD normalized flow) with $\beta = 0$ can be expressed as

$$\frac{\Phi^* - \Phi^n}{k} = -(D + E)\Phi^n, \quad \Phi^{n+1} = \frac{\Phi^*}{\|\Phi^*\|}, \qquad n = 0, 1, \cdots . \qquad (10.49)$$

It is easy to see that D and G are symmetric positive definite matrices. Furthermore D is also an M-matrix and $\rho(D) = \left(1 + \cos \frac{\pi}{M}\right)/h^2 < 2/h^2$ and $\rho(G) = e^{-k\mu_1^2/2} < 1$. Applying the theorem 10.12 and remarks 10.13, 10.14 and 10.15, we have

Theorem 10.16: *Suppose $V \geq 0$ in Ω and $\beta = 0$. We have that*

(i). The BEFD normalized flow (10.37) is energy diminishing and monotone for any $k > 0$.

(ii). The TSSP normalized flow (10.42) is energy diminishing for any $k > 0$.

(iii). The CNFD normalized flow (10.44) is energy diminishing and monotone provided that

$$k \leq \frac{2}{2/h^2 + \max_j V(x_j)} = \frac{2h^2}{2 + h^2 \max_j V(x_j)}. \qquad (10.50)$$

(iv). The FEFD normalized flow (10.45) is energy diminishing and monotone provided that

$$k \leq \frac{1}{2/h^2 + \max_j V(x_j)} = \frac{h^2}{2 + h^2 \max_j V(x_j)}. \qquad (10.51)$$

For nonlinear case, i.e. $\beta > 0$, we only analyze the *energy* between two steps of BEFD flow (10.37). In this case, consider

$$\frac{\tilde{\Phi}^{n+1} - \Phi^n}{k} = -\left(D + E + \beta F(\Phi^n)\right)\tilde{\Phi}^{n+1}, \qquad \Phi^{n+1} = \frac{\tilde{\Phi}^{n+1}}{\|\tilde{\Phi}^{n+1}\|}. \qquad (10.52)$$

Lemma 10.17: *Suppose $V \geq 0$, $\beta > 0$ and $\|\Phi^n\| = 1$. Then for the flow (10.52), we have*

$$\tilde{E}\left(\tilde{\Phi}^{n+1}\right) \leq \tilde{E}\left(\Phi^n\right), \qquad \tilde{E}_{\Phi^n}\left(\Phi^{n+1}\right) \leq \tilde{E}_{\Phi^n}\left(\Phi^n\right) \tag{10.53}$$

where

$$\tilde{E}\left(\Phi\right) = (\Phi, (D + E + \beta F(\Phi))\Phi) = \Phi^T(D + E)\Phi + \beta \sum_{j=1}^{M-1} \phi_j^4, \tag{10.54}$$

$$\tilde{E}_{\Phi^n}\left(\Phi\right) = (\Phi, (D + E + \beta F(\Phi^n))\Phi)$$

$$= \Phi^T(D + E)\Phi + \beta \sum_{j=1}^{M-1} \phi_j^2 \left(\phi_j^n\right)^2. \tag{10.55}$$

Proof: Combining (10.52), (10.27) and Theorem 10.12, we have

$$\left(\tilde{\Phi}^{n+1}, (D + E + \beta F(\Phi^n))\tilde{\Phi}^{n+1}\right) \leq \frac{\left(\tilde{\Phi}^{n+1}, (D + E + \beta F(\Phi^n))\tilde{\Phi}^{n+1}\right)}{\left(\tilde{\Phi}^{n+1}, \tilde{\Phi}^{n+1}\right)}$$

$$\leq \frac{(\Phi^n, (D + E + \beta F(\Phi^n))\Phi^n)}{(\Phi^n, \Phi^n)} = \tilde{E}\left(\Phi^n\right). \tag{10.56}$$

Similar to the proof of (10.22), we have

$$\sum_{j=1}^{M-1} \left(\phi_j^n\right)^2 \left(\tilde{\phi}_j^{n+1}\right)^2 \geq \sum_{j=1}^{M-1} \left(\tilde{\phi}_j^{n+1}\right)^4. \tag{10.57}$$

The required result (10.53) is a combination of (10.57), and (10.56). $\qquad\square$

10.8. *Numerical results*

Here we report the ground state solutions in BEC with different potentials by the method BEFD. Due to the ground state solution $\phi_g(\mathbf{x}) \geq 0$ for $\mathbf{x} \in \Omega$ in non-rotating BEC [87], in our computations, the initial condition (10.3) is always chosen such that $\phi_0(\mathbf{x}) \geq 0$ and decays to zero sufficiently fast as $|\mathbf{x}| \to \infty$. We choose an appropriately large interval, rectangle and box in 1d, 2d and 3d, respectively, to avoid that the homogeneous periodic boundary condition (10.36) introduce a significant (aliasing) error relative to the whole space problem. To quantify the ground state solution $\phi_g(\mathbf{x})$, we define the radius mean square

$$\alpha_{\mathrm{rms}} = \|\alpha\phi_g\|_{L^2(\Omega)} = \sqrt{\int_\Omega \alpha^2 \phi_g^2(\mathbf{x})\, d\mathbf{x}}, \qquad \alpha = x, y, \text{ or } z. \tag{10.58}$$

Example 1 Ground state solution of 1d BEC with harmonic oscillator potential

$$V(x) = \frac{x^2}{2}, \qquad \phi_0(x) = \frac{1}{(\pi)^{1/4}} e^{-x^2/2}, \quad x \in \mathbb{R}.$$

The CNGF (10.15)-(10.16) with $d = 1$ is solved on $\Omega = [-16, 16]$ with mesh size $h = 1/8$ and time step $k = 0.1$ by using BEFD. The steady state solution is reached when $\max \left| \Phi^{n+1} - \Phi^n \right| < \varepsilon = 10^{-6}$. Fig. 1 shows the ground state solution $\phi_g(x)$ and energy evolution for different β. Tab. 1 displays the values of $\phi_g(0)$, radius mean square x_{rms}, energy $E(\phi_g)$ and chemical potential μ_g.

β	$\phi_g(0)$	x_{rms}	$E(\phi_g)$	$\mu_g = \mu_\beta(\phi_g)$
0	0.7511	0.7071	0.5000	0.5000
3.1371	0.6463	0.8949	1.0441	1.5272
12.5484	0.5301	1.2435	2.2330	3.5986
31.371	0.4562	1.6378	3.9810	6.5587
62.742	0.4067	2.0423	6.2570	10.384
156.855	0.3487	2.7630	11.464	19.083
313.71	0.3107	3.4764	18.171	30.279
627.42	0.2768	4.3757	28.825	48.063
1254.8	0.2467	5.5073	45.743	76.312

Tab. 1: Maximum value of the wave function $\phi_g(0)$, root mean square size x_{rms}, energy $E(\phi_g)$ and ground state chemical potential μ_g verus the interaction coefficient β in 1d.

The results in Fig. 1. and Tab. 1. agree very well with the ground state solutions of BEC obtained by directly minimizing the energy functional [5].

Example 2 Ground state solution of BEC in 2d. Two cases are considered:

I. With a harmonic oscillator potential [5, 8, 44]*, i.e.*

$$V(x, y) = \frac{1}{2} \left(\gamma_x^2 x^2 + \gamma_y^2 y^2 \right).$$

II. With a harmonic oscillator potential and a potential of a stirrer corresponding a far-blue detuned Gaussian laser beam [27] *which is used to generate vortices in BEC* [27]*, i.e.*

$$V(x, y) = \frac{1}{2} \left(\gamma_x^2 x^2 + \gamma_y^2 y^2 \right) + w_0 e^{-\delta((x-r_0)^2 + y^2)}.$$

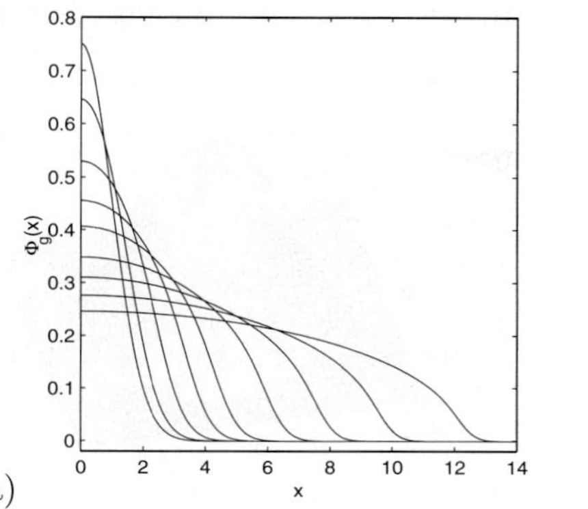

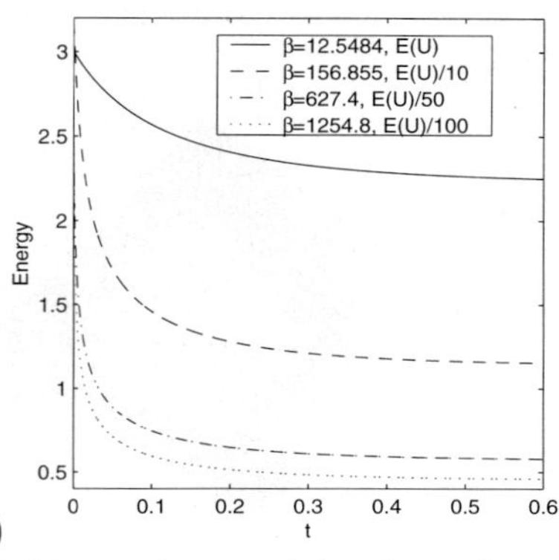

a)   b)

Fig. 1: Ground state solution ϕ_g in Example 1. (a). For $\beta = 0, 3.1371, 12.5484, 31.371, 62.742, 156.855, 313.71, 627.42, 1254.8$ (with decreasing peak). (b). Energy evolution for different β.

The initial condition is chosen as

$$\phi_0(x, y) = \frac{(\gamma_x \gamma_y)^{1/4}}{\pi^{1/2}} e^{-(\gamma_x x^2 + \gamma_y y^2)/2}.$$

For the case I, we choose $\gamma_x = 1$, $\gamma_y = 4$, $w_0 = \delta = r_0 = 0$, $\beta = 200$ and solve the problem by BEFD on $\Omega = [-8, 8] \times [-4, 4]$ with mesh size $h_x = 1/8$, $h_y = 1/16$ and time step $k = 0.1$. We get the following results from the ground state solution ϕ_g:

$$x_{\text{rms}} = 2.2734, \quad y_{\text{rms}} = 0.6074, \quad \phi_g^2(0) = 0.0808,$$

$$E(\phi_g) = 11.1563, \quad \mu_g = 16.3377.$$

For case II, we choose $\gamma_x = 1$, $\gamma_y = 1$, $w_0 = 4$, $\delta = r_0 = 1$, $\beta = 200$ and solve the problem by TSSP on $\Omega = [-8, 8]^2$ with mesh size $h = 1/8$ and time step $k = 0.001$. We get the following results from the ground state solution ϕ_g:

$$x_{\text{rms}} = 1.6951, \quad y_{\text{rms}} = 1.7144, \quad \phi_g^2(0) = 0.034,$$

$$E(\phi_g) = 5.8507, \quad \mu_g = 8.3269.$$

In addition, Fig. 2 shows surface plots of the ground state solution ϕ_g.

Example 3 Ground state solution of BEC in 3d. Two cases are considered:

I. With a harmonic oscillator potential [5, 8, 44], *i.e.*

$$V(x, y, z) = \frac{1}{2} \left(\gamma_x^2 x^2 + \gamma_y^2 y^2 + \gamma_z^2 z^2 \right).$$

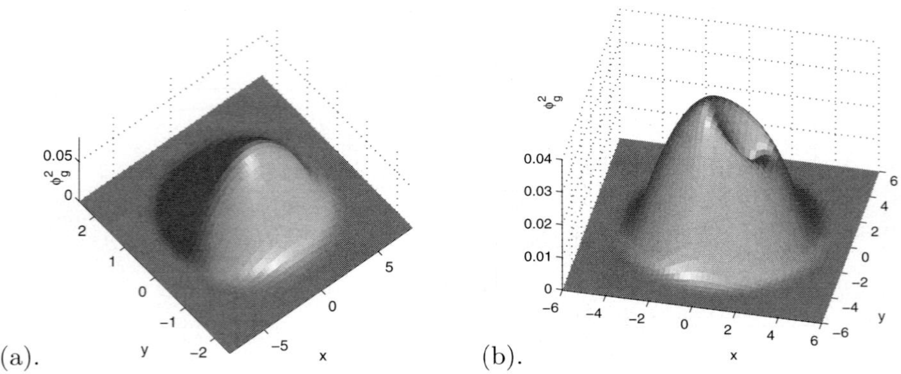

(a).   (b).

Fig. 2: Ground state solutions ϕ_g^2 in Example 2, case I (a), and case II (b).

II. With a harmonic oscillator potential and a potential of a stirrer corresponding a far-blue detuned Gaussian laser beam [27] *which is used to generate vortex in BEC* [27], i.e.

$$V(x, y, z) = \frac{1}{2} \left(\gamma_x^2 x^2 + \gamma_y^2 y^2 + \gamma_z^2 z^2 \right) + w_0 e^{-\delta((x-r_0)^2 + y^2)}.$$

The initial condition is chosen as

$$\phi_0(x, y, z) = \frac{(\gamma_x \gamma_y \gamma_z)^{1/4}}{\pi^{3/4}} e^{-(\gamma_x x^2 + \gamma_y y^2 + \gamma_z z^2)/2}.$$

For case I, we choose $\gamma_x = 1$, $\gamma_y = 2$, $\gamma_z = 4$, $w_0 = \delta = r_0 = 0$, $\beta = 200$ and solve the problem by TSSP on $\Omega = [-8, 8] \times [-6, 6] \times [-4, 4]$ with mesh size $h_x = \frac{1}{8}$, $h_y = \frac{3}{32}$, $h_z = \frac{1}{16}$ and time step $k = 0.001$. The ground state solution ϕ_g gives:

$$x_{\text{rms}} = 1.67, \quad y_{\text{rms}} = 0.87, \quad z_{\text{rms}} = 0.49,$$

$$\phi_g^2(0) = 0.052, \quad E(\phi_g) = 8.33, \quad \mu_g = 11.03.$$

For case II, we choose $\gamma_x = 1$, $\gamma_y = 1$, $\gamma_z = 2$, $w_0 = 4$, $\delta = r_0 = 1$, $\beta = 200$ and solve the problem by TSSP on $\Omega = [-8, 8]^3$ with mesh size $h = \frac{1}{8}$ and time step $k = 0.001$. The ground state solution ϕ_g gives:

$$x_{\text{rms}} = 1.37, \quad y_{\text{rms}} = 1.43, \quad z_{\text{rms}} = 0.70,$$

$$\phi_g^2(0) = 0.025, \quad E(\phi_g) = 5.27, \quad \mu_g = 6.71.$$

Furthermore, Fig. 3 shows surface plots of the ground state solution $\phi_g^2(x, 0, z)$. BEFD gives similar results with $k = 0.1$.

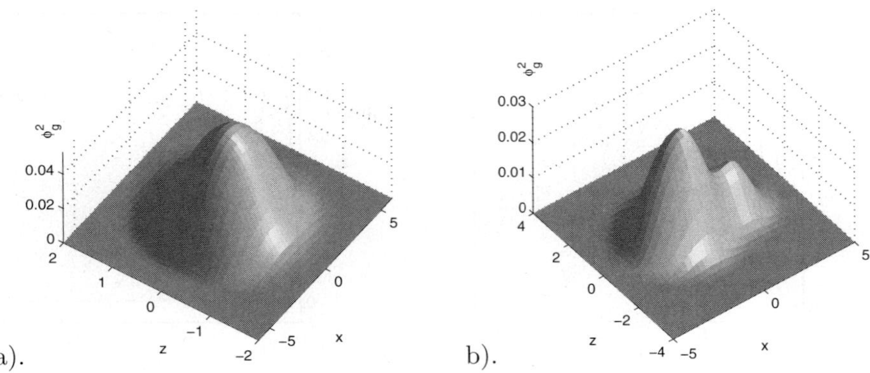

a).   b).

Fig. 3: Ground state solutions $\phi_g^2(x, 0, z)$ in Example 3. (a). For case I. (b). For case II.

Example 4 2d central vortex states in BEC, i.e.

$$V(x, y) = V(r) = \frac{1}{2}\left(\frac{m^2}{r^2} + r^2\right), \quad \phi_0(x, y) = \phi_0(r) = \frac{1}{\sqrt{\pi m!}}\, r^m\, e^{-r^2/2},\; 0 \le r.$$

The CNGF (10.15)-(10.16) is solved in polar coordinate with $\Omega = [0, 8]$ with mesh size $h = \frac{1}{64}$ and time step $k = 0.1$ by using BEFD. Fig. 4a shows the ground state solution $\phi_g(r)$ with $\beta = 200$ for different index of the central vortex m. Tab. 2 displays the values of $\phi_g(0)$, radius mean square $r_{\rm rms}$, energy $E(\phi_g)$ and chemical potential μ_g.

Index m	$\phi_g(0)$	$r_{\rm rms}$	$E(\phi_g)$	$\mu_g = \mu_\beta(\phi_g)$
1	0.0000	2.4086	5.8014	8.2967
2	0.0000	2.5258	6.3797	8.7413
3	0.0000	2.6605	7.0782	9.3160
4	0.0000	2.8015	7.8485	9.9772
5	0.0000	2.9438	8.6660	10.6994
6	0.0000	3.0848	9.5164	11.4664

Tab. 2: Numerical results for 2d central vortex states in BEC.

Example 5. The first excited state solution of BEC in 1d with a harmonic oscillator potential, i.e.

$$V(x) = \frac{x^2}{2}, \qquad \phi_0(x) = \frac{\sqrt{2}}{(\pi)^{1/4}}\, x\, e^{-x^2/2}, \quad x \in \mathbb{R}.$$

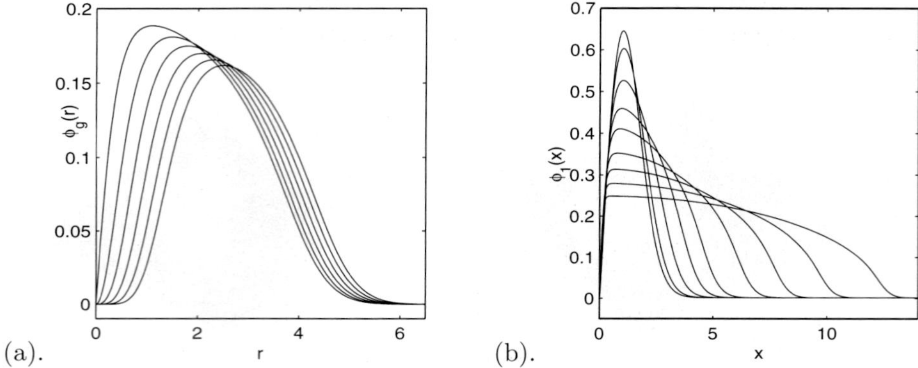

(a). (b).

Fig. 4: (a). 2d central vortex states $\phi_g(r)$ in Example 4. $\beta = 200$ for $m = 1$ to 6 (with decreasing peak). (b). First excited state solution $\phi_1(x)$ (an odd function) in Example 5. For $\beta = 0$, 3.1371, 12.5484, 31.371, 62.742, 156.855, 313.71, 627.42, 1254.8 (with decreasing peak).

The CNGF (10.15)-(10.16) with $d = 1$ is solved on $\Omega = [-16, 16]$ with mesh size $h = 1/64$ and time step $k = 0.1$ by using BEFD. Fig. 4b shows the first excited state solution $\phi_1(x)$ for different β. Tab. 3 displays the radius mean square $x_{\mathrm{rms}} = \|x\phi_1\|_{L^2(\Omega)}$, ground state and first excited state energies $E(\phi_g)$ and $E(\phi_1)$, ratio $E(\phi_1)/E(\phi_g)$, chemical potentials $\mu_g = \mu_\beta(\phi_g)$ and $\mu_1 = \mu_\beta(\phi_1)$, ratio μ_1/μ_g.

β	x_{rms}	$E(\phi_g)$	$E(\phi_1)$	$\dfrac{E(\phi_1)}{E(\phi_g)}$	μ_g	μ_1	$\dfrac{\mu_1}{\mu_g}$
0	1.2247	0.500	1.500	3.000	0.500	1.500	3.000
3.1371	1.3165	1.044	1.941	1.859	1.527	2.357	1.544
12.5484	1.5441	2.233	3.037	1.360	3.598	4.344	1.207
31.371	1.8642	3.981	4.743	1.192	6.558	7.279	1.110
62.742	2.2259	6.257	6.999	1.119	10.38	11.089	1.068
156.855	2.8973	11.46	12.191	1.063	19.08	19.784	1.037
313.71	3.5847	18.17	18.889	1.040	30.28	30.969	1.023
627.42	4.4657	28.82	29.539	1.025	48.06	48.733	1.014
1254.8	5.5870	45.74	46.453	1.016	76.31	76.933	1.008

Tab. 3: Numerical results for the first excited state solution in 1d in Example 5.

From the results in Tab. 3 and Fig. 4b, we can see that the BEFD can be applied directly to compute the first excited states in BEC. Furthermore, we have

$$\lim_{\beta \to +\infty} \frac{E(\phi_1)}{E(\phi_g)} = 1, \qquad \lim_{\beta \to +\infty} \frac{\mu_1}{\mu_g} = 1.$$

These results are confirmed with the results in [5] where the ground and first excited states are computed by directly minimizing the energy functional through the finite element discretization.

11. Numerical methods for dynamics of NLSE

In this section we present time-splitting sine pseudospectral (TSSP) methods for the problem (4.1), (4.2) with/without external driven field with homogeneous Dirichlet boundary conditions. For the simplicity of notation we shall introduce the method for the case of one space dimension ($d = 1$). Generalizations to $d > 1$ are straightforward for tensor product grids and the results remain valid without modifications. For $d = 1$, the problem with an external driven field becomes

$$i\partial_t \psi = -\frac{1}{2}\partial_{xx}\psi + V(x)\psi + W(x,t)\psi + \beta|\psi|^2\psi, \quad a < x < b, \ t > 0, \quad (11.1)$$

$$\psi(x, t = 0) = \psi_0(x), \quad a \le x \le b, \quad \psi(a,t) = \psi(b,t) = \mathbf{0}, \quad t \ge 0; \quad (11.2)$$

where $W(x,t)$ is an external driven field. Typical external driven fields used in physical literatures include a far-blued detuned Gaussian laser beam stirrer [27]

$$W(\mathbf{x}, t) = W_s(t) \exp\left[-\left(\frac{|\mathbf{x} - \mathbf{x}_s(t)|^2}{w_s/2}\right)\right], \quad (11.3)$$

with W_s the height, w_s the width, and $\mathbf{x}_s(t)$ the position of the stirrer; or a Delta-kicked potential [77]

$$W(x,t) = K\cos(kx) \sum_{n=-\infty}^{\infty} \delta(t - n\tau), \quad (11.4)$$

with K the kick strength, k the wavenumber, τ the time interval between kicks, and $\delta(\tau)$ is the Dirac delta function.

11.1. *General high-order split-step method*

As preparatory steps, we begin by introducing the general high-order split-step method [53] for a general evolution equation

$$i\,\partial_t u = f(u) = A\,u + B\,u, \quad (11.5)$$

where $f(u)$ is a nonlinear operator and the splitting $f(u) = Au + Bu$ can be quite arbitrary, in particular, A and B do not need to commute. For a given time step $k = \Delta t > 0$, let $t_n = n\,k$, $n = 0, 1, 2, \ldots$ and u^n be the approximation of $u(t_n)$. A second-order symplectic time integrator (cf. [111]) for (11.5) is as follows:

$$u^{(1)} = e^{-ik\,A/2}\,u^n;$$
$$u^{(2)} = e^{-ik\,B}\,u^{(1)};$$
$$u^{n+1} = e^{-ik\,A/2}\,u^{(2)}. \tag{11.6}$$

A fourth-order symplectic time integrator (cf. [120]) for (11.5) is as follows:

$$u^{(1)} = e^{-i2w_1 k\,A}\,u^n;$$
$$u^{(2)} = e^{-i2w_2 k\,B}\,u^{(1)};$$
$$u^{(3)} = e^{-i2w_3 k\,A}\,u^{(2)};$$
$$u^{(4)} = e^{-i2w_4 k\,B}\,u^{(3)};$$
$$u^{(5)} = e^{-i2w_3 k\,A}\,u^{(4)};$$
$$u^{(6)} = e^{-i2w_2 k\,B}\,u^{(5)};$$
$$u^{n+1} = e^{-i2w_1 k\,A}\,u^{(6)}; \tag{11.7}$$

where

$$w_1 = 0.33780\ 17979\ 89914\ 40851, \quad w_2 = 0.67560\ 35959\ 79828\ 81702,$$
$$w_3 = -0.08780\ 17979\ 89914\ 40851, \quad w_4 = -0.85120\ 71979\ 59657\ 63405.$$

11.2. *Fourth-order TSSP for GPE without external driving field*

We choose the spatial mesh size $h = \Delta x > 0$ with $h = (b - a)/M$ for M an even positive integer, and let $x_j := a + j\,h$, $j = 0, 1, \cdots, M$. Let ψ_j^n be the approximation of $\psi(x_j, t_n)$ and ψ^n be the solution vector at time $t = t_n = nk$ with components ψ_j^n.

We now rewrite the GPE (11.1) without external driven field, i.e. $W(\mathbf{x}, t) \equiv 0$, in the form of (11.5) with

$$A\psi = V(x)\psi(x, t) + \beta|\psi(x, t)|^2\psi(x, t), \qquad B\psi = -\frac{1}{2}\partial_{xx}\psi(x, t). \tag{11.8}$$

Thus, the key for an efficient implementation of (11.7) is to solve efficiently the following two subproblems:

$$i\,\partial_t\psi(x, t) = B\psi = -\frac{1}{2}\partial_{xx}\psi, \tag{11.9}$$

and

$$i\,\partial_t\psi(x,t) = V(x)\psi(x,t) + \beta|\psi(x,t)|^2\psi(x,t). \tag{11.10}$$

Equation (11.9) will be discretized in space by the sine pseudospectral method and integrated in time *exactly*. For $t \in [t_n, t_{n+1}]$, the ODE (11.10) leaves $|\psi|$ invariant in t [11, 10] and therefore becomes

$$i\psi_t(x,t) = V(x)\psi(x,t) + \beta|\psi(x,t_n)|^2\psi(x,t) \tag{11.11}$$

and thus can be integrated *exactly*.

From time $t = t_n$ to $t = t_{n+1}$, we combine the splitting steps via the fourth-order split-step method and obtain a fourth-order time-splitting sine-spectral (**TSSP4**) method for the GPE (10.34). The detailed method is given by

$$\psi_j^{(1)} = e^{-i2w_1 k(V(x_j)+\beta|\psi_j^n|^2)}\,\psi_j^n,$$

$$\psi_j^{(2)} = \sum_{l=1}^{M-1} e^{-iw_2 k\mu_l^2}\,\widehat{\psi}_l^{(1)}\,\sin(\mu_l(x_j - a)),$$

$$\psi_j^{(3)} = e^{-i2w_3 k(V(x_j)+\beta|\psi_j^{(2)}|^2)}\,\psi_j^{(2)},$$

$$\psi_j^{(4)} = \sum_{l=1}^{M-1} e^{-iw_4 k\mu_l^2}\,\widehat{\psi}_l^{(3)}\,\sin(\mu_l(x_j - a)), \qquad j = 1, 2, \cdots, M - 1,$$

$$\psi_j^{(5)} = e^{-i2w_3 k(V(x_j)+\beta|\psi_j^{(4)}|^2)}\,\psi_j^{(4)},$$

$$\psi_j^{(6)} = \sum_{l=1}^{M-1} e^{-iw_2 k\mu_l^2}\,\widehat{\psi}_l^{(5)}\,\sin(\mu_l(x_j - a)),$$

$$\psi_j^{n+1} = e^{-i2w_1 k(V(x_j)+\beta|\psi_j^{(6)}|^2)}\,\psi_j^{(6)}, \tag{11.12}$$

where $\widehat{U}_l$, the sine-transform coefficients of a complex vector $U = (U_0, U_1, \cdots, U_M)$ with $U_0 = U_M = \mathbf{0}$, are defined as

$$\mu_l = \frac{\pi l}{b - a}, \quad \widehat{U}_l = \frac{2}{M} \sum_{j=1}^{M-1} U_j\,\sin(\mu_l(x_j - a)),\ l = 1, 2, \cdots, M-1, \tag{11.13}$$

with

$$\psi_j^0 = \psi(x_j, 0) = \psi_0(x_j), \qquad j = 0, 1, 2, \cdots, M. \tag{11.14}$$

Note that the only time discretization error of TSSP4 is the splitting error, which is fourth order in k for any fixed mesh size $h > 0$.

This scheme is explicit, time reversible, just as the IVP for the GPE. Also, a main advantage of the time-splitting method is its time-transverse invariance, just as it holds for the GPE itself. If a constant α is added to the potential V_1, then the discrete wave functions ψ_j^{n+1} obtained from TSSP4 get multiplied by the phase factor $e^{-i\alpha(n+1)k}$, which leaves the discrete quadratic observables unchanged. This property does not hold for finite difference schemes.

11.3. *Second-order TSSP for GPE with external driving field*

We now rewrite the GPE (11.1) with an external driven field

$$A\psi = -\frac{1}{2}\partial_{xx}\psi(x,t),$$
$$B\psi = V(x)\psi(x,t) + W(x,t)\psi(x,t) + \beta|\psi(x,t)|^2\psi(x,t). \quad (11.15)$$

Due to the external driven field could be vary complicated, e.g. it may be a Delta-function [77], here we only use a second-order split-step scheme in time discretization. More precisely, from time $t = t_n$ to $t = t_{n+1}$, we proceed as follows:

$$\psi_j^* = \sum_{l=1}^{M-1} e^{-ik\mu_l^2/4}\,\widehat{(\psi^n)}_l\,\sin(\mu_l(x_j - a)), \qquad j = 1, 2, \cdots, M - 1,$$

$$\psi_j^{**} = \exp\left[-ik(V(x_j) + \beta|\psi_j^n|^2) - i\int_{t_n}^{t_{n+1}} W(x_j,t)dt\right]\psi_j^*,$$

$$\psi_j^{n+1} = \sum_{l=1}^{M-1} e^{-ik\mu_l^2/4}\,\widehat{(\psi^{**})}_l\,\sin(\mu_l(x_j - a)). \qquad (11.16)$$

Remark 11.1: If the integral in (11.16) could not be evaluated analytically, one can use numerical quadrature to evaluate, e.g.

$$\int_{t_n}^{t_{n+1}} W(x_j,t)dt \approx \frac{k}{6}\left[W(x_j,t_n) + 4W(x_j,t_n + k/2) + W(x_j,t_{n+1})\right].$$

11.4. *Stability*

Let $U = (U_0, U_1, \cdots, U_M)^T$ with $U_0 = U_M = \mathbf{0}$, $f(x)$ a homogeneous periodic function on the interval $[a,b]$, and let $\|\cdot\|_{l^2}$ be the usual discrete

l^2-norm on the interval (a, b), i.e.,

$$\|U\|_{l^2} = \sqrt{\frac{b-a}{M} \sum_{j=1}^{M-1} |U_j|^2}, \qquad \|f\|_{l^2} = \sqrt{\frac{b-a}{M} \sum_{j=1}^{M-1} |f(x_j)|^2}. \quad (11.17)$$

For the *stability* of the second-order time-splitting spectral approximations TSSP2 (11.16) and fourth-order scheme (11.12), we have the following lemma, which shows that the total charge is conserved.

Lemma 11.2: *The second-order time-splitting sine pseudospectral scheme (11.16) and fourth-order scheme (11.12) are unconditionally stable. In fact, for every mesh size $h > 0$ and time step $k > 0$,*

$$\|\psi^n\|_{l^2} = \|\psi^0\|_{l^2} = \|\psi_0\|_{l^2}, \qquad n = 1, 2, \cdots \quad (11.18)$$

Proof: For the scheme TSSP2 (11.16), noting (10.43) and (11.17), one has

$$\frac{1}{b-a}\|\psi^{n+1}\|_{l^2}^2 = \frac{1}{M} \sum_{j=1}^{M-1} |\psi_j^{n+1}|^2$$

$$= \frac{1}{M} \sum_{j=1}^{M-1} \left| \sum_{l=1}^{M-1} e^{-ik\mu_l^2/4} \widehat{(\psi^{**})}_l \ \sin(\mu_l(x_j - a)) \right|^2$$

$$= \frac{1}{M} \sum_{j=1}^{M-1} \left| \sum_{l=1}^{M-1} e^{-ik\mu_l^2/4} \widehat{(\psi^{**})}_l \ \sin(jl\pi/M) \right|^2$$

$$= \frac{1}{2} \sum_{l=1}^{M-1} \left| e^{-ik\mu_l^2/4} \widehat{(\psi^{**})}_l \right|^2 = \frac{1}{2} \sum_{l=1}^{M-1} \left| \widehat{(\psi^{**})}_l \right|^2. \quad (11.19)$$

Plugging (10.43) into (11.19), we obtain

$$\frac{1}{b-a}\|\psi^{n+1}\|_{l^2}^2 = \frac{1}{2} \sum_{l=1}^{M-1} \left| \frac{2}{M} \sum_{j=1}^{M-1} \psi_j^{**} \ \sin(\mu_l(x_j - a)) \right|^2$$

$$= \frac{1}{2} \sum_{l=1}^{M-1} \left| \frac{2}{M} \sum_{j=1}^{M-1} \psi_j^{**} \ \sin(lj\pi/M) \right|^2 = \frac{1}{M} \sum_{j=1}^{M-1} |\psi_j^{**}|^2$$

$$= \frac{1}{M} \sum_{j=1}^{M-1} \left| \exp\left[-ik(V(x_j) + \beta|\psi_j^n|^2) - i \int_{t_n}^{t_{n+1}} W(x_j, t)dt \right] \psi_j^* \right|^2$$

$$= \frac{1}{M} \sum_{j=1}^{M-1} |\psi_j^*|^2 = \frac{1}{M} \sum_{j=1}^{M-1} |\psi_j^n|^2 = \frac{1}{b-a}\|\psi^n\|_{l^2}^2. \quad (11.20)$$

Here we used the identities

$$\sum_{j=1}^{M-1} \sin(lj\pi/M)\sin(kj\pi/M) = \begin{cases} 0, & k = l, \\ M/2, & k \neq l. \end{cases} \qquad (11.21)$$

For the scheme TSSP4 (11.12), using (10.43), (11.17) and (11.21), we get similarly

$$\frac{1}{b-a}\|\psi^{n+1}\|_{l^2}^2 = \frac{1}{M}\sum_{j=0}^{M-1}|\psi_j^{n+1}|^2$$

$$= \frac{1}{M}\sum_{j=0}^{M-1}\left|e^{-i2w_1 k(V(x_j)+\beta|\psi_j^{(6)}|^2)}\,\psi_j^{(6)}\right|^2$$

$$= \frac{1}{M}\sum_{j=0}^{M-1}\left|\psi_j^{(6)}\right|^2$$

$$= \frac{1}{M}\sum_{j=0}^{M-1}\left|\sum_{l=1}^{M-1}e^{-iw_2 k\mu_l^2}\,\widehat{\psi}_l^{(5)}\,\sin(\mu_l(x_j-a))\right|^2$$

$$= \frac{1}{M}\sum_{j=0}^{M-1}\left|\sum_{l=1}^{M-1}e^{-iw_2 k\mu_l^2}\,\widehat{\psi}_l^{(5)}\,\sin(lj\pi/M)\right|^2$$

$$= \frac{1}{2}\sum_{l=1}^{M-1}\left|e^{-iw_2 k\mu_l^2}\,\widehat{\psi}_l^{(5)}\right|^2 = \frac{1}{2}\sum_{l=1}^{M-1}\left|\widehat{\psi}_l^{(5)}\right|^2. \qquad (11.22)$$

Plugging (10.43)) into (11.22), we have

$$\frac{1}{b-a}\|\psi^{n+1}\|_{l^2}^2 = \frac{1}{2}\sum_{l=1}^{M-1}\left|\frac{2}{M}\sum_{j=1}^{M-1}\psi_j^{(5)}\,\sin(\mu_l(x_j-a))\right|^2$$

$$= \frac{1}{2}\sum_{l=1}^{M-1}\left|\frac{2}{M}\sum_{j=1}^{M-1}\psi_j^{(5)}\,\sin(jl\pi/M)\right|^2 = \frac{1}{M}\sum_{j=1}^{M-1}\left|\psi_j^{(5)}\right|^2$$

$$= \frac{1}{M}\sum_{j=1}^{M-1}\left|\psi_j^{(4)}\right|^2 = \frac{1}{M}\sum_{j=1}^{M-1}\left|\psi_j^{(3)}\right|^2 = \frac{1}{M}\sum_{j=1}^{M-1}\left|\psi_j^{(2)}\right|^2$$

$$= \frac{1}{M}\sum_{j=1}^{M-1}\left|\psi_j^{(1)}\right|^2 = \frac{1}{M}\sum_{j=1}^{M-1}\left|\psi_j^n\right|^2$$

$$= \frac{1}{b-a}\|\psi^n\|_{l^2}^2. \qquad (11.23)$$

Thus the equality (11.18) can be obtained from (11.19) for the scheme TSSP2 and (11.22) for the scheme TSSP4 by induction. $\qquad\square$

11.5. *Crank-Nicolson finite difference method (CNFD)*

Another scheme used to disretize the NLSE (11.1) is the *Crank-Nicolson finite difference method (CNFD)*. In this method one uses the Crank-Nicolson scheme for time derivative and the second order central difference scheme for spatial derivative. The detailed method is:

$$i\frac{\psi_j^{n+1} - \psi_j^n}{k} = -\frac{1}{4h^2}\left(\psi_{j+1}^{n+1} - 2\psi_j^{n+1} + \psi_{j-1}^{n+1} + \psi_{j+1}^n - 2\psi_j^n + \psi_{j-1}^n\right)$$
$$+\frac{V(x_j)}{2}\left(\psi_j^{n+1} + \psi_j^n\right) + \frac{\beta}{2}\left(|\psi_j^{n+1}|^2 + |\psi_j^n|^2\right)\left(\psi_j^{n+1} + \psi_j^n\right),$$
$$j = 1, 2, \cdots, M-1, \qquad n = 0, 1, \cdots, \qquad (11.24)$$
$$\psi_0^{n+1} = \psi_M^{n+1} = 0, \qquad n = 0, 1, \cdots,$$
$$\psi_j^0 = \psi_0(x_j), \qquad j = 0, 1, 2, \cdots, M.$$

11.6. *Numerical results*

In this subsection we present numerical results to confirm spectral accuracy in space and fourth order accuracy in time of the numerical method (11.12), and then apply it to study time-evolution of condensate width in 1D, 2D and 3D.

Example 6 1d Gross-Pitaevskii equation, i.e. in (4.1) we choose $d = 1$ and $\gamma_x = 1$. The initial condition is taken as

$$\psi_0(x) = \frac{1}{\pi^{1/4}} e^{-x^2/2}, \qquad x \in \mathbb{R}.$$

We solve on the interval $[-32, 32]$, i.e. $a = -32$ and $b = 32$ with homogeneous Dirichlet boundary condition (11.2). We compute a numerical solution by using TSSP4 with a very fine mesh, e.g. $h = \frac{1}{128}$, and a very small time step, e.g. $k = 0.0001$, as the 'exact' solution ψ. Let $\psi^{h,k}$ denote the numerical solution under mesh size h and time step k.

First we test the spectral accuracy of TSSP4 in space. In order to do so, for each fixed β_1, we solve the problem with different mesh size h but a very small time step, e.g. $k = 0.0001$, such that the truncation error from time discretization is negligible comparing to that from space discretization. Tab. 4 shows the errors $\|\psi(t) - \psi^{h,k}(t)\|_{l^2}$ at $t = 2.0$ with $k = 0.0001$ for different β_1 and h.

mesh	$h = 1$	$h = \frac{1}{2}$	$h = \frac{1}{4}$	$h = \frac{1}{8}$	$h = \frac{1}{16}$
$\beta_1 = 10$	0.2745	1.081E-2	1.805E-6	3.461E-11	
$\beta_1 = 20\sqrt{2}$	1.495	0.1657	7.379E-4	7.588E-10	
$\beta_1 = 80$	1.603	1.637	6.836E-2	3.184E-5	3.47E-11

Tab. 4: Spatial error analysis: Error $\|\psi(t) - \psi^{h,k}(t)\|_{l^2}$ at $t = 2.0$ with $k = 0.0001$ in Example 6.

Then we test the fourth-order accuracy of TSSP4 in time. In order to do so, for each fixed β_1, we solve the problem with different time step k but a very fine mesh, e.g. $h = \frac{1}{64}$, such that the truncation error from space discretization is negligible comparing to that from time discretization. Tab. 5 shows the errors $\|\psi(t) - \psi^{h,k}(t)\|_{l^2}$ at $t = 2.0$ with $h = \frac{1}{64}$ for different β_1 and k.

time step	$k = \frac{1}{20}$	$k = \frac{1}{40}$	$k = \frac{1}{80}$	$k = \frac{1}{160}$	$k = \frac{1}{320}$
$\delta = 10.0$	1.261E-4	8.834E-6	5.712E-7	3.602E-8	2.254E-9
$\delta = 20\sqrt{2}$	1.426E-3	9.715E-5	6.367E-6	4.034E-7	2.529E-8
$\delta = 80$	4.375E-2	1.693E-3	8.982E-5	5.852E-6	3.706E-7

Tab. 5: Temporal error analysis: $\|\psi(t) - \psi^{h,k}(t)\|_{l^2}$ at $t = 2.0$ with $h = \frac{1}{64}$ in Example 6.

As shown in Tabs. 4&5, spectral order accuracy for spatial derivatives and fourth-order accuracy for time derivative of TSSP4 are demonstrated numerically for 1d GPE, respectively. Another issue is how to choose mesh size h and time step k in the strong repulsive interaction regime or semi-classical regime, i.e. $\beta_d \gg 1$, in order to get "correct" physical observables. In fact, after a rescaling in (4.1) under the normalization (4.3): $\mathbf{x} \to \varepsilon^{-1/2}\mathbf{x}$ and $\psi \to \varepsilon^{d/4}\psi$ with $\varepsilon = \beta_d^{-2/(d+2)}$, then the GPE (4.1) can be rewritten as

$$i\varepsilon\, \partial_t \psi(\mathbf{x}, t) = -\frac{\varepsilon^2}{2}\nabla^2\psi + V_d(\mathbf{x})\psi + |\psi|^2\psi, \quad \mathbf{x} \in \mathbb{R}^d. \qquad (11.25)$$

As demonstrated in [10, 11], the meshing strategy to capture 'correct' phys-

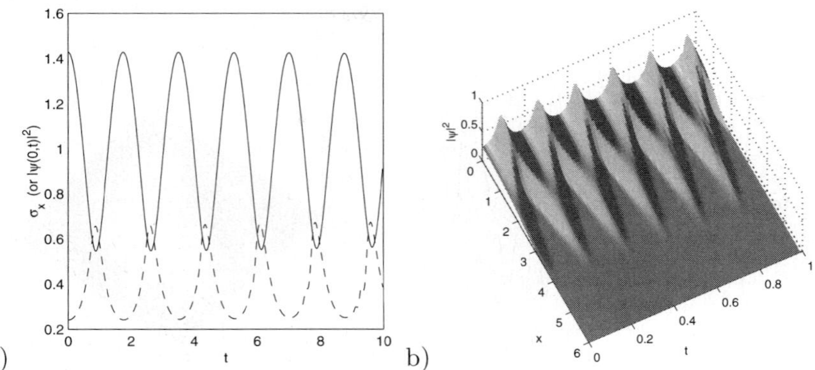

a)   b)

Fig. 5: Numerical results for Example 7:   a) Condensate width σ_x ('—') and central density $|\psi(0,t)|^2$ ('- - -'). b) Evolution of the density function $|\psi|^2$.

ical observables for this this problem is

$$h = O(\varepsilon), \qquad k = O(\varepsilon).$$

Thus the admissible meshing strategy for the GPE with strong repulsive interaction is

$$h = O(\varepsilon) = O\left(1/\beta_d^{2/(d+2)}\right), \quad k = O(\varepsilon) = O\left(1/\beta_d^{2/(d+2)}\right), \qquad d = 1, 2, 3.$$

$$(11.26)$$

Example 7 1d Gross-Pitaevskii equation, i.e. in (4.1) we choose $d = 1$. The initial condition is taken as the ground-state solution of (4.1) under $d = 1$ with $\gamma_x = 1$ and $\beta_1 = 20.0$ [6, 5], i.e. initially the condensate is assumed to be in its ground state. When $t = 0$, we double the trap frequency by setting $\gamma_x = 2$.

We solve this problem on the interval $[-12, 12]$ under mesh size $h = \frac{3}{64}$ and time step $k = 0.005$ with homogeneous Dirichlet boundary condition. Fig. 5 plots the condensate width and central density $|\psi(0,t)|^2$ as functions of time, as well as evolution of the density $|\psi|^2$ in space-time. One can see from this figure that the sudden change in the trap potential leads to oscillations in the condensate width and the peak value of the wave function. Note that the condensate width contracts in an oscillatory way (cf. Fig. 5a), which agrees with the analytical results in (4.43).

Example 8 2d Gross-Pitaevskii equation, i.e. in (4.1) we choose $d = 2$. The initial condition is taken as the ground-state solution of (4.1) under

                                   Weizhu Bao

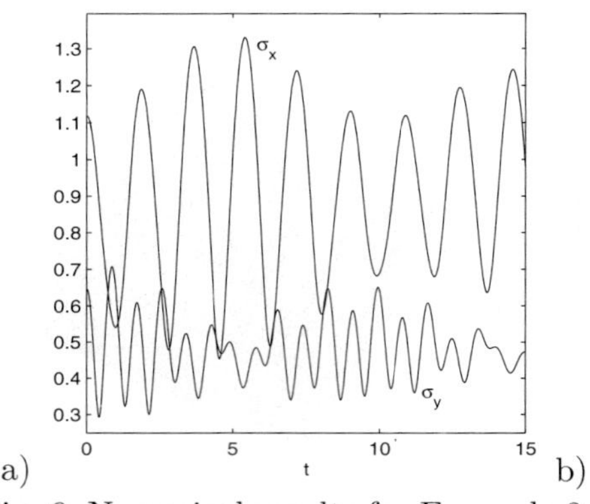
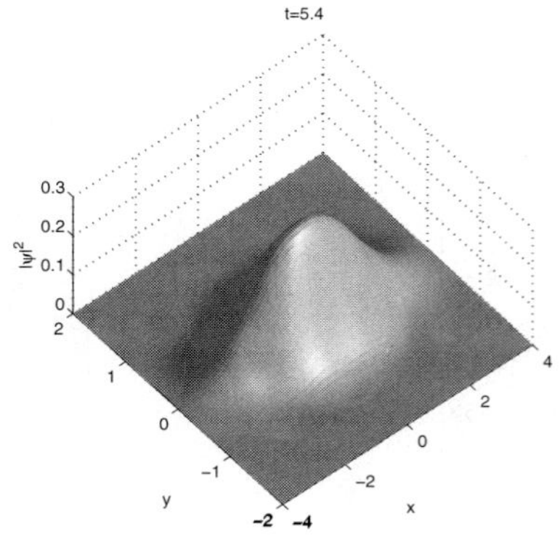

Fig. 6: Numerical results for Example 8:   a) Condensate width. b) Surface plot of the density $|\psi|^2$ at $t = 5.4$.

$d = 2$ with $\gamma_x = 1$, $\gamma_y = 2$ and $\beta_2 = 20.0$ [6, 5], i.e. initially the condensate is assumed to be in its ground state. When at $t = 0$, we double the trap frequency by setting $\gamma_x = 2$ and $\gamma_y = 4$.

We solve this problem on $[-8, 8]^2$ under mesh size $h = \frac{1}{32}$ and time step $k = 0.005$ with homogeneous Dirichlet boundary condition. Fig. 6 shows the condensate widths σ_x and σ_y as functions of time and the surface of the density $|\psi|^2$ at time $t = 5.4$. Fig. 7 the contour plots of the density $|\psi|^2$ at different times. Again, the sudden change in the trap potential leads to oscillations in the condensate width. Due to $\gamma_y = 2\gamma_x$, the oscillation frequency of σ_y is roughly double that of σ_x and the amplitudes of σ_x are larger than those of σ_y in general (cf. Fig. 6a). Again this agrees with the analytical results in (4.43).

Example 9 3d Gross-Pitaevskii equation, i.e. in (4.1) we choose $d = 3$. We present computations for two cases:

Case I. Intermediate ratio between trap frequencies along different axis (data for ^{87}Rb used in JILA [3]). The initial condition is taken as the ground-state solution of (4.1) under $d = 3$ with $\gamma_x = \gamma_y = 1$, $\gamma_z = 4$ and $\beta_3 = 37.62$ [6, 5]. When at $t = 0$, we four times the trap frequency by setting $\gamma_x = \gamma_y = 4$ and $\gamma_z = 16$.

Case II. High ratio between trap frequencies along different axis (data for ^{23}Na used in MIT (group of Ketterle) [38]). The initial condition is taken as the ground-state solution of (4.1) under $d = 3$ with $\gamma_x = \gamma_y = \frac{360}{3.5}$,

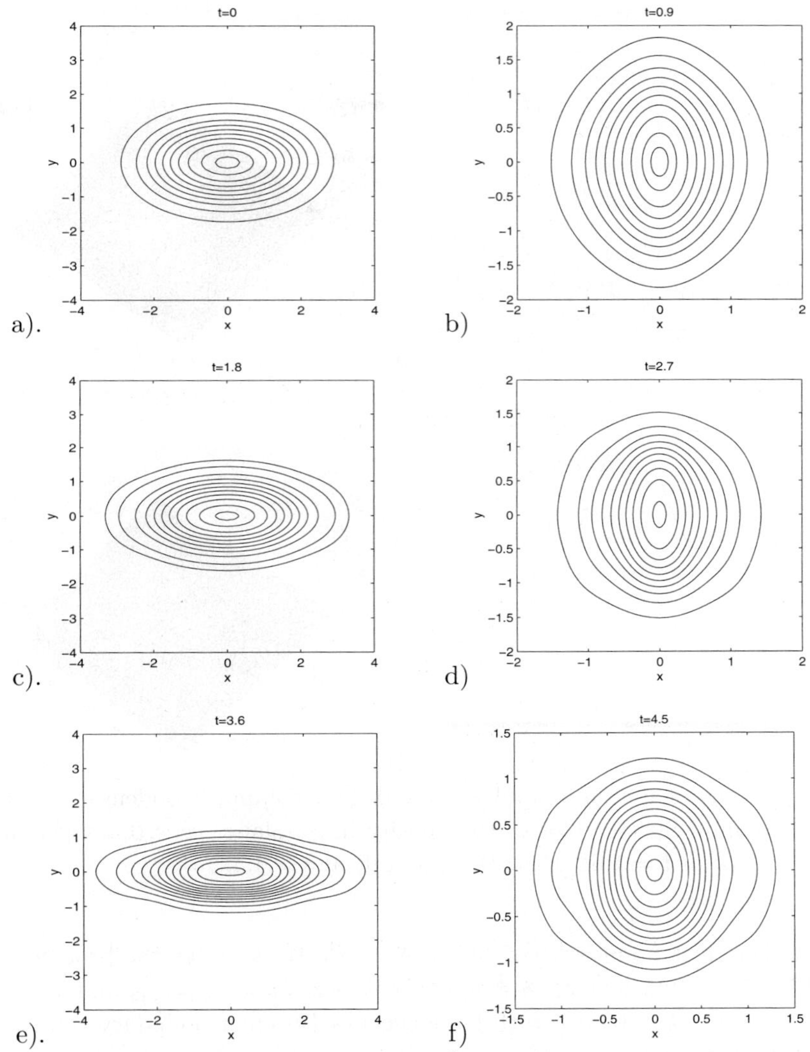

Fig. 7: Contour plots of the density $|\psi|^2$ at different times in Example 8.
a). $t = 0$, b). $t = 0.9$, c). $t = 1.8$, d). $t = 2.7$, e). $t = 3.6$, f). $t = 4.5$.

$\gamma_z = 1$ and $\beta_3 = 3.083$ [6, 5]. When at $t = 0$, we double the trap frequency by setting $\gamma_x = \gamma_y = \frac{720}{3.5}$ and $\gamma_z = 2$.

For case I, we solve the problem on $[-6, 6] \times [-6, 6] \times [-3, 3]$ under mesh size $h_x = h_y = \frac{3}{32}$ and $h_z = \frac{3}{64}$, and time step $k = 0.0025$ with homoge-

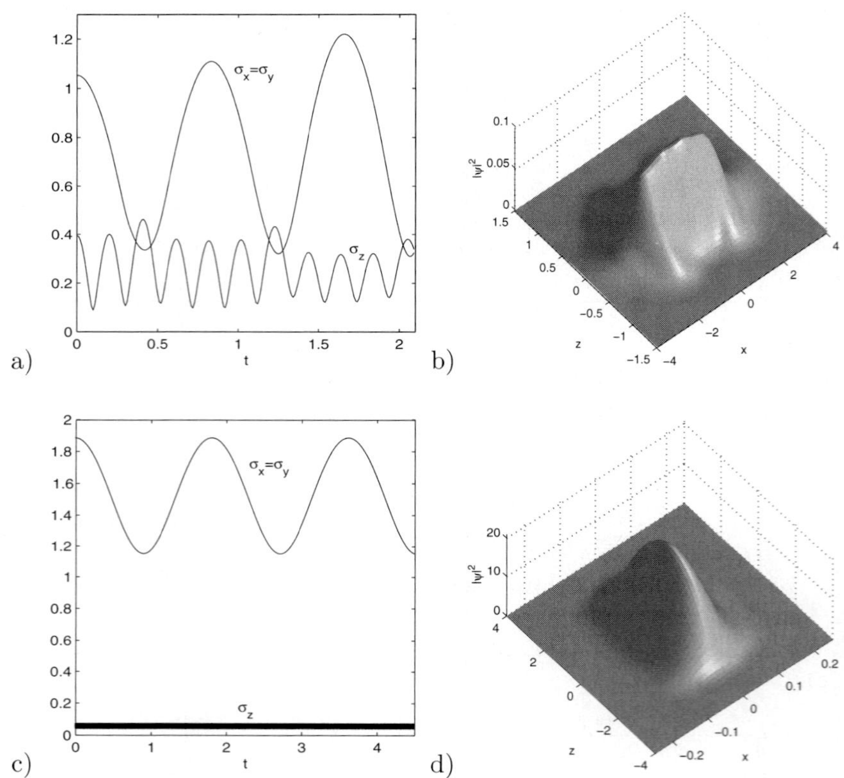

Fig. 8: Numerical results for Example 9: Left column: Condensate width; right column: Surface plot of the density in xz-plane, $|\psi(x,0,z,t)|^2$. Case I: a) and b) at $t = 1.64$. Case II: c) and d) at $t = 4.5$.

neous Dirichlet boundary condition. For case II, we solve the problem on $[-0.5, 0.5] \times [-0.5, 0.5] \times [-8, 8]$ under mesh size $h_x = h_y = \frac{1}{128}$ and $h_z = \frac{1}{8}$, and time step $k = 0.0005$ with homogeneous Dirichlet boundary condition along their boundaries.

Fig. 8 shows the condensate widths $\sigma_x = \sigma_y$ and σ_z as functions of time, as well as the surface of the density in xz-plane $|\psi(x,0,z,t)|^2$. Similar phenomena in case I in 3d is observed as those in Example 8 which is in 2d (cf. Fig. 6a). The ratio between the condensate widths increases with increasing the ratio between trap frequencies along different axis, i.e. it becomes more difficult to excite oscillations for large trap frequencies. In case II, the curves of the condensate widths are very well separated. This behavior is one of the basic assumptions allowing the reduction of GPE to 2d and 1d in the cases one or two of the trap frequencies are much larger than the others [85, 5, 8].

12. Derivation of the vector Zakharov system

In this section, we derive VZS from the two-fluid model [113] for ion-electron dynamics in plasma physics. Here we will use a more formal approach based on the multiple-scale modulation analysis. Following from [113], we will consider a plasma as two interpenetrating fluids, an electron fluid and an ion fluid, and denote the mass, number density (number of particles per unit volume) and velocity of the electrons (respectively of the ions), by m_e, $N_e(\mathbf{x}, t)$ and $\mathbf{v}_e(\mathbf{x}, t)$ (respectively m_i, $N_i(\mathbf{x}, t)$ and $\mathbf{v}_i(\mathbf{x}, t)$). The continuity equations for these fluids read

$$\partial_t N_e + \nabla \cdot (N_e \mathbf{v}_e) = 0, \tag{12.1}$$

$$\partial_t N_i + \nabla \cdot (N_i \mathbf{v}_i) = 0, \qquad \mathbf{x} \in \mathbb{R}^3, \qquad t > 0 \tag{12.2}$$

and the momentum equations read

$$m_e \, N_e (\partial_t \mathbf{v}_e + \mathbf{v}_e \cdot \nabla \mathbf{v}_e) = -\nabla p_e - e \, N_e \left(\mathcal{E} + \frac{1}{c} \mathbf{v}_e \times \mathcal{B} \right), \tag{12.3}$$

$$m_i \, N_i (\partial_t \mathbf{v}_i + \mathbf{v}_i \cdot \nabla \mathbf{v}_i) = -\nabla p_i + e \, N_i \left(\mathcal{E} + \frac{1}{c} \mathbf{v}_i \times \mathcal{B} \right), \tag{12.4}$$

where $-e$ and e represent the charge of the electron and the ions assumed to reduce to protons, respectively; p_e and p_i are the pressure. For small fluctuations, we write $\nabla p_e = \gamma_e \, T_e \, \nabla N_e$ and $\nabla p_i = \gamma_i \, T_i \, \nabla N_i$, where γ_e and γ_i denote the specific heat ratios of the electrons and the ions and T_e and T_i their respective temperatures in energy units. The electric field $\mathcal{E}$ and magnetic field $\mathcal{B}$ are provided by the Maxwell equations

$$-\frac{1}{c} \partial_t \mathcal{E} + \nabla \times \mathcal{B} = \frac{4\pi}{c} \mathbf{j}, \qquad \nabla \cdot \mathcal{E} = 4\pi \rho, \tag{12.5}$$

$$\frac{1}{c} \partial_t \mathcal{B} + \nabla \times \mathcal{E} = 0, \qquad \nabla \cdot \mathcal{B} = 0, \tag{12.6}$$

where $\rho = -e(N_e - N_i)$ and $\mathbf{j} = -e(N_e \mathbf{v}_e - N_i \mathbf{v}_i)$ are the densities of total charge and total current, respectively.

Equations (12.5) and (12.6) yield

$$\frac{1}{c^2} \partial_{tt} \mathcal{E} + \nabla \times (\nabla \times \mathcal{E}) + \frac{4\pi}{c^2} \partial_t \mathbf{j} = 0, \tag{12.7}$$

and using equations (12.1)-(12.4), we have

$$\partial_t \mathbf{j} = e(\nabla \cdot (N_e \mathbf{v}_e)\mathbf{v}_e + N_e \mathbf{v}_e \cdot \nabla \mathbf{v}_e + \frac{1}{m_e}\nabla p_e + \frac{eN_e}{m_e}(\mathcal{E} + \frac{1}{c}\mathbf{v}_e \times \mathcal{B}))$$

$$-e(\nabla \cdot (N_i \mathbf{v}_i)\mathbf{v}_i + N_i \mathbf{v}_i \cdot \nabla \mathbf{v}_i + \frac{1}{m_i}\nabla p_i + \frac{eN_i}{m_i}(\mathcal{E} + \frac{1}{c}\mathbf{v}_i \times \mathcal{B})). \quad (12.8)$$

In order to get VZS from the two-fluid model just mentioned, as in [113], we consider a long-wavelength small-amplitude Langmuir oscillation of the form

$$\mathcal{E} = \frac{\varepsilon}{2}(\mathbf{E}(\mathbf{X}, T)e^{-i\omega_e t} + c.c.) + \varepsilon^2 \hat{\mathbf{E}}(\mathbf{X}, T) + \cdots, \quad (12.9)$$

where the complex amplitude $\mathbf{E}$ depends on the slow variables $\mathbf{X} = \varepsilon \mathbf{x}$ and $T = \varepsilon^2 t$, the notation $c.c.$ stands for the complex conjugate and $\hat{\mathbf{E}}(\mathbf{X}, T)$ denotes the mean complex amplitude. It induces fluctuations for the density and velocity of the electrons and of the ions whose dynamical time will be seen to be $\tau = \varepsilon t$, thus shorter than T. We write

$$N_e = N_0 + \frac{\varepsilon^2}{2}(\tilde{N}_e(\mathbf{X}, \tau)e^{-i\omega_e t} + c.c.) + \varepsilon^2 \hat{N}_e(\mathbf{X}, \tau) + \cdots, \quad (12.10)$$

$$N_i = N_0 + \frac{\varepsilon^2}{2}(\tilde{N}_i(\mathbf{X}, \tau)e^{-i\omega_e t} + c.c.) + \varepsilon^2 \hat{N}_i(\mathbf{X}, \tau) + \cdots, \quad (12.11)$$

$$\mathbf{v}_e = \frac{\varepsilon}{2}(\tilde{\mathbf{v}}_e(\mathbf{X}, \tau)e^{-i\omega_e t} + c.c.) + \varepsilon^2 \hat{\mathbf{v}}_e(\mathbf{X}, \tau) + \cdots, \quad (12.12)$$

$$\mathbf{v}_i = \frac{\varepsilon}{2}(\tilde{\mathbf{v}}_i(\mathbf{X}, \tau)e^{-i\omega_e t} + c.c.) + \varepsilon^2 \hat{\mathbf{v}}_i(\mathbf{X}, \tau) + \cdots, \quad (12.13)$$

where N_0 is the unperturbed plasma density.

From the momentum equation (12.3), considering the leading order and noting that the magnetic field $\mathcal{B}$ is of order ε^2, we can easily get

$$m_e N_e \left(i\omega_e \tilde{\mathbf{v}}_e \frac{\varepsilon}{2}e^{-i\omega_e t}\right) = e N_e \left(\mathbf{E}\frac{\varepsilon}{2}e^{-i\omega_e t}\right),$$

thus the amplitude of the electron velocity oscillations is given by

$$\tilde{\mathbf{v}}_e = -\frac{ie}{m_e \omega_e}\mathbf{E}. \quad (12.14)$$

Neglecting the velocity oscillations of the ions due to their large mass, we take

$$\tilde{\mathbf{v}}_i = 0. \quad (12.15)$$

Applying (12.14) and (12.15) into the continuity equations (12.1) and (12.2), at the order of ε^2, we have

$$-i\omega_e \tilde{N}_e \frac{\varepsilon^2}{2} e^{-i\omega_e t} + N_0 \nabla \cdot \tilde{\mathbf{v}}_e \frac{\varepsilon^2}{2} e^{-i\omega_e t} = 0,$$

thus the density fluctuations are obtained as

$$\tilde{N}_e = -i\frac{N_0}{\omega_e}\nabla \cdot \tilde{\mathbf{v}}_e = -\frac{eN_0}{m_e\omega_e^2}\nabla \cdot \mathbf{E}, \tag{12.16}$$

$$\tilde{N}_i = 0. \tag{12.17}$$

At leading order, the equation for the electric field (12.7) with $\mathbf{j} = -e(N_e\mathbf{v}_e - N_i\mathbf{v}_i)$ becomes

$$-\frac{1}{c^2}\omega_e^2\mathbf{E}\frac{\varepsilon}{2}e^{-i\omega_e t} + \frac{4\pi}{c^2}i\omega_e eN_0\tilde{\mathbf{v}}_e\frac{\varepsilon}{2}e^{-i\omega_e t} = 0,$$

from which, with (12.14), we finally get the electron plasma frequency

$$\omega_e = \sqrt{\frac{4\pi e^2 N_0}{m_e}}. \tag{12.18}$$

At the order of ε^3, if no large-scale magnetic field is generated, then the equation (12.7) with (12.8) implies that

$$-2i\frac{\omega_e}{c^2}\partial_T\mathbf{E}\frac{\varepsilon^3}{2}e^{-i\omega_e t} + \nabla \times (\nabla \times \mathbf{E})\frac{\varepsilon^3}{2}e^{-i\omega_e t}$$
$$-\frac{4\pi e^2 N_0\gamma_e T_e}{c^2 m_e^2\omega_e^2}\nabla(\nabla \cdot \mathbf{E})\frac{\varepsilon^3}{2}e^{-i\omega_e t} + \frac{4\pi e^2 \hat{N}_e\mathbf{E}}{c^2 m_e}\frac{\varepsilon^3}{2}e^{-i\omega_e t} = 0,$$

and thus

$$-2i\frac{\omega_e}{c^2}\partial_T\mathbf{E} + \nabla \times (\nabla \times \mathbf{E}) - \frac{\gamma_e T_e}{m_e c^2}\nabla(\nabla \cdot \mathbf{E}) + \frac{4\pi e^2}{c^2 m_e}\hat{N}_e\mathbf{E} = 0, \tag{12.19}$$

where, resulting from (12.15) and (12.17), the contribution of the ions is negligible.

We rewrite the amplitude equation (12.19) as

$$i\partial_T\mathbf{E} - \frac{c^2}{2\omega_e}\nabla \times (\nabla \times \mathbf{E}) + \frac{3v_e^2}{2\omega_e}\nabla(\nabla \cdot \mathbf{E}) = \frac{\omega_e}{2}\frac{\hat{N}_e}{N_0}\mathbf{E}, \tag{12.20}$$

where the electron thermal velocity v_e is defined by

$$v_e = \sqrt{\frac{T_e}{m_e}} \tag{12.21}$$

and γ_e is taken to be 3 [113].

It is seen from (12.3), (12.4) and (12.15) that the mean electron velocity $\hat{\mathbf{v}}_e$ and the mean ion velocity $\hat{\mathbf{v}}_i$ satisfy

$$m_e\left(\partial_\tau \hat{\mathbf{v}}_e + \frac{1}{4}(\tilde{\mathbf{v}}_e \cdot \nabla \tilde{\mathbf{v}}_e^* + \tilde{\mathbf{v}}_e^* \cdot \nabla \tilde{\mathbf{v}}_e)\right) = -\frac{\gamma_e T_e}{N_0}\nabla \hat{N}_e - e\hat{\mathcal{E}}, \qquad (12.22)$$

$$m_i \partial_\tau \hat{\mathbf{v}}_i = -\frac{\gamma_i T_i}{N_0}\nabla \hat{N}_i + e\hat{\mathcal{E}}, \qquad (12.23)$$

where

$$\frac{1}{4}(\tilde{\mathbf{v}}_e \cdot \nabla \tilde{\mathbf{v}}_e^* + \tilde{\mathbf{v}}_e^* \cdot \nabla \tilde{\mathbf{v}}_e) = \frac{e^2}{4m_e^2\omega_e^2}\nabla|\mathbf{E}|^2, \qquad (12.24)$$

and $\overline{\tilde{\mathbf{v}}_e}$ denotes the conjugate of $\tilde{\mathbf{v}}_e$ and $m_e\partial_\tau\hat{\mathbf{v}}_e$ is negligible because of the small mass of the electron. Furthermore, $\hat{\mathcal{E}}$ denotes the leading contribution (of order ε^3) of the mean electron field. We thus replace (12.22) by

$$\frac{e^2}{4m_e\omega_e^2}\nabla|\mathbf{E}|^2 = -\frac{\gamma_e T_e}{N_0}\nabla \hat{N}_e - e\hat{\mathcal{E}}. \qquad (12.25)$$

The system is closed by using the quasi-neutrality of the plasma in the form

$$\hat{N}_e = \hat{N}_i, \qquad (12.26)$$

$$\hat{\mathbf{v}}_e = \hat{\mathbf{v}}_i, \qquad (12.27)$$

which we denote by N and $\mathbf{v}$, respectively. Then from the continuity equations, one gets

$$\partial_\tau N + N_0 \nabla \cdot \mathbf{v} = 0. \qquad (12.28)$$

Adding (12.25) to (12.23) and noting (12.28), we have

$$\partial_\tau \mathbf{v} = -\frac{c_s^2}{N_0}\nabla N - \frac{1}{16\pi m_i N_0}\nabla|\mathbf{E}|^2, \qquad (12.29)$$

with the speed of sound c_s,

$$c_s^2 = \eta\frac{T_e}{m_i}, \qquad \eta = \frac{\gamma_e T_e + \gamma_i T_i}{T_e}. \qquad (12.30)$$

Finally, we obtain the VZS [113] from equations (12.20), (12.28) and (12.29) as

$$i\partial_T \mathbf{E} - \frac{c^2}{2\omega_e}\nabla \times (\nabla \times \mathbf{E}) + \frac{3v_e^2}{2\omega_e}\nabla(\nabla \cdot \mathbf{E}) = \frac{\omega_e}{2}\frac{N}{N_0}\mathbf{E}, \qquad (12.31)$$

$$\varepsilon^2\partial_{TT} N - c_s^2 \triangle N = \frac{1}{16\pi m_i}\triangle|\mathbf{E}|^2. \qquad (12.32)$$

This VZS governs the coupled dynamics of the electric-field amplitude and of the low-frequency density fluctuations of the ions and describes the dynamics of the complex envelope of the electric field oscillations near the electron plasma frequency and the slow variations of the density perturbations.

In order to obtain a dimensionless form of the system (12.31)-(12.32), we define the normalized variables

$$t' = \frac{2\eta}{3}\mu_m\,\omega_e\,T, \qquad \mathbf{x}' = \frac{2}{3}\,(\eta\mu_m)^{1/2}\,\frac{\mathbf{X}}{\zeta_d}, \tag{12.33}$$

$$N' = \frac{3}{4\eta}\frac{1}{\mu_m}\frac{N}{N_0}, \qquad \mathbf{E}' = \frac{1}{\eta}\frac{1}{\mu_m^{1/2}}\left(\frac{3}{64\pi N_0 T_e}\right)^{1/2}\mathbf{E}. \tag{12.34}$$

with

$$\zeta_d = \sqrt{\frac{T_e}{4\pi e^2 N_0}}, \qquad \mu_m = \frac{m_e}{m_i}, \tag{12.35}$$

where ζ_d is the Debye length and μ_m is the ratio of the electron to the ion masses. Then defining

$$a = \frac{c^2}{3v_e^2} = \frac{c^2}{3\omega_e^2\zeta_d^2} \tag{12.36}$$

and plugging (12.33)-(12.34) into (12.31)-(12.32), and then removing all primes, we get the following dimensionless vector Zakharov system in three dimension

$$i\partial_t\mathbf{E} - a\nabla\times(\nabla\times\mathbf{E}) + \nabla(\nabla\cdot\mathbf{E}) = N\,\mathbf{E}, \tag{12.37}$$

$$\varepsilon^2\partial_{tt}N - \triangle N = \triangle|\mathbf{E}|^2, \qquad \mathbf{x}\in\mathbb{R}^3, \quad t > 0. \tag{12.38}$$

In fact, the equation (12.37) is equalivent to

$$i\partial_t\mathbf{E} + a\triangle\mathbf{E} + (1-a)\nabla(\nabla\cdot\mathbf{E}) = N\,\mathbf{E}. \tag{12.39}$$

13. Generalization and simplification of ZS

The VZS (12.37), (12.38) can be easily generalized to a physical situation when the dispersive waves interact with $\mathcal{M}$ different acoustic modes, e.g. in a multi-component plasma, which may be described by the following VZSM [113, 72, 73]:

$$i\,\partial_t\mathbf{E} + a\,\triangle\mathbf{E} + (1-a)\,\nabla(\nabla\cdot\mathbf{E}) - \mathbf{E}\sum_{J=1}^{\mathcal{M}}N_J = 0, \; \mathbf{x}\in\mathbb{R}^d, \; t > 0, \tag{13.1}$$

$$\varepsilon_J^2\,\partial_{tt}N_J - \triangle N_J + \nu_J\,\triangle|\mathbf{E}|^2 = 0, \quad J = 1,\cdots,\mathcal{M}; \tag{13.2}$$

where the real unknown function N_J is the Jth-component deviation of the ion density from its equilibrium value, $\varepsilon_J > 0$ is a parameter inversely proportional to the acoustic speed of the Jth-component, and ν_J are real constants.

The VZSM (13.1), (13.2) is time reversible and time transverse invariant, and preserves the following three conserved quantities. They are the wave energy

$$D^{VZSM} = \int_{\mathbb{R}^d} |\mathbf{E}(\mathbf{x},t)|^2 \, dx, \tag{13.3}$$

the momentum

$$\mathbf{P}^{VZSM} = \int_{\mathbb{R}^d} \left[\frac{i}{2} \sum_{j=1}^{d} (E_j \, \nabla E_j^* - E_j^* \, \nabla E_j) - \sum_{J=1}^{\mathcal{M}} \frac{\varepsilon_J^2}{\nu_J} N_J \mathbf{V_J} \right] dx \tag{13.4}$$

and the Hamiltonian

$$H^{VZSM} = \int_{\mathbb{R}^d} \left[a \, \|\nabla \mathbf{E}\|_{l^2}^2 + (1-a)|\nabla \cdot \mathbf{E}|^2 + \sum_{J=1}^{\mathcal{M}} N_J |\mathbf{E}|^2 \right.$$
$$\left. - \frac{1}{2} \sum_{J=1}^{\mathcal{M}} \left(\frac{\varepsilon_J^2}{\nu_J} |\mathbf{V}_J|^2 + \frac{1}{\nu_J} N_J^2 \right) \right] dx; \tag{13.5}$$

where the flux vector $\mathbf{V_J} = ((v_J)_1, \cdots, (v_J)_d)^T$ for the Jth-component is introduced through the equations

$$\partial_t N_J = -\nabla \cdot \mathbf{V}_J, \quad \partial_t \mathbf{V}_J = -\frac{1}{\varepsilon_J^2} \nabla (N_J - \nu_J |\mathbf{E}|^2), \qquad J = 1, \cdots, \mathcal{M}. \tag{13.6}$$

13.1. *Reduction from VZSM to GVZS*

In the VZSM (13.1)-(13.2), if we choose $\mathcal{M} = 2$, and assume that $1/\varepsilon_2^2 \gg 1/\varepsilon_1^2$, i.e. the acoustic speed of the second component is much faster than the first component, then formally the fast nondispersive component N_2 can be excluded by means of the relation

$$N_2 = \nu_2 \, |\mathbf{E}|^2 + \varepsilon_2^2 \, \triangle^{-1} \partial_{tt} N_2 \approx \nu_2 \, |\mathbf{E}|^2 + O(\varepsilon_2^2), \qquad \text{when } \varepsilon_2 \to 0. \tag{13.7}$$

Plugging (13.7) into (13.1), then the VZSM (13.1), (13.2) is reduced to GVZS with $N = N_1$, $\nu = \nu_1$, $\varepsilon = \varepsilon_1$, $\lambda = -\nu_2$ and $\alpha = 1$:

$$i \, \partial_t \mathbf{E} + a \, \triangle \mathbf{E} + (1-a) \, \nabla(\nabla \cdot \mathbf{E}) - \alpha \, N \, \mathbf{E} + \lambda \, |\mathbf{E}|^2 \mathbf{E} = 0, \tag{13.8}$$
$$\varepsilon^2 \partial_{tt} N - \triangle N + \nu \triangle |\mathbf{E}|^2 = 0, \qquad \mathbf{x} \in \mathbb{R}^d, \quad t > 0. \tag{13.9}$$

The GVZS (13.8), (13.9) is time reversible, time transverse invariant and preserves the following three conserved quantities, i.e. the wave energy, momentum and Hamiltonian:

$$D^{GVZS} = \int_{\mathbb{R}^d} |\mathbf{E}(\mathbf{x},t)|^2 \, d\mathbf{x}, \tag{13.10}$$

$$\mathbf{P}^{GVZS} = \int_{\mathbb{R}^d} \left[\frac{i}{2} \sum_{j=1}^{d} \left(E_j \, \nabla E_j^* - E_j^* \, \nabla E_j \right) - \frac{\alpha \varepsilon^2}{\nu} N\mathbf{V} \right] \, d\mathbf{x}, \tag{13.11}$$

$$H^{GVZS} = \int_{\mathbb{R}^d} \left[a \, \|\nabla \mathbf{E}\|_{l^2}^2 + (1-a)|\nabla \cdot \mathbf{E}|^2 + \alpha N|\mathbf{E}|^2 - \frac{\lambda}{2}|\mathbf{E}|^4 \right.$$
$$\left. - \frac{\alpha}{2\nu}N^2 - \frac{\alpha \varepsilon^2}{2\nu}|\mathbf{V}|^2 \right] \, d\mathbf{x}; \tag{13.12}$$

where the flux vector $\mathbf{V} = (v_1, \cdots, v_d)^T$ is introduced through the equations

$$\partial_t N = -\nabla \cdot \mathbf{V}, \qquad \partial_t \mathbf{V} = -\frac{1}{\varepsilon^2}\nabla(N - \nu|\mathbf{E}|^2). \tag{13.13}$$

In the case of $\mathcal{M} = 2$, $\nu = \nu_1$ and $\varepsilon = \varepsilon_1$, $N = N_1$ and $\mathbf{V} = \mathbf{V}_1$ in (13.4) and (13.5), and $\lambda = -\nu_2$, $\alpha = 1$ in (13.11), (13.12), letting $\varepsilon_2 \to 0$ and noting (13.7), we get formally quadratic convergence rate of the momentum and Hamiltonian from VZSM to GVZS in the 'subsonic limit' regime of the second component, i.e., $0 < \varepsilon_2 \ll 1$:

$$\mathbf{P}^{VZSM} = \int_{\mathbb{R}^d} \left[\frac{i}{2} \sum_{j=1}^{d} \left(E_j \, \nabla E_j^* - E_j^* \, \nabla E_j \right) - \frac{\varepsilon_1^2}{\nu_1}N_1\mathbf{V} \right] \, d\mathbf{x}$$
$$- \frac{\varepsilon_2^2}{\nu_2} \int_{\mathbb{R}^d} N_2\mathbf{V_2} \, d\mathbf{x}$$
$$\approx \mathbf{P}^{GVZS} + O(\varepsilon_2^2), \tag{13.14}$$

$$H^{VZSM} = \int_{\mathbb{R}^d} \left[a \, \|\nabla \mathbf{E}\|_{l^2}^2 + (1-a)|\nabla \cdot \mathbf{E}|^2 + N_1|\mathbf{E}|^2 - \frac{1}{2\nu_1}N_1^2 \right.$$
$$\left. - \frac{\varepsilon_1^2}{2\nu_1}|\mathbf{V_1}|^2 \right] \, d\mathbf{x} + \int_{\mathbb{R}^d} \left[N_2|\mathbf{E}|^2 - \frac{1}{2\nu_2}N_2^2 - \frac{\varepsilon_2^2}{2\nu_2}|\mathbf{V_2}|^2 \right] \, d\mathbf{x}$$
$$\approx H^{GVZS} + O(\varepsilon_2^2). \tag{13.15}$$

Choosing $a = 1$, $\alpha = 1$, $\nu = -1$ and $\lambda = 0$ in the GVZS (13.8)-(13.9), it collapses to the standard VZS [113]

$$i \, \partial_t \mathbf{E} + \triangle \mathbf{E} - N\mathbf{E} = 0, \qquad \mathbf{x} \in \mathbb{R}^d, \quad t > 0, \tag{13.16}$$

$$\varepsilon^2 \, \partial_{tt} N - \triangle N - \triangle |\mathbf{E}|^2 = 0. \tag{13.17}$$

13.2. *Reduction from GVZS to GZS*

In the case when $E_2 = \cdots = E_d = 0$ and $a = 1$ in the GVZS (13.8), (13.9), it reduces to the scalar GZS [113, 15],

$$i\,\partial_t E + \triangle E - \alpha\, N\, E + \lambda\, |E|^2 E = 0, \qquad \mathbf{x} \in \mathbb{R}^d, \quad t > 0, \qquad (13.18)$$

$$\varepsilon^2\,\partial_{tt} N - \triangle N + \nu\,\triangle |E|^2 = 0. \qquad (13.19)$$

The GZS (13.18), (13.19) is time reversible, time transverse invariant and conserved the following wave energy, momentum and Hamiltonian:

$$D^{GZS} = \int_{\mathbb{R}^d} |E(\mathbf{x}, t)|^2 \, d\mathbf{x}, \qquad (13.20)$$

$$\mathbf{P}^{GZS} = \int_{\mathbb{R}^d} \left[\frac{i}{2}\,(E\nabla E^* - E^*\nabla E) - \frac{\varepsilon^2 \alpha}{\nu} N\mathbf{V} \right] d\mathbf{x}, \qquad (13.21)$$

$$H^{GZS} = \int_{\mathbb{R}^d} \left[|\nabla E|^2 + \alpha N|E|^2 - \frac{\lambda}{2}|E|^4 - \frac{\alpha}{2\nu}N^2 - \frac{\alpha\varepsilon^2}{2\nu}|\mathbf{V}|^2 \right] d\mathbf{x}; \qquad (13.22)$$

where the flux vector $\mathbf{V} = (v_1, \cdots, v_d)^T$ is introduced through the equations

$$N_t = -\nabla \cdot \mathbf{V}, \qquad \mathbf{V}_t = -\frac{1}{\varepsilon^2}\nabla(N - \nu|E|^2). \qquad (13.23)$$

Choosing $\alpha = 1$, $\nu = -1$, $\varepsilon = 1$ and $\lambda = 0$ in the GZS (13.18)-(13.19), it collapses to the standard ZS [113, 15, 121]. When $\lambda \neq 0$, a cubic nonlinear term is added to the standard ZS.

Proof of the conservation laws in GZS: Multiplying (13.18) by $\overline{E}$, the conjugate of E, we get

$$iE_t\, E^* + E^*\,\triangle E - \alpha N\,|E|^2 + \lambda\,|E|^4 = 0. \qquad (13.24)$$

Then calculating the conjugate of (13.24) and multiplying it by E, one finds

$$-iE_t^*\, E + E\,\triangle E^* - \alpha N\,|E|^2 + \lambda\,|E|^4 = 0. \qquad (13.25)$$

Subtracting (13.25) from (13.24) and then multiplying both sides by $-i$, one gets

$$E_t\, E^* + E_t^*E + i(E\,\triangle E^* - E^*\,\triangle E) = 0. \qquad (13.26)$$

Integrating over $\mathbb{R}^d$, integration by parts, (13.26) leads to the conservation of the wave energy

$$\frac{d}{dt}D^{GZS} = \frac{d}{dt}\int_{\mathbb{R}^d} |E(\mathbf{x}, t)|^2 \, d\mathbf{x} = 0.$$

From (13.21), noting (13.23), (13.18), one has the conservation of the momentum

$$\frac{d}{dt}\mathbf{P}^{GZS} = \frac{i}{2}\int_{\mathbb{R}^d}(E_t\nabla E^* + E\nabla E_t^* - E^*\nabla E_t - E_t^*\nabla E)\,d\mathbf{x}$$

$$-\frac{\varepsilon^2\alpha}{\nu}\int_{\mathbb{R}^d}(N_t\mathbf{V} + N\mathbf{V}_t)\,d\mathbf{x}$$

$$= i\int_{\mathbb{R}^d}(E_t\nabla E^* - E_t^*\nabla E)\,d\mathbf{x} - \frac{\varepsilon^2\alpha}{\nu}\int_{\mathbb{R}^d}(N_t\mathbf{V} + N\mathbf{V}_t)\,d\mathbf{x}$$

$$= i\int_{\mathbb{R}^d}\nabla E^*(i\,\triangle\,E - i\alpha NE + i\lambda|E|^2E)\,d\mathbf{x}$$

$$-\frac{\varepsilon^2\alpha}{\nu}\int_{\mathbb{R}^d}(N_t\mathbf{V} + N\mathbf{V}_t)\,d\mathbf{x}$$

$$-i\int_{\mathbb{R}^d}\nabla E(-i\,\triangle\,E^* + i\alpha NE^* - i\lambda|E|^2E^*)\,d\mathbf{x}$$

$$= \alpha\int_{\mathbb{R}^d}N\nabla|E|^2\,d\mathbf{x} + \frac{\varepsilon^2\alpha}{\nu}\int_{\mathbb{R}^d}\mathbf{V}\nabla\cdot\mathbf{V}\,d\mathbf{x}$$

$$+\frac{\alpha}{\nu}\int_{\mathbb{R}^d}\nabla(N - \nu|E|^2)N\,d\mathbf{x} = 0.$$

Noting (13.23), (13.19) and multiplying (13.18) by E_t^*, the conjugate of E_t, we write it

$$\mathcal{T} = \int_{\mathbb{R}^d}\left[i|E_t|^2 + E_t^*\,\triangle\,E - \alpha NEE_t^* + \lambda|E|^2EE_t^*\right]\,d\mathbf{x} = 0. \quad (13.27)$$

Then the real part of $\mathcal{T}$ is

$$0 = Re(\mathcal{T}) = Re\int_{\mathbb{R}^d}\left[E_t^*\,\triangle\,E - \alpha NEE_t^* + \lambda|E|^2EE_t^*\right]\,d\mathbf{x}$$

$$= Re\int_{\mathbb{R}^d}\left[-\nabla E\nabla E_t^* - \frac{\alpha}{2}(N|E|^2)_t + \frac{\alpha}{2}N_t|E|^2 + \frac{\lambda}{4}(|E|^4)_t\right]\,d\mathbf{x}$$

$$= -\frac{1}{2}\int_{\mathbb{R}^d}\left[\left(|\nabla E|^2 + \alpha N|E|^2 - \frac{\lambda}{2}|E|^4\right)_t + \frac{\alpha}{2}N_t|E|^2\right]\,d\mathbf{x}$$

$$= -\frac{1}{2}\int_{\mathbb{R}^d}\left[\left(|\nabla E|^2 + \alpha N|E|^2 - \frac{\lambda}{2}|E|^4\right)_t - \frac{\alpha}{2}|E|^2\nabla\cdot\mathbf{V}\right]\,d\mathbf{x}$$

$$= -\frac{1}{2}\int_{\mathbb{R}^d}\left[\left(|\nabla E|^2 + \alpha N|E|^2 - \frac{\lambda}{2}|E|^4\right)_t + \frac{\alpha}{2}\nabla|E|^2\cdot\mathbf{V}\right]\,d\mathbf{x}$$

$$= -\frac{1}{2}\int_{\mathbb{R}^d}\left[\left(|\nabla E|^2 + \alpha N|E|^2 - \frac{\lambda}{2}|E|^4\right)_t + \frac{\alpha}{2}\left(\frac{\varepsilon^2}{\nu}\mathbf{V}_t + \frac{1}{\nu}\nabla N\right)\cdot\mathbf{V}\right]\,d\mathbf{x}.$$

210 *Weizhu Bao*

Thus we have,

$$
\begin{aligned}
0 &= -\frac{1}{2}\int_{\mathbb{R}^d}\left(|\nabla E|^2 + \alpha N|E|^2 - \frac{\lambda}{2}|E|^4\right)\,d\mathbf{x} \\
&\quad + \frac{\alpha\varepsilon^2}{2\nu}\int_{\mathbb{R}^d}\frac{1}{2}(|\mathbf{V}|^2)_t\,d\mathbf{x} - \frac{\alpha}{2\nu}\int_{\mathbb{R}^d}N\nabla\cdot\mathbf{V}\,d\mathbf{x} \\
&= -\frac{1}{2}\int_{\mathbb{R}^d}\partial_t\left(|\nabla E|^2 + \alpha N|E|^2 - \frac{\lambda}{2}|E|^4\right)\,d\mathbf{x} \\
&\quad + \frac{\alpha\varepsilon^2}{2\nu}\int_{\mathbb{R}^d}\frac{1}{2}(|\mathbf{V}|^2)_t\,d\mathbf{x} + \frac{\alpha}{2\nu}\int_{\mathbb{R}^d}N\partial_t N\,d\mathbf{x} \\
&= -\frac{1}{2}\int_{\mathbb{R}^d}\partial_t\left(|\nabla E|^2 + \alpha N|E|^2 - \frac{\lambda}{2}|E|^4 - \frac{\alpha}{2\nu}N^2 - \frac{\alpha\varepsilon^2}{2\nu}|\mathbf{V}|^2)\right)\,d\mathbf{x},
\end{aligned}
$$

which implies the conservation of Hamiltonian

$$
\frac{d}{dt}H^{GZS} = 0.
$$

13.3. *Reduction from GVZS to VNLS*

In the "subsonic limit", i.e. $\varepsilon \to 0$ in GVZS (13.8), (13.9), which corresponds to that the density fluctuations are assumed to follow adiabatically the modulation of the Langmuir wave, it collapses to the VNLS equation. In fact, letting $\varepsilon \to 0$ in (13.9), we get formally

$$
N = \nu\,|\mathbf{E}|^2 + \varepsilon^2\,\triangle^{-1}\,\partial_{tt}N = \nu\,|\mathbf{E}|^2 + O(\varepsilon^2), \qquad \text{when } \varepsilon \to 0. \quad (13.28)
$$

Plugging (13.28) into (13.8), we obtain formally the VNLS:

$$
i\,\partial_t\mathbf{E} + a\,\triangle\mathbf{E} + (1-a)\,\nabla(\nabla\cdot\mathbf{E}) + (\lambda - \alpha\nu)|\mathbf{E}|^2\mathbf{E} = 0, \quad \mathbf{x} \in \mathbb{R}^d,\ t > 0. \quad (13.29)
$$

The VNLS (13.29) is time reversible, time transverse invariant and preserves the following wave energy, momentum and Hamiltonian:

$$
D^{VNLS} = \int_{\mathbb{R}^d}|\mathbf{E}(\mathbf{x},t)|^2\,d\mathbf{x}, \tag{13.30}
$$

$$
\mathbf{P}^{VNLS} = \int_{\mathbb{R}^d}\frac{i}{2}\sum_{j=1}^{d}\left(E_j\,\nabla E_j^* - E_j^*\,\nabla E_j\right)\,d\mathbf{x}, \tag{13.31}
$$

$$
H^{VNLS} = \int_{\mathbb{R}^d}\left[a\,\|\nabla\mathbf{E}\|_{l^2}^2 + (1-a)|\nabla\cdot\mathbf{E}|^2 + \frac{\alpha\nu - \lambda}{2}|\mathbf{E}|^4\right]\,d\mathbf{x}. \tag{13.32}
$$

Letting $\varepsilon \to 0$ in (13.11), (13.12), noting (13.28), we get formally quadratic convergence rate of the momentum and Hamiltonian from GVZS to VNLS in the 'subsonic limit' regime, i.e., $0 < \varepsilon \ll 1$:

$$\mathbf{P}^{GVZS} = \int_{\mathbb{R}^d} \frac{i}{2} \sum_{j=1}^{d} \left(E_j \, \nabla E_j^* - E_j^* \, \nabla E_j \right) \, d\mathbf{x} - \frac{\alpha\varepsilon^2}{\nu} \int_{\mathbb{R}^d} N \mathbf{V} \, d\mathbf{x}$$

$$\approx \mathbf{P}^{VNLS} + O(\varepsilon^2),$$

$$H^{GVZS} = \int_{\mathbb{R}^d} \left[a \, \|\nabla \mathbf{E}\|_{l^2}^2 + (1-a)|\nabla \cdot \mathbf{E}|^2 + \frac{\alpha\nu - \lambda}{2}|\mathbf{E}|^4 \right] \, d\mathbf{x}$$

$$- \frac{\alpha\varepsilon^2}{2\nu} \int_{\mathbb{R}^d} |\mathbf{V}|^2 \, d\mathbf{x}$$

$$\approx H^{VNLS} + O(\varepsilon^2).$$

13.4. *Reduction from GZS to NLSE*

Similarly, in the "subsonic limit", i.e. $\varepsilon \to 0$ in GZS (13.18), (13.19), it collpases to the well-known NLSE with a cubic nonlinearity. In fact, letting $\varepsilon \to 0$ in (13.19), we get formally

$$N = \nu \, |E|^2 + \varepsilon^2 \, \triangle^{-1} \, \partial_{tt} N = \nu \, |E|^2 + O(\varepsilon^2), \qquad \text{when } \varepsilon \to 0. \quad (13.33)$$

Plugging (13.33) into (13.18), we obtain formally the NLS equation:

$$i \, E_t + \triangle \, E + (\lambda - \alpha\nu)|E|^2 \, E = 0, \qquad \mathbf{x} \in \mathbb{R}^d, \quad t > 0. \quad (13.34)$$

The NLSE (13.34) is time reversible, time transverse invariant, and preserves the following wave energy, momentum and Hamiltonian:

$$D^{NLS} = \int_{\mathbb{R}^d} |E(\mathbf{x}, t)|^2 \, d\mathbf{x}, \qquad (13.35)$$

$$\mathbf{P}^{NLS} = \int_{\mathbb{R}^d} \left[\frac{i}{2} \left(E\nabla E^* - E^*\nabla E \right) \right] \, d\mathbf{x}, \qquad (13.36)$$

$$H^{NLS} = \int_{\mathbb{R}^d} \left[|\nabla E|^2 + \frac{\alpha\nu - \lambda}{2}|E|^4 \right] \, d\mathbf{x}. \qquad (13.37)$$

Similarly, letting $\varepsilon \to 0$ in (13.21), (13.22), noting (13.33), we get formally the quadratic convergence rate of the momentum and Hamiltonian from GZS to NLSE in the 'subsonic limit' regime, i.e., $0 < \varepsilon \ll 1$:

$$\mathbf{P}^{GZS} = \int_{\mathbb{R}^d} \frac{i}{2} \left(E\nabla E^* - E^*\nabla E \right) \, d\mathbf{x} - \frac{\varepsilon^2\alpha}{\nu} \int_{\mathbb{R}^d} N\mathbf{V} \, d\mathbf{x}$$

$$\approx \mathbf{P}^{NLS} + O(\varepsilon^2), \qquad (13.38)$$

$$H^{GZS} = \int_{\mathbb{R}^d} \left[|\nabla E|^2 + \frac{\alpha\nu - \lambda}{2}|E|^4 \right] \, d\mathbf{x} - \frac{\alpha\varepsilon^2}{2\nu} \int_{\mathbb{R}^d} |\mathbf{V}|^2 \, d\mathbf{x}$$

$$\approx H^{NLS} + O(\varepsilon^2). \qquad (13.39)$$

13.5. *Add a linear damping term to arrest blowup*

When $d \geq 2$ and initial Hamiltonian $H^{GZS} < 0$ in the GZS (13.18), (13.19), mathematically, it will blowup in finite time [113]. However, the physical quantities modeled by E and N do not become infinite which implies the validity of (13.18), (13.19) breaks down near singularity. Additional physical mechanisms, which were initially small, become important near the singular point and prevent the formation of singularity. In order to arrest blowup, in physical literatures, a small linear damping (absorption) term is introduced into the GZS [64]:

$$i\,\partial_t E + \triangle E - \alpha\,N\,E + \lambda\,|E|^2 E + i\,\gamma\,E = 0, \tag{13.40}$$

$$\varepsilon^2\,\partial_{tt}N - \triangle N + \nu\,\triangle\,|E|^2 = 0, \qquad \mathbf{x} \in \mathbb{R}^d, \quad t > 0; \tag{13.41}$$

where $\gamma > 0$ is a damping parameter. The decay rate of the wave energy D^{GZS} of the damped GZS (13.40), (13.41) is

$$D^{GZS}(t) = \int_{\mathbb{R}^d} |E(\mathbf{x},t)|^2\,d\mathbf{x} = e^{-2\gamma t} \int_{\mathbb{R}^d} |E(\mathbf{x},0)|^2\,d\mathbf{x}$$

$$= e^{-2\gamma t} D^{GZS}(0), \qquad t \geq 0. \tag{13.42}$$

Similarly, when $d \geq 2$ and initial Hamiltonian $H^{GVZS} < 0$ in the GVZS (13.8), (13.9) (or $H^{VZSM} < 0$ in the VZSM (13.1), (13.2)), mathematically, it will blowup in finite time too. In order to arrest blowup, in physical literatures, a small linear damping (absorption) term is introduced into the GVZS (or VZSM):

$$i\,\partial_t \mathbf{E} + a\,\triangle\,\mathbf{E} + (1-a)\,\nabla(\nabla \cdot \mathbf{E}) - \alpha\,N\,\mathbf{E} + \lambda\,|\mathbf{E}|^2\mathbf{E} + i\,\gamma\,\mathbf{E} = 0, \tag{13.43}$$

$$\varepsilon^2\partial_{tt}N - \triangle N + \nu\,\triangle\,|\mathbf{E}|^2 = 0, \qquad \mathbf{x} \in \mathbb{R}^d, \quad t > 0; \tag{13.44}$$

where $\gamma > 0$ is a damping parameter. The decay rate of the wave energy D^{GVZS} of the damped GVZS (13.43), (13.44) is

$$D^{GVZS}(t) = \int_{\mathbb{R}^d} |\mathbf{E}(\mathbf{x},t)|^2\,d\mathbf{x} = e^{-2\gamma t} \int_{\mathbb{R}^d} |\mathbf{E}(\mathbf{x},0)|^2\,d\mathbf{x}$$

$$= e^{-2\gamma t} D^{GVZS}(0), \qquad t \geq 0. \tag{13.45}$$

14. Well-posedness of ZS

Based on the conservation laws, the wellposedness for the standard ZS (13.16)-(13.17) were proven [113, 112, 23, 24]

Theorem 14.1: *In one dimension, for initial conditions, $E^0 \in H^p(\mathbb{R})$, $N^0 \in H^{p-1}(\mathbb{R})$, and $N^{(1)} \in H^{p-2}(\mathbb{R})$ with $p \leq 3$, there exists a unique solution $E \in L^\infty(\mathbb{R}^+, H^p(\mathbb{R}))$, $N \in L^\infty(\mathbb{R}^+, H^{p-1}(\mathbb{R}))$ for (13.16)-(13.17).*

Theorem 14.2: *In dimensions 2 and 3, for initial conditions $E^0 \in H^p(\mathbb{R}^d)$, $N^0 \in H^{p-1}(\mathbb{R}^d)$, and $N^{(1)} \in H^{p-2}(\mathbb{R}^d)$ with $p \leq 3$, there exists a unique solution $E \in L^\infty([0, T^*), H^p(\mathbb{R}^d))$, $N \in L^\infty([0, T^*), H^{p-1}(\mathbb{R}^d))$ for (13.16)-(13.17), where time T^* depends on the initial conditions.*

15. Plane wave and soliton wave solutions of ZS

In one spatial dimension (1D), the GZS (13.40)- (13.41) collapses to

$$i\, E_t + E_{xx} - \alpha N\, E + \lambda |E|^2\, E + i\gamma\, E = 0, \ a < x < b, \ t > 0, \tag{15.1}$$

$$\varepsilon^2 N_{tt} - N_{xx} + \nu(|E|^2)_{xx} = 0, \qquad a < x < b, \quad t > 0, \tag{15.2}$$

which admits plane wave and soliton wave solutions.

Firstly, it is instructive to examine some explicit solutions to (15.1) and (15.2). The well-known plane wave solutions [97] can be given in the following form:

$$N(x, t) = d, \qquad a < x < b, \quad t \geq 0, \tag{15.3}$$

$$E(x, t) = \begin{cases} c\, e^{i\left(\frac{2\pi r x}{b-a} - \omega_1 t\right)}, & \gamma = 0, \\ c\, e^{-\gamma t} e^{i\left(\frac{2\pi r x}{b-a} - \omega_2 t - \frac{\lambda c^2}{2\gamma}(e^{-2\gamma t} - 1)\right)}, & \gamma \neq 0, \end{cases} \tag{15.4}$$

where r is an integer, c, d are constants and

$$\omega_1 = \alpha d + \frac{4\pi^2 r^2}{(b-a)^2} - \lambda c^2, \qquad \omega_2 = \alpha d + \frac{4\pi^2 r^2}{(b-a)^2}.$$

Secondly, as is well known, the standard ZS is not exactly integrable. Therefore the generalized ZS cannot be exactly integrable, either. However, it has exact one-soliton solutions to (15.1) and (15.2) for $\gamma = 0$ [72] for $x \in \mathbb{R}$ and $t \geq 0$:

$$E_s(x, t; \eta, V, \varepsilon, \nu) = \left[\frac{\lambda}{2} - \frac{\alpha\nu}{2\varepsilon^2}(1/\varepsilon^2 - V^2)^{-1}\right]^{-1/2} U_s, \tag{15.5}$$

$$U_s \equiv 2i\eta\, \text{sech}[2\eta(x - Vt)] \exp\left[iVx/2 + i(4\eta^2 - V^2/4)t + i\Phi_0\right], \tag{15.6}$$

$$N_s(x, t; \eta, V, \varepsilon, \nu) = \frac{\nu}{\varepsilon^2}(1/\varepsilon^2 - V^2)^{-1}|E_s|^2, \tag{15.7}$$

where η and V are the soliton's amplitude and velocity, respectively, and Φ_0 is a trivial phase constant.

Finally, we will consider the periodic soliton solution with a period L in 1d of the standard ZS, that is, $d = 1$, $\varepsilon = 1$, $\alpha = 1$, $\lambda = 0$, $\gamma = 0$ and

$\nu = -1$ in (13.40)-(13.41). The analytic solution of the ZS (15.1)-(15.2) was derived [100] and used to test different numerical methods for the ZS in [100, 28]. The solution can be written as

$$E_s(x,t;v,E_{\max}) = F(x - vt)\exp[i\phi(x - ut)], \qquad (15.8)$$

$$N_s(x,t;v,E_{\max}) = G(x - vt), \qquad (15.9)$$

where

$$F(x - vt) = E_{\max} \cdot dn(w,q), \quad G(x - vt) = \frac{|F(x - vt)|^2}{v^2 - 1} + N_0,$$

$$w = \frac{E_{\max}}{\sqrt{(2(1 - v^2))}} \cdot (x - vt), \quad q = \frac{\sqrt{(E_{\max}^2 - E_{\min}^2)}}{E_{\max}},$$

$$\phi = v/2, \ \frac{v}{2}L = 2\pi m, \ m = 1, 2, 3 \cdots, \ u = \frac{v}{2} + \frac{2N_0}{v} - \frac{E_{\max}^2 + E_{\min}^2}{v(1 - v^2)},$$

$$L = \frac{2\sqrt{2(1 - v^2)}}{E_{\max}} K(q) = \frac{2\sqrt{2(1 - v^2)}}{E_{\max}} K'\left(\frac{E_{\min}}{E_{\max}}\right),$$

with $dn(w,q)$ a Jacobian elliptic function, L the period of the Jacobian elliptic functions, K and K' the complete elliptic integrals of the first kind satisfying $K(q) = K'\left(\sqrt{1 - q^2}\right)$, and N_0 chosen such that

$$\langle N_s \rangle = \frac{1}{L} \int_0^L N_s(x,t)\,dx = 0.$$

16. Time-splitting spectral method for GZS

In this section we present new numerical methods for the GZS (13.40), (13.41). For simplicity of notations, we shall introduce the method in one space dimension $(d = 1)$ of the GZS with periodic boundary conditions. Generalizations to $d > 1$ are straightforward for tensor product grids and the results remain valid without modifications. For $d = 1$, the problem becomes

$$i\,\partial_t E + \partial_{xx} E - \alpha N\,E + \lambda|E|^2\,E + i\gamma\,E = 0, \ a < x < b, \ t > 0, \quad (16.1)$$

$$\varepsilon^2\partial_{tt} N - \partial_{xx}(N - \nu\,|E|^2) = 0, \qquad a < x < b, \quad t > 0, \qquad (16.2)$$

$$E(x,0) = E^{(0)}(x), \ N(x,0) = N^{(0)}(x), \ \partial_t N(x,0) = N^{(1)}(x), \qquad (16.3)$$

$$E(a,t) = E(b,t), \qquad \partial_x E(a,t) = \partial_x E(b,t), \qquad t \geq 0, \qquad (16.4)$$

$$N(a,t) = N(b,t), \qquad \partial_x N(a,t) = \partial_x N(b,t), \qquad t \geq 0. \qquad (16.5)$$

Moreover, we supplement (16.1)-(16.5) by imposing the compatibility condition

$$E^{(0)}(a) = E^{(0)}(b), \qquad N^{(0)}(a) = N^{(0)}(b),$$

$$N^{(1)}(a) = N^{(1)}(b), \qquad \int_a^b N^{(1)}(x)\,dx = 0. \tag{16.6}$$

As is well known, the GZS has the following property

$$D^{GZS}(t) = \int_a^b |E(x,t)|^2\,dx = e^{-2\gamma t} \int_a^b |E^{(0)}(x)|^2\,dx$$

$$= e^{-2\gamma t} D^{GZS}(0), \quad t \geq 0. \tag{16.7}$$

When $\gamma = 0$, $D^{GZS}(t) \equiv D^{GZS}(0)$, i.e., it is an invariant of the GZS [28]. When $\gamma > 0$, it decays to 0 exponentially. Furthermore, the GZS also has the following properties for $t \geq 0$

$$\int_a^b \partial_t N(x,t)\,dx = 0, \qquad \int_a^b N(x,t)\,dx = \int_a^b N^{(0)}(x)\,dx = \text{const.} \tag{16.8}$$

In some cases, the boundary conditions (16.4) and (16.5) may be replaced by

$$E(a,t) = E(b,t) = 0, \qquad N(a,t) = N(b,t) = 0, \qquad t \geq 0. \tag{16.9}$$

We choose the spatial mesh size $h = \triangle x > 0$ with $h = (b-a)/M$ for M being an even positive integer, the time step $k = \triangle t > 0$ and let the grid points and the time step be

$$x_j := a + j\,h, \qquad j = 0, 1, \cdots, M; \qquad t_m := m\,k, \qquad m = 0, 1, 2, \cdots .$$

Let E_j^m and N_j^m be the approximations of $E(x_j, t_m)$ and $N(x_j, t_m)$, respectively. Furthermore, let E^m and N^m be the solution vector at time $t = t_m = mk$ with components E_j^m and N_j^m, respectively.

From time $t = t_m$ to $t = t_{m+1}$, the first NLS-type equation (16.1) is solved in two splitting steps. One solves first

$$i\,\partial_t E + \partial_{xx} E = 0, \tag{16.10}$$

for the time step of length k, followed by solving

$$i\,\partial_t E = \alpha N\,E - \lambda |E|^2\,E - i\gamma\,E, \tag{16.11}$$

for the same time step. Equation (16.10) will be discretized in space by the Fourier spectral method and integrated in time *exactly*. For $t \in [t_m, t_{m+1}]$, multiplying (16.11) by $\overline{E}$, we get

$$i\,\partial_t E\,E^* = \alpha N|E|^2 - \lambda |E|^4 - i\gamma |E|^2. \tag{16.12}$$

Then calculating the conjugate of the ODE (16.11) and multiplying it by E, one finds

$$-i\,\partial_t E^*\,E = \alpha N|E|^2 - \lambda|E|^4 + i\gamma|E|^2. \tag{16.13}$$

Subtracting (16.13) from (16.12) and then multiplying both sides by $-i$, one gets

$$\partial_t(|E(x,t)|^2) = \partial_t E(x,t)E(x,t)^* + \partial_t E(x,t)^* E(x,t)$$
$$= -2\gamma|E(x,t)|^2 \tag{16.14}$$

and therefore

$$|E(x,t)|^2 = e^{-2\gamma(t-t_m)}|E(x,t_m)|^2, \quad t_m \le t \le t_{m+1}. \tag{16.15}$$

Substituting (16.15) into (16.11), we obtain

$$i\partial_t E(x,t) = \alpha N(x,t)E(x,t) - \lambda e^{-2\gamma(t-t_m)}|E(x,t_m)|^2 E(x,t)$$
$$-i\gamma E(x,t). \tag{16.16}$$

Integrating (16.16) from t_m to t_{m+1}, and then approximating the integral of N on $[t_m, t_{m+1}]$ via the trapezoidal rule, one obtains

$$E(x,t_{m+1}) = e^{-i\int_{t_m}^{t_{m+1}}[\alpha N(x,\tau)-\lambda e^{-2\gamma(\tau-t_m)}|E(x,t_m)|^2-i\gamma]\,d\tau}\,E(x,t_m)$$

$$\approx \begin{cases} e^{-ik[\alpha(N(x,t_m)+N(x,t_{m+1}))/2-\lambda|E(x,t_m)|^2]}\,E(x,t_m), & \gamma = 0, \\ e^{-\gamma k-i[k\alpha(N(x,t_m)+N(x,t_{m+1}))/2+\lambda|E(x,t_m)|^2(e^{-2\gamma k}-1)/2\gamma]}\,E(x,t_m), & \gamma \ne 0. \end{cases}$$

16.1. *Crank-Nicolson leap-frog time-splitting spectral discretizations (CN-LF-TSSP) for GZS*

The second wave-type equation (16.2) in the GZS is discretized by pseudo-spectral method for spatial derivatives, and then applying Crank-Nicolson /leap-frog for linear/nonlinear terms for time derivatives:

$$\varepsilon^2 \frac{N_j^{m+1} - 2N_j^m + N_j^{m-1}}{k^2} - D_{xx}^f\Big[(\beta N^{m+1} + (1-2\beta)N^m + \beta N^{m-1})$$

$$-\nu|E^m|^2\Big]_{x=x_j} = 0, \quad j = 0,\cdots,M, \quad m = 1,2,\cdots, \tag{16.17}$$

where $0 \le \beta \le 1$ is a constant, D_{xx}^f, a spectral differential operator approximation of ∂_{xx}, is defined as

$$D_{xx}^f U\big|_{x=x_j} = -\sum_{l=-M/2}^{M/2-1} \mu_l^2\,\widetilde{U}_l\,e^{i\mu_l(x_j-a)} \tag{16.18}$$

and $\widetilde{U}_l$, the Fourier coefficients of a vector $U = (U_0, U_1, U_2, \cdots, U_M)^T$ with $U_0 = U_M$, are defined as

$$\mu_l = \frac{2\pi l}{b-a}, \quad \widetilde{U}_l = \frac{1}{M} \sum_{j=0}^{M-1} U_j \, e^{-i\mu_l(x_j-a)}, \quad l = -\frac{M}{2}, \cdots, \frac{M}{2} - 1.$$

$$(16.19)$$

When $\beta = 0$ in (16.17), the discretization (16.17) to the wave-type equation (16.2) is *explicit* and was used in [15, 114]. When $0 < \beta \le 1$, the discretization is *implicit*, but can be solved *explicitly*. In fact, suppose

$$N_j^m = \sum_{l=-M/2}^{M/2-1} (\widetilde{N^m})_l \, e^{i\mu_l(x_j-a)}, \qquad j = 0, \cdots, M; \quad m = 0, 1, \cdots,$$

$$(16.20)$$

Plugging (16.20) into (16.17), using the orthogonality of the Fourier basis, we obtain for $m \ge 1$

$$\varepsilon^2 \frac{(\widetilde{N^{m+1}})_l - 2(\widetilde{N^m})_l + (\widetilde{N^{m-1}})_l}{k^2} + \mu_l^2 \Big[\beta(\widetilde{N^{m+1}})_l + (1-2\beta)(\widetilde{N^m})_l$$

$$+ \beta(\widetilde{N^{m-1}})_l - \nu(\widetilde{|E^m|^2})_l \Big] = 0, \ l = -\frac{M}{2}, \cdots, \frac{M}{2} - 1. \qquad (16.21)$$

Solving the above equation, we get

$$(\widetilde{N^{m+1}})_l = \left(2 - \frac{k^2\mu_l^2}{\varepsilon^2 + \beta k^2\mu_l^2} \right) (\widetilde{N^m})_l - (\widetilde{N^{m-1}})_l + \frac{\nu k^2\mu_l^2}{\varepsilon^2 + \beta k^2\mu_l^2} (\widetilde{|E^m|^2})_l,$$

$$l = -M/2, \cdots, M/2 - 1; \qquad m = 1, 2, \cdots. \quad (16.22)$$

From time $t = t_m$ to $t = t_{m+1}$, we combine the splitting steps via the standard Strang splitting for $m \ge 0$:

$$N_j^{m+1} = \sum_{l=-M/2}^{M/2-1} (\widetilde{N^{m+1}})_l \, e^{i\mu_l(x_j-a)}, \qquad\qquad (16.23)$$

$$E_j^* = \sum_{l=-M/2}^{M/2-1} e^{-ik\mu_l^2/2} (\widetilde{E^m})_l \, e^{i\mu_l(x_j-a)},$$

$$E_j^{**} = \begin{cases} e^{-ik[\alpha(N_j^m+N_j^{m+1})/2 - \lambda|E_j^*|^2]} E_j^*, & \gamma = 0, \\ e^{-\gamma k - i[k\alpha(N_j^m+N_j^{m+1})/2 + \lambda|E_j^*|^2(e^{-2\gamma k}-1)/2\gamma]} E_j^*, & \gamma \ne 0, \end{cases}$$

$$E_j^{m+1} = \sum_{l=-M/2}^{M/2-1} e^{-ik\mu_l^2/2} (\widetilde{E^{**}})_l \, e^{i\mu_l(x_j-a)}, \ 0 \le j \le M - 1; \qquad (16.24)$$

where $(\widetilde{N^{m+1}})_l$ is given in (16.22) for $m > 0$ and (16.27) for $m = 0$. The initial conditions (16.3) are discretized as

$$E_j^0 = E^{(0)}(x_j), \ N_j^0 = N^{(0)}(x_j), \ \frac{N_j^1 - N_j^{-1}}{2k} = N_j^{(1)}, \quad j = 0, 1, 2, \cdots, M-1, \tag{16.25}$$

where

$$N_j^{(1)} = \begin{cases} N^{(1)}(x_j), & 0 \le j \le M - 2, \\ -\sum_{l=0}^{M-2} N^{(1)}(x_l), & j = M - 1. \end{cases} \tag{16.26}$$

This implies that

$$(\widetilde{N^1})_l = \left(1 - \frac{k^2 \mu_l^2}{2(\varepsilon^2 + \beta k^2 \mu_l^2)} \right) (\widetilde{N^{(0)}})_l + k \, (\widetilde{N^{(1)}})_l$$

$$+ \frac{\nu k^2 \mu_l^2}{2(\varepsilon^2 + \beta k^2 \mu_l^2)} (\widetilde{|E^{(0)}|^2})_l, \ l = -\frac{M}{2}, \cdots, \frac{M}{2} - 1. \tag{16.27}$$

This type of discretization for the initial condition (16.3) is equivalent to use the trapezoidal rule for the periodic function $N^{(1)}$ and such that (16.8) is satisfied in discretized level. The discretization error converges to 0 exponentially fast as the mesh size h goes to 0.

Note that the spatial discretization error of the method is of spectral-order accuracy in h and time discretization error is demonstrated to be second-order accuracy in k from our numerical results.

16.2. *Phase space analytical solver + time-splitting spectral discretizations (PSAS-TSSP)*

Another way to discretize the second wave-type equation (16.2) in GZS is by pseudo-spectral method for spatial derivatives, and then solving the ODEs in phase space analytically under appropriate chosen transmission conditions between different time intervals. From time $t = t_m$ to $t = t_{m+1}$, assume

$$N(x,t) = \sum_{l=-M/2}^{M/2-1} \tilde{N}_l^m(t) \, e^{i\mu_l(x-a)}, \quad a \le x \le b, \ t_m \le t \le t_{m+1}. \tag{16.28}$$

Plugging (16.28) into (16.2) and noticing the orthogonality of the Fourier series, we get the following ODEs for $m \geq 0$:

$$\varepsilon^2 \, \frac{d^2 \widetilde{N}_l^m(t)}{d\,t^2} + \mu_l^2 \left[\widetilde{N}_l^m(t) - \nu \, \widetilde{(|E(t_m)|^2)}_l \right] = 0, \quad t_m \leq t \leq t_{m+1}, \quad (16.29)$$

$$\widetilde{N}_l^m(t_m) = \begin{cases} \widetilde{(N^{(0)})}_l, & m = 0, \\ \widetilde{N}_l^{m-1}(t_m), & m > 0, \end{cases} \quad l = -M/2, \cdots, M/2 - 1. \quad (16.30)$$

For each fixed l $(-M/2 \leq l \leq M/2 - 1)$, Eq. (16.29) is a second-order ODE. It needs two initial conditions such that the solution is unique. When $m = 0$ in (16.29), (16.30), we have the initial condition (16.30) and we can pose the other initial condition for (16.29) due to the initial condition (16.3) for the GZS (16.1)-(16.5):

$$\frac{d}{dt} \widetilde{N}_l^0(t_0) = \frac{d}{dt} \widetilde{N}_l^0(0) = \widetilde{(N^{(1)})}_l, \quad l = -M/2, \cdots, M/2 - 1. \quad (16.31)$$

Then the solution of (16.29), (16.30) with $m = 0$ and (16.31) is:

$$\widetilde{N}_l^0(t) = \begin{cases} \widetilde{(N^{(0)})}_0 + t \, \widetilde{(N^{(1)})}_0, & l = 0, \\[2mm] \left[\widetilde{(N^{(0)})}_l - \nu \widetilde{(|E^{(0)}|^2)}_l \right] \cos(\mu_l \, t/\varepsilon) + \nu \, \widetilde{(|E^{(0)}|^2)}_l & \\[2mm] \quad + \frac{\varepsilon}{\mu_l} \widetilde{(N^{(1)})}_l \sin(\mu_l \, t/\varepsilon), & l \neq 0, \end{cases} \quad (16.32)$$

$$0 \leq t \leq t_1, \quad l = -M/2, \cdots, M/2 - 1.$$

But when $m > 0$, we only have one initial condition (16.30). One **can't** simply pose the continuity between $\frac{d}{dt} \widetilde{N}_l^m(t)$ and $\frac{d}{dt} \widetilde{N}_l^{m-1}(t)$ across the time $t = t_m$ due to the last term in (16.29) is usually different in two adjacent time intervals $[t_{m-1}, t_m]$ and $[t_m, t_{m+1}]$, i.e. $\widetilde{(|E(t_{m-1})|^2)}_l \neq \widetilde{(|E(t_m)|^2)}_l$. Since our goal is to develop explicit scheme and we need linearize the nonlinear term in (16.2) in our discretization (16.29), in general,

$$\frac{d}{dt} \widetilde{N}_l^{m-1}(t_m^-) = \lim_{t \to t_m^-} \frac{d}{dt} \widetilde{N}_l^{m-1}(t) \neq \lim_{t \to t_m^+} \frac{d}{dt} \widetilde{N}_l^m(t) = \frac{d}{dt} \widetilde{N}_l^m(t_m^+), \quad (16.33)$$

$$m = 1, \cdots, \quad l = -M/2, \cdots, M/2 - 1.$$

Unfortunely, we don't know the jump $\frac{d}{dt} \widetilde{N}_l^m(t_m^+) - \frac{d}{dt} \widetilde{N}_l^{m-1}(t_m^-)$ across the time $t = t_m$. In order to get a unique solution of (16.29), (16.30) for $m > 0$, here we pose an additional condition:

$$\widetilde{N}_l^m(t_{m-1}) = \widetilde{N}_l^{m-1}(t_{m-1}), \quad l = -M/2, \cdots, M/2 - 1. \quad (16.34)$$

The condition (16.34) is equivalent to pose the solution $\widetilde{N}_l^m(t)$ on the time interval $[t_m, t_{m+1}]$ of (16.29), (16.30) is also continuity at the time $t = t_{m-1}$. After a simple computation, we get the solution of (16.29), (16.30) and (16.34) for $m > 0$:

$$\widetilde{N}_l^m(t) = \begin{cases} \widetilde{N}_0^{m-1}(t_m) + \frac{t-t_m}{k}\left[\widetilde{N}_0^{m-1}(t_m) - \widetilde{N}_0^{m-1}(t_{m-1})\right], & l = 0, \\[2ex] \left[\widetilde{N}_l^{m-1}(t_m) - \nu\,(\widetilde{|E^m|^2})_l\right]\cos(\mu_l(t-t_m)/\varepsilon) & \\ \quad + \nu\,(\widetilde{|E^m|^2})_l + \frac{\sin(\mu_l(t-t_m)/\varepsilon)}{\sin(k\mu_l/\varepsilon)}\Big[\widetilde{N}_l^{m-1}(t_m)\cos(k\mu_l/\varepsilon) & \\ \quad - \widetilde{N}_l^{m-1}(t_{m-1}) + \nu\left[1 - \cos(k\mu_l/\varepsilon)\right](\widetilde{|E^m|^2})_l\Big], & l \neq 0, \end{cases}$$

$$t_m \leq t \leq t_{m+1}, \quad l = -M/2, \cdots, M/2 - 1.$$

(16.35)

From time $t = t_m$ to $t = t_{m+1}$, we combine the splitting steps via the standard Strang splitting for $m \geq 0$:

$$N_j^{m+1} = \sum_{l=-M/2}^{M/2-1} \widetilde{N}_l^m(t_{m+1})\, e^{i\mu_l(x_j-a)}, \tag{16.36}$$

$$E_j^* = \sum_{l=-M/2}^{M/2-1} e^{-ik\mu_l^2/2}(\widetilde{E^m})_l\, e^{i\mu_l(x_j-a)},$$

$$E_j^{**} = \begin{cases} e^{-ik[\alpha(N_j^m + N_j^{m+1})/2 - \lambda|E_j^*|^2]}\, E_j^*, & \gamma = 0, \\[1.5ex] e^{-\gamma k - i[k\alpha(N_j^m + N_j^{m+1})/2 + \lambda|E_j^*|^2(e^{-2\gamma k}-1)/2\gamma]}\, E_j^*, & \gamma \neq 0, \end{cases}$$

$$E_j^{m+1} = \sum_{l=-M/2}^{M/2-1} e^{-ik\mu_l^2/2}(\widetilde{E^{**}})_l\, e^{i\mu_l(x_j-a)}, \quad 0 \leq j \leq M - 1; \tag{16.37}$$

where

$$\widetilde{N}_l^m(t_{m+1}) = \begin{cases} (\widetilde{N^{(0)}})_0 + k\,(\widetilde{N^{(1)}})_0, & l = 0,\ m = 0, \\[2ex] (\widetilde{N^{(0)}})_l\cos(k\mu_l/\varepsilon) + \frac{\varepsilon}{\mu_l}(\widetilde{N^{(1)}})_l\sin(k\mu_l/\varepsilon)\ l \neq 0,\ m = 0, \\ \quad + \nu\left[1 - \cos(k\mu_l/\varepsilon)\right](\widetilde{|E^{(0)}|^2})_l, \\[2ex] 2\widetilde{N}_l^{m-1}(t_m)\cos(k\mu_l/\varepsilon) - \widetilde{N}_l^{m-1}(t_{m-1}) & m \geq 1, \\ \quad + 2\nu\left[1 - \cos(k\mu_l/\varepsilon)\right](\widetilde{|E^m|^2})_l, \end{cases} \tag{16.38}$$

The initial conditions (16.3) are discretized as

$$E_j^0 = E^{(0)}(x_j), \quad N_j^0 = N^{(0)}(x_j), \quad (\partial_t N)_j^0 = N_j^{(1)}, \quad j = 0, 1, 2, \cdots, M - 1. \tag{16.39}$$

Note that the spatial discretization error of the above method is again of spectral-order accuracy in h and time discretization error is demonstrated to be second-order accuracy in k from our numerical results.

16.3. *Properties of the numerical methods*

(1). Plane wave solution: If the initial data in (16.3) is chosen as

$$E^{(0)}(x) = c\, e^{i2\pi lx/(b-a)}, \quad N^{(0)}(x) = d, \quad N^{(1)}(x) = 0, \quad a \le x \le b,$$
$$\tag{16.40}$$

where l is an integer and c, d are constants, then the GZS (16.1)-(16.5) admits the plane wave solution [97]

$$N(x,t) = d, \qquad a < x < b, \quad t \ge 0, \tag{16.41}$$

$$E(x,t) = \begin{cases} c\, e^{i\left(\frac{2\pi lx}{b-a} - \omega_1 t\right)}, & \gamma = 0, \\[2mm] c\, e^{-\gamma t} e^{i\left(\frac{2\pi lx}{b-a} - \omega_2 t - \frac{\lambda c^2}{2\gamma}(e^{-2\gamma t}-1)\right)}, & \gamma \ne 0. \end{cases} \tag{16.42}$$

where

$$\omega_1 = \alpha d + \frac{4\pi^2 l^2}{(b-a)^2} - \lambda c^2, \qquad \omega_2 = \alpha d + \frac{4\pi^2 l^2}{(b-a)^2}.$$

It is easy to see that in this case our numerical methods CN-LF-TSSP (16.23), (16.24) and PAAS-TSSP (16.36), (16.37) give exact results provided that $M \ge 2(|l| + 1)$.

(2). Time transverse invariant: A main advantage of CN-LF-TSSP and PAAS-TSSP is that if a constant r is added to the initial data $N^0(x)$ in (16.3) when $\gamma = 0$ in (16.1), then the discrete functions N_j^{m+1} obtained from (16.23) or (16.36) get added by r and E_j^{m+1} obtained from (16.24) or (16.37) get multiplied by the phase factor $e^{-ir(m+1)k}$, which leaves the discrete function $|E_j^{m+1}|^2$ unchanged. This property also holds for the exact solution of GZS, but does not hold for the finite difference schemes proposed in [62, 28] and the spectral method proposed in [100].

(3). Conservation: Let $U = (U_0, U_1, \cdots, U_M)^T$ with $U_0 = U_M$, $f(x)$ a periodic function on the interval $[a, b]$, and let $\| \cdot \|_{l^2}$ be the usual discrete l^2-norm on the interval (a, b), i.e.,

$$\|U\|_{l^2} = \sqrt{\frac{b-a}{M} \sum_{j=0}^{M-1} |U_j|^2}, \qquad \|f\|_{l^2} = \sqrt{\frac{b-a}{M} \sum_{j=0}^{M-1} |f(x_j)|^2}. \tag{16.43}$$

Then we have

Theorem 16.1: *The CN-LF-TSSP (16.23), (16.24) and PSAS-TSSP (16.36), (16.37) for GZS possesses the following properties (in fact, they are the discretized version of (16.7) and (16.8)):*

$$\|E^m\|_{l^2}^2 = e^{-2\gamma t_m}\|E^0\|_{l^2}^2 = e^{-2\gamma t_m}\|E^{(0)}\|_{l^2}^2, \quad m = 0,1,2,\cdots, \tag{16.44}$$

$$\frac{b-a}{M}\sum_{j=0}^{M-1}\frac{N_j^{m+1}-N_j^m}{k} = 0, \qquad m = 0,1,2,\cdots \tag{16.45}$$

and

$$\frac{b-a}{M}\sum_{j=0}^{M-1} N_j^m = \frac{b-a}{M}\sum_{j=0}^{M-1} N_j^0 = \frac{b-a}{M}\sum_{j=0}^{M-1} N^{(0)}(x_j), \quad m \geq 0. \tag{16.46}$$

Proof: From (16.43), (16.37) and (16.19), using the orthogonality of the discrete Fourier series and noticing the Pasavel equality, we have

$$\frac{M}{b-a}\|E^{m+1}\|_{l^2}^2 = \sum_{j=0}^{M-1}|E_j^{m+1}|^2 = M\sum_{l=-M/2}^{M/2-1}\left|e^{-ik\mu_l^2/2}(\widetilde{E^{**}})_l\right|^2$$

$$= M\sum_{l=-M/2}^{M/2-1}|(\widetilde{E^{**}})_l|^2 = \sum_{j=0}^{M-1}|E_j^{**}|^2$$

$$= e^{-2\gamma k}\sum_{j=0}^{M-1}|E_j^*|^2 = e^{-2\gamma k}\sum_{j=0}^{M-1}|E_j^m|^2$$

$$= e^{-2\gamma k}\frac{M}{b-a}\|E^m\|_{l^2}^2, \quad m \geq 0. \tag{16.47}$$

Thus (16.44) is obtained from (16.47) by induction. The equalities (16.45) and (16.46) can be obtained in a similar way.

(4). Unconditional stability: By the standard Von Neumann analysis for (16.23) and (16.36), noting (16.44), we get PSAS-TSSP and CN-LF-TSSP with $1/4 \leq \beta \leq 1$ are unconditionally stable, and CN-LF-TSSP with $0 \leq \beta < 1/4$ is conditionally stable with stability constraint $k \leq \frac{2h\varepsilon}{\pi\sqrt{d(1-4\beta)}}$ in d-dimensions ($d = 1,2,3$). In fact, for PSAS-TSSP (16.36), (16.37), setting $(\widetilde{|E^m|^2})_l = 0$ and plugging $\widetilde{N}_l^m(t_{m+1}) = \mu\widetilde{N}_l^{m-1}(t_m) = \mu^2\widetilde{N}_l^{m-1}(t_{m-1})$ into (16.38) with $|\mu|$ the amplification factor, we obtain the characteristic equation:

$$\mu^2 - 2\cos(k\mu_l/\varepsilon)\mu + 1 = 0. \tag{16.48}$$

This implies

$$\mu = \cos(k\mu_l/\varepsilon) \pm i\,\sin(k\mu_l/\varepsilon). \qquad (16.49)$$

Thus the amplification factor

$$G_l = |\mu| = \sqrt{\cos^2(k\mu_l/\varepsilon) + \sin^2(k\mu_l/\varepsilon)} = 1, \qquad l = -M/2, \cdots M/2 - 1.$$

This together with (16.44) imply that PSAS-TSSP is unconditionally stable. Similarly for CN-LF-TSSP (16.23), (16.24), noting (16.22), we have the characteristic equation:

$$\mu^2 - \left(2 - \frac{k^2\mu_l^2}{\varepsilon^2 + \beta k^2\mu_l^2}\right)\mu + 1 = 0. \qquad (16.50)$$

This implies

$$\mu = 1 - \frac{k^2\mu_l^2}{2(\varepsilon^2 + \beta k^2\mu_l^2)} \pm \sqrt{\left(1 - \frac{k^2\mu_l^2}{2(\varepsilon^2 + \beta k^2\mu_l^2)}\right)^2 - 1}. \qquad (16.51)$$

When $1/4 \le \beta \le 1$, we have

$$\left|1 - \frac{k^2\mu_l^2}{2(\varepsilon^2 + \beta k^2\mu_l^2)}\right| \le 1, \qquad k > 0, \quad l = -M/2, \cdots M/2 - 1.$$

Thus

$$\mu = 1 - \frac{k^2\mu_l^2}{2(\varepsilon^2 + \beta k^2\mu_l^2)} \pm i\sqrt{1 - \left(1 - \frac{k^2\mu_l^2}{2(\varepsilon^2 + \beta k^2\mu_l^2)}\right)^2}. \qquad (16.52)$$

This implies the amplification factor

$$G_l = |\mu| = \sqrt{\left(1 - \frac{k^2\mu_l^2}{2(\varepsilon^2 + \beta k^2\mu_l^2)}\right)^2 + 1 - \left(1 - \frac{k^2\mu_l^2}{2(\varepsilon^2 + \beta k^2\mu_l^2)}\right)^2}$$
$$= 1, \qquad l = -M/2, \cdots M/2 - 1.$$

This together with (16.44) imply that CN-LF-TSSP with $1/4 \le \beta \le 1/2$ is unconditionally stable. On the other hand, when $0 \le \beta < 1/4$, we need the stability condition

$$\left|1 - \frac{k^2\mu_l^2}{2(\varepsilon^2 + \beta k^2\mu_l^2)}\right| \le 1 \Longrightarrow k \le \min_{-M/2 \le l \le M/2-1} \frac{2\varepsilon}{\sqrt{(1 - 4\beta)\mu_l^2}} = \frac{2h\varepsilon}{\pi\sqrt{1 - 4\beta}}.$$

This together with (16.44) imply that CN-LF-TSSP with $0 \le \beta < 1/4$ is conditionally stable in one dimension with stability condition

$$k \le \frac{2h\varepsilon}{\pi\sqrt{1 - 4\beta}}. \qquad (16.53)$$

All above stability results are confirmed by our numerical experiments.
(5). ε-resolution in the 'subsonic limit' regime ($0 < \varepsilon \ll 1$): As our
numerical results suggest: The meshing strategy (or ε-resolution) which
guarantees good numerical approximations of our new numerical methods
PSAS-TSSP and CN-LF-TSSP with $1/4 \le \beta \le 1/2$ in the 'subsonic limit'
regime, i.e. $0 < \varepsilon \ll 1$, is: i). for initial data with $O(\varepsilon)$-wavelength: $h = O(\varepsilon)$
and $k = O(\varepsilon)$; ii). for initial data with $O(1)$-wavelength: $h = O(1)$ and
$k = O(1)$. Where the meshing strategy for CN-LF-TSSP with $0 \le \beta < 1/4$
is: $h = O(\varepsilon)$ & $k = O(h\varepsilon) = O(\varepsilon^2)$; $h = O(1)$ & $k = O(\varepsilon)$, respectively.

Remark 16.2: If the periodic boundary conditions (16.4) and (16.5) are
replaced by the homogeneous Dirichlet boundary condition (16.9), then
the Fourier basis used in the above algorithm can be replaced by the sine
basis [15] or the algorithm in section 4 for VZSM. Similarly, if homogeneous
Neumann conditions are used, then cosine series can be applied in designing
the algorithm.

16.4. *Extension TSSP to GVZS*

The idea to construct the numerical methods CN-LF-TSSP and PSAS-
TSSP for GZS (16.1)-(16.5) can be easily extended to the VZSM [114] in
three dimensions for $\mathcal{M}$ different acoustic modes in a box $\Omega = [a_1, b_1] \times
[a_2, b_2] \times [a_3, b_3]$ with homogeneous Dirichlet boundary conditions:

$$i\partial_t \mathbf{E} + a \,\triangle \mathbf{E} + (1-a)\,\nabla(\nabla \cdot \mathbf{E}) - \alpha \mathbf{E} \sum_{J=1}^{\mathcal{M}} N_J + \lambda|\mathbf{E}|^2\mathbf{E} + i\gamma\mathbf{E} = 0, \quad (16.54)$$

$$\varepsilon_J^2 \partial_{tt} N_J - \triangle N_J + \nu_J \triangle |\mathbf{E}|^2 = 0, \quad J = 1, \cdots, \mathcal{M}, \qquad \mathbf{x} \in \Omega,\ t > 0, \quad (16.55)$$

$$\mathbf{E}(\mathbf{x},0) = \mathbf{E}^{(0)}(\mathbf{x}),\ N_J(\mathbf{x},0) = N_J^{(0)}(\mathbf{x}),\ \partial_t N_J(\mathbf{x},0) = N_J^{(1)}(\mathbf{x}),\ \mathbf{x} \in \Omega, \quad (16.56)$$

$$\mathbf{E}(\mathbf{x},t) = \mathbf{0},\ N_J(\mathbf{x},t) = 0 \qquad (J = 1, \cdots, \mathcal{M}), \qquad \mathbf{x} \in \partial\Omega; \quad (16.57)$$

where $\mathbf{x} = (x,y,z)^T$ and $\mathbf{E}(\mathbf{x},t) = (E_1(\mathbf{x},t), E_2(\mathbf{x},t), E_3(\mathbf{x},t))^T$. Moreover,
we supplement (16.54)-(16.57) by imposing the compatibility condition

$$\mathbf{E}^{(0)}(\mathbf{x}) = \mathbf{0}, \quad N_J^{(0)}(\mathbf{x}) = N_J^{(1)}(\mathbf{x}) = 0, \quad \mathbf{x} \in \partial\Omega,\ J = 1, \cdots, \mathcal{M}. \quad (16.58)$$

In some cases, the homogeneous Dirichlet boundary condition (16.57)
may be replaced by periodic boundary conditions:

With periodic boundary conditions for $\mathbf{E}$, $N_J(J = 1, \cdots, \mathcal{M})$ on $\partial\Omega$.
$$(16.59)$$

We choose the spatial mesh sizes $h_1 = \frac{b_1 - a_1}{M_1}$, $h_2 = \frac{b_2 - a_2}{M_2}$ and $h_3 = \frac{b_3 - a_3}{M_3}$
in x-, y- and z-direction respectively, with M_1, M_2 and M_3 given positive

integers; the time step $k = \triangle t > 0$. Denote grid points and time steps as

$$x_j := a_1 + jh_1, \; j = 0, 1, \cdots, M_1; \quad y_p := a_2 + ph_2, \; p = 0, 1, \cdots, M_2;$$

$$z_s := a_3 + sh_3, \; s = 0, 1, \cdots, M_3; \quad t_m := mk, \; m = 0, 1, 2, \cdots.$$

Let $\mathbf{E}_{j,p,s}^m$ and $(N_J)_{j,p,s}^m$ be the approximations of $\mathbf{E}(x_j, y_p, z_s, t_m)$ and $N_J(x_j, y_p, z_s, t_m)$, respectively.

For simplicity, here we only extend PSAS-TSSP from GZS (16.1)-(16.5) to VZSM (16.54)-(16.57) with homogeneous Dirichlet conditions. For periodic boundary conditions (16.59) or extension of CN-LF-TSSP can be done in a similar way. Following the idea of constructing PSAS-TSSP for GZS and the TSSP for VZSM in [114] here we only present the numerical algorithm. From time $t = t_m$ to $t = t_{m+1}$, the PSAS-TSSP method for VZSM (16.54)-(16.57) reads:

$$(N_J)_{j,p,s}^{m+1} = \sum_{(l,g,r) \in \mathcal{N}} \widetilde{(N_J)}_{l,g,r}^m (t_{m+1}) \sin\left(\frac{lj\pi}{M_1}\right) \sin\left(\frac{pg\pi}{M_2}\right) \sin\left(\frac{sr\pi}{M_3}\right),$$

$$\mathbf{E}_{j,p,s}^* = \sum_{(l,g,r) \in \mathcal{N}} B_{l,g,r}(k/2) \, \widetilde{(\mathbf{E}^m)}_{l,g,r} \sin\left(\frac{lj\pi}{M_1}\right) \sin\left(\frac{pg\pi}{M_2}\right) \sin\left(\frac{sr\pi}{M_3}\right),$$

$$\mathbf{E}_{j,p,s}^{**} = \begin{cases} e^{-ik[\alpha \sum_{J=1}^{M} ((N_J)_{j,p,s}^m + (N_J)_{j,p,s}^{m+1})/2 - \lambda |\mathbf{E}_{j,p,s}^*|^2]} \, \mathbf{E}_{j,p,s}^*, & \gamma = 0, \\[2mm] e^{-\gamma k - i[k\alpha \sum_{J=1}^{M} ((N_J)_{j,p,s}^m + (N_J)_{j,p,s}^{m+1})/2 + \lambda |\mathbf{E}_{j,p,s}^*|^2 (e^{-2\gamma k} - 1)/2\gamma]} \, \mathbf{E}_{j,p,s}^*, & \gamma \neq 0, \end{cases}$$

$$\mathbf{E}_{j,p,s}^{m+1} = \sum_{(l,g,r) \in \mathcal{N}} B_{l,g,r}(k/2) \, \widetilde{(\mathbf{E}^{**})}_{l,g,r} \sin\left(\frac{lj\pi}{M_1}\right) \sin\left(\frac{pg\pi}{M_2}\right) \sin\left(\frac{sr\pi}{M_3}\right),$$

where

$$\mathcal{N} = \{(l,g,r) \mid 1 \leq l \leq M_1 - 1, \; 1 \leq g \leq M_2 - 1, \; 1 \leq r \leq M_3 - 1\}.$$

$$\widetilde{(N_J)}_{l,g,r}^m (t_{m+1}) = \begin{cases} \widetilde{(N_J^{(0)})}_{0,0,0} + k \, \widetilde{(N_J^{(1)})}_{0,0,0}, & R_{l,g,r} = 0, m = 0, \\[3mm] \widetilde{(N_J^{(0)})}_{l,g,r} \cos(kR_{l,g,r}/\varepsilon_J) & R_{l,g,r} \neq 0, m = 0, \\ \quad + \frac{\varepsilon_J}{R_{l,g,r}} \widetilde{(N_J^{(1)})}_{l,g,r} \sin(kR_{l,g,r}/\varepsilon_J) & \\ \quad + \nu_J \left[1 - \cos(kR_{l,g,r}/\varepsilon_J)\right] \widetilde{(|\mathbf{E}^{(0)}|^2)}_{l,g,r}, & \\[3mm] 2\widetilde{(N_J)}_{l,g,r}^{m-1}(t_m) \cos(kR_{l,g,r}/\varepsilon_J) & m \geq 1, \\ \quad + 2\nu_J \left[1 - \cos(kR_{l,g,r}/\varepsilon_J)\right] \widetilde{(|\mathbf{E}^m|^2)}_{l,g,r} & \\ \quad - \widetilde{(N_J)}_{l,g,r}^{m-1}(t_{m-1}), & \end{cases}$$

and

$$B_{l,g,r}(\tau) = \begin{cases} I_3, & l = g = r = 0, \\[2em] e^{-ia\tau R_{l,g,r}^2}\left[I_3 + \dfrac{e^{-i(1-a)\tau R_{l,g,r}^2} - 1}{R_{l,g,r}^2} A_{l,g,r}\right], & \text{otherwise,} \end{cases}$$

with

$$R_{l,g,r}^2 = \kappa_l^2 + \zeta_g^2 + \eta_r^2, \quad A_{l,g,r} = \begin{pmatrix} \kappa_l^2 & \kappa_l\zeta_g & \kappa_l\eta_r \\ \kappa_l\zeta_g & \zeta_g^2 & \zeta_g\eta_r \\ \kappa_l\eta_r & \zeta_g\eta_r & \eta_r^2 \end{pmatrix} = \begin{pmatrix} \kappa_l \\ \zeta_g \\ \eta_r \end{pmatrix} \begin{pmatrix} \kappa_l & \zeta_g & \eta_r \end{pmatrix};$$

where I_3 is the 3×3 identity matrix, and $\widetilde{\mathbf{U}}_{l,g,r}$, the sine-transform coefficients, are defined as

$$\widetilde{\mathbf{U}}_{l,g,r} = \frac{8}{M_1 M_2 M_3} \sum_{j=1}^{M_1-1} \sum_{p=1}^{M_2-1} \sum_{s=1}^{M_3-1} \mathbf{U}_{j,p,s} \, \sin\left(\frac{lj\pi}{M_1}\right) \sin\left(\frac{pg\pi}{M_2}\right) \sin\left(\frac{sr\pi}{M_3}\right),$$

$$(16.60)$$

with

$$\kappa_l = \frac{\pi l}{b_1 - a_1}, \quad l = 1, \ldots, M_1 - 1, \qquad \zeta_g = \frac{\pi g}{b_2 - a_2}, \quad g = 1, \ldots, M_2 - 1,$$

$$\eta_r = \frac{\pi r}{b_3 - a_3}, \qquad r = 1, \ldots, M_3 - 1.$$

The initial conditions (16.56) are discretized as

$$\mathbf{E}_{j,p,s}^0 = \mathbf{E}^{(0)}(x_j, y_p, z_s),$$

$$(N_J)_{j,p,s}^0 = N_J^{(0)}(x_j, y_p, z_s), \quad j = 0, \cdots, M_1, \ p = 0, \cdots, M_2, \ s = 0, \cdots, M_3,$$

$$(\partial_t N_J)_{j,p,s}^0 = N_J^{(1)}(x_j, y_p, z_s), \qquad J = 1, \cdots, \mathcal{M}.$$

The properties of the numerical method for GZS in section 3 are still valid here.

17. Crank-Nicolson finite difference (CNFD) method for GZS

Another method for the GZS (13.40)-(13.41) is to use centered finite difference for spatial derivatives and Crank-Nicolson for time derivative. For simplicity of notations , here we only present the CNFD method for the standard ZS [28] with homogeneous Dirichlet boundary condition (16.9), i.e., in (16.1)-(16.2) with $\varepsilon = 1$, $\nu = -1$, $\alpha = 1$, $\lambda = 0$ and $\gamma = 0$:

$$i\frac{E_j^{m+1} - E_j^m}{k} + \frac{1}{2}\left(\frac{E_{j+1}^{m+1} - 2E_j^{m+1} + E_{j-1}^{m+1}}{h^2} + \frac{E_{j+1}^m - 2E_j^m + E_{j-1}^m}{h^2}\right)$$

$$= \frac{1}{4}(N_j^m + N_j^{m+1})(E_j^{m+1} + E_j^m), \qquad j = 1, 2, \cdots, M - 1,$$

$$\frac{N_j^{m+1} - 2N_j^m + N_j^{m-1}}{k^2} - \theta\frac{N_{j+1}^{m+1} - 2N_j^{m+1} + N_{j-1}^{m+1}}{h^2}$$

$$- \theta\frac{N_{j+1}^{m-1} - 2N_j^{m-1} + N_{j-1}^{m-1}}{h^2} - (1 - 2\theta)\frac{N_{j+1}^m - 2N_j^m + N_{j-1}^m}{h^2}$$

$$= \frac{|E_{j+1}^m|^2 - 2|E_j^m|^2 + |E_{j-1}^m|^2}{h^2},$$

$$E_0^{m+1} = E_M^{m+1} = 0, \qquad N_0^{m+1} = N_0^{m+1} = 0, \qquad m = 0, 1 \cdots.$$

where $0 \leq \theta \leq 1$ is a parameter. The initial conditions are discretized as:

$$E_j^0 = E^0(x_j), \qquad N_j^0 = N^0(x_j), \qquad j = 0, 1, \cdots, M, \qquad (17.1)$$

$$N_j^1 = N_j^0 + kN^1(x_j) + \frac{k^2}{2}\left[\frac{N_{j+1}^0 - 2N_j^0 + N_{j-1}^0}{h^2}\right.$$

$$\left. + \frac{|E_{j+1}^0|^2 - 2|E_j^0|^2 + |E_{j-1}^0|^2}{h^2}\right]. \qquad (17.2)$$

When $\theta = 0$, the discretization (17.1) for wave-type equation is explicit; when $\theta > 0$, it is implicit but can be solved explicitly when periodic boundary conditions are applied. Generalization of the method to GZS are straightforward. In our computations in next subsection, we choose $\theta = 0.5$.

Remark 17.1: In [62, 63], convergence and error estimate of the CNFD discretization (17.1), (17.1) are proved.

18. Numerical results of GZS

In this section, we present numerical results of GZS with a solitary wave solution in one dimension to compare the accuracy, stability and ε-resolution of different methods. We also present numerical examples solitary-wave collisions in one dimension GZS.

In our computation, the initial conditions for (16.3) are always chosen such that $|E^0|$, N^0 and $N^{(1)}$ decay to zero sufficiently fast as $|\mathbf{x}| \to \infty$. We always compute on a domain, which is large enough such that the periodic boundary conditions do not introduce a significant aliasing error relative to the problem in the whole space.

Example 10 The standard ZS with a solitary-wave solution, i.e., we choose $d = 1$, $\alpha = 1$, $\lambda = 0$, $\gamma = 0$ and $\nu = -1$ in (13.40)-(13.41). The well-known solitary-wave solution (15.5)-(15.7) of the ZS in this case is given in [97, 73]

$$E(x,t) = \sqrt{2B^2(1 - \varepsilon^2 C^2)} \, \mathrm{sech}(B(x - Ct)) \, e^{i[(C/2)x - ((C/2)^2 - B^2)t]}, \quad (18.1)$$

$$N(x,t) = -2B^2 \, \mathrm{sech}^2(B(x - Ct)), \quad -\infty < x < \infty, \quad t \geq 0, \quad (18.2)$$

where B, C are constants. The initial condition is taken as

$$E^{(0)}(x) = E(x,0), \; N^{(0)}(x) = N(x,0), \; N^{(1)}(x,0) = \partial_t N(x,0), \quad (18.3)$$

where $E(x,0)$, $N(x,0)$ and $\partial_t N(x,0)$ are obtained from (18.1), (18.2) by setting $t = 0$.

We present computations for two different regimes of the acoustic speed, i.e. $1/\varepsilon$:

Case I. $O(1)$-acoustic speed, i.e. we choose $\varepsilon = 1$, $B = 1$, $C = 0.5$ in (18.1), (18.2). Here we test the spatial and temporal discretization errors, conservation of the conserved quantities as well as the stability constraint of different numerical methods. We solve the problem on the interval [-32, 32], i.e., $a = -32$ and $b = 32$ with periodic boundary conditions. Let $E_{h,k}$ and $N_{h,k}$ be the numerical solution of (16.1), (16.5) in one dimension with the initial condition (18.3) by using a numerical method with mesh size h and time step k. To quantify the numerical methods, we define the error functions as

$$e_1 = \|E(\cdot,t) - E_{h,k}(t)\|_{l^2}, \qquad e_2 = \|N(\cdot,t) - N_{h,k}(t)\|_{l^2},$$

$$e = \frac{\|E(\cdot,t) - E_{h,k}(t)\|_{l^2}}{\|E(\cdot,t)\|_{l^2}} + \frac{\|N(\cdot,t) - N_{h,k}(t)\|_{l^2}}{\|N(\cdot,t)\|_{l^2}}$$

$$= \frac{e_1}{\|E(\cdot,t)\|_{l^2}} + \frac{e_2}{\|N(\cdot,t)\|_{l^2}}$$

and evaluate the conserved quantities D^{GZS}, P^{GZS} and H^{GZS} by using the numerical solution, i.e. replacing E and N by their numerical counterparts $E_{h,k}$ and $N_{h,k}$ respectively, in (13.20)-(13.22).

First, we test the discretization error in space. In order to do this, we choose a very small time step, e.g., $k = 0.0001$ such that the error from time discretization is negligible comparing to the spatial discretization error, and solve the ZS with different methods under different mesh sizes h. Tab. 6

lists the numerical errors of e_1 and e_2 at $t = 2.0$ with different mesh sizes h for different numerical methods.

	Mesh	$h = 1.0$	$h = \frac{1}{2}$	$h = \frac{1}{4}$
PSAS-TSSP	e_1	9.810E-2	1.500E-4	8.958E-9
	e_2	0.143	1.168E-3	6.500E-8
CN-LF-TSSP	e_1	9.810E-2	1.500E-4	7.409E-9
($\beta = 0$)	e_2	0.143	1.168E-3	3.904E-8
CN-LF-TSSP	e_1	9.810E-2	1.500E-4	8.628E-9
($\beta = 1/4$)	e_2	0.143	1.168E-3	6.521E-8
CN-LF-TSSP	e_1	9.810E-2	1.500E-4	1.098E-8
($\beta = 1/2$)	e_2	0.143	1.168E-3	6.326E-8
CNFD	e_1	0.491	0.120	2.818E-2
	e_2	0.889	0.209	4.726E-2

Tab. 6: Spatial discretization error analysis: e_1, e_2 at time $t=2$ under $k = 0.0001$.

Secondly, we test the discretization error in time. Tab. 7 shows the numerical errors of e_1 and e_2 at $t = 2.0$ under different time steps k and mesh sizes h for different numerical methods.

Thirdly, we test the conservation of conserved quantities. Tab. 8 presents the quantities and numerical errors at different times with mesh size $h = \frac{1}{8}$ and time step $k = 0.0001$ for different numerical methods.

Case II: 'Subsonic limit' regime, i.e. $0 < \varepsilon \ll 1$, we choose $B = 1$ and $C = 1/2\varepsilon$ in (18.1), (18.2). Here we test the ε-resolution of different numerical methods. We solve the problem on the interval [-8, 120], i.e., $a = -8$ and $b = 120$ with periodic boundary conditions. Fig. 9 shows the numerical results of PSAS-TSSP at $t = 1$ when we choose the meshing strategy $h = O(\varepsilon)$ and $k = O(\varepsilon)$: $\mathcal{T}_0 = (\varepsilon_0, h_0, k_0) = (0.125, 0.5, 0.04)$, $\mathcal{T}_0/4$, $\mathcal{T}_0/16$; and $h = O(\varepsilon)$ and $k = 0.04$-independent of ε: $\mathcal{T}_0 = (\varepsilon_0, h_0) = (0.125, 0.5)$, $\mathcal{T}_0/4$, $\mathcal{T}_0/16$. CN-LF-TSSP with $\beta = 1/4$ or $\beta = 1/2$ gives similar numerical results at the same meshing strategies, where CN-LF-TSSP with $\beta = 0$ gives correct numerical results at meshing strategy $h = O(\varepsilon)$ and $k = O(\varepsilon^2)$ and incorrect results at $h = O(\varepsilon)$ and $k = O(\varepsilon)$ [15]. Furthermore, our additional numerical experiments confirm that PSAS-TSSP and CN-LF-TSSP with $1/4 \leq \beta \leq 1/2$ are unconditionally stable and CN-LF-TSSP with $\beta = 0$ is stable under the stability constraint (16.53).

From Tabs. 6-8 and Fig. 9, we can draw the following observations:

                  Weizhu Bao

	h	Error	$k = \frac{1}{100}$	$k = \frac{1}{400}$	$k = \frac{1}{1600}$	$k = \frac{1}{6400}$
PSAS-TSSP	$\frac{1}{4}$	e_1	4.968E-5	3.109E-6	1.944E-7	1.226E-8
		e_2	1.225E-4	7.664E-6	4.797E-7	3.871E-8
	$\frac{1}{8}$	e_1	4.968E-5	3.109E-6	1.944E-7	1.172E-8
		e_2	1.225E-4	7.664E-6	4.797E-7	3.157E-8
CN-LF-TSSP	$\frac{1}{4}$	e_1	4.829E-5	3.022E-6	1.888E-7	1.156E-8
($\beta = 0$)		e_2	1.032E-4	6.456E-6	4.041E-7	3.673E-8
	$\frac{1}{8}$	e_1	4.829E-5	3.022E-6	1.888E-7	1.100E-8
		e_2	1.032E-4	6.456E-6	4.043E-7	2.946E-8
CN-LF-TSSP	$\frac{1}{4}$	e_1	5.679E-5	3.556E-6	2.224E-7	1.425E-8
($\beta = 1/4$)		e_2	1.623E-4	1.015E-5	6.351E-7	4.970E-8
	$\frac{1}{8}$	e_1	5.679E-5	3.556E-6	2.224E-7	1.377E-8
		e_2	1.623E-4	1.015E-5	6.351E-7	4.356E-8
CN-LF-TSSP	$\frac{1}{4}$	e_1	7.468E-5	4.678E-6	2.924E-7	1.868E-8
($\beta = 1/2$)		e_2	2.232E-4	1.396E-5	8.732E-7	6.360E-8
	$\frac{1}{8}$	e_1	7.468E-5	4.678E-6	2.924E-7	1.841E-8
		e_2	2.232E-4	1.396E-5	8.732E-7	5.942E-8
CNFD	$\frac{1}{4}$	e_1	0.802	3.480E-2	2.855E-2	2.820E-2
		e_2	0.674	9.012-2	5.005E-2	4.743E-2
	$\frac{1}{8}$	e_1	0.809	1.753E-2	7.363E-3	6.961E-3
		e_2	0.656	5.491E-2	1.427E-2	1.167E-2

Tab. 7: Time discretization error analysis: e_1, e_2 at time $t=2$.

In $O(1)$-acoustic speed regime, our new methods PSAS-TSSP and CN-LF-TSSP with $\beta = 1/2$ or $1/4$ give similar results as the old method, i.e. CN-LF-TSSP with $\beta = 0$, proposed in [15]: they are of spectral order accuracy in space discretization and second-order accuracy in time, conserve D^{GZS} exactly and P^{GZS}, H^{GZS} very well (up to 8 digits). However, they are improved in two aspects: (i) They are unconditionally stable where the old method is conditionally stable under the stability condition $k \le \frac{2h\varepsilon}{\pi\sqrt{d(1-4\beta)}}$ in d-dimensions ($d = 1, 2$ or 3); (ii) In the 'subsonic limit' regime for initial data with $O(\varepsilon)$-wavelength, i.e. $0 < \varepsilon \ll 1$, the ε-resolution of our new methods is improved to $h = O(\varepsilon)$ and $k = O(\varepsilon)$, where the old

	Time	D^{GZS}	P^{GZS}	H^{GZS}
PSAS-TSSP	1.0	3.0000000000	3.41181556	0.510202736
	2.0	3.0000000000	3.41181562	0.510202765
$\beta = 0$	1.0	3.0000000000	3.41181557	0.510202736
	2.0	3.0000000000	3.41181562	0.510202766
$\beta = 1/4$	1.0	3.0000000000	3.41181556	0.510202740
	2.0	3.0000000000	3.41181564	0.510202779
$\beta = 1/2$	1.0	3.0000000000	3.41181556	0.510202737
	2.0	3.0000000000	3.41181562	0.510202768
CNFD	1.0	3.0000000000	3.394829741	0.510115589
	2.0	3.0000000000	3.394791238	0.510076710

Tab. 8: Conserved quantities analysis: $k = 0.0001$ and $h = \frac{1}{8}$.

method required $h = O(\varepsilon)$ and $k = O(\varepsilon h) = O(\varepsilon^2)$. Thus in the following, we only present numerical results by PSAS-TSSP. In fact, CN-LF-TSSP with $1/4 \le \beta \le 1$ gives similar numerical results at the same mesh size and time step for all the following numerical examples.

Example 11 Soliton-soliton collisions in one dimension GZS, i.e., we choose $d = 1$, $\varepsilon = 1$, $\alpha = -2$ and $\gamma = 0$ in (13.40)-(13.41). We use the family of one-soliton solutions (15.5)-(15.7) in [72] to test our new numerical method PSAS-TSSP.

The initial data is chosen as

$$E(x,0) = E_s(x + p, 0, \eta_1, V_1, \varepsilon, \nu) + E_s(x - p, 0, \eta_2, V_2, \varepsilon, \nu),$$
$$N(x,0) = N_s(x + p, 0, \eta_1, V_1, \varepsilon, \nu) + N_s(x - p, 0, \eta_2, V_2, \varepsilon, \nu),$$
$$\partial_t N(x,0) = \partial_t N_s(x + p, 0, \eta_1, V_1, \varepsilon, \nu) + \partial_t N_s(x - p, 0, \eta_2, V_2, \varepsilon, \nu),$$

where $x = \mp p$ are initial locations of the two solitons.

In all the numerical simulations reported in this example, we set $\lambda = 2$, and $\Phi_0 = 0$. We only simulated the symmetric collisions, i.e., the collisions of solitons with equal amplitudes $\eta_1 = \eta_2 = \eta$ and opposite velocities $V_1 = -V_2 \equiv V$. Here, we present computations for two cases:

I. Collision between solitons moving with the subsonic velocities, $V < 1/\varepsilon = 1$, i.e. we take $\nu = 0.2$, $\eta = 0.3$ and $V = 0.5$;

II. Collision between solitons in the transonic regime, $V > 1/\varepsilon = 1$, i.e. we take $\nu = 2.0$, $\eta = 0.3$ and $V = 3.0$.

We solve the problem on the interval [-128,128], i.e., a = -128 and b = 128 with mesh size $h = \frac{1}{4}$ and time step $k = 0.005$. We take p = 10. Fig.

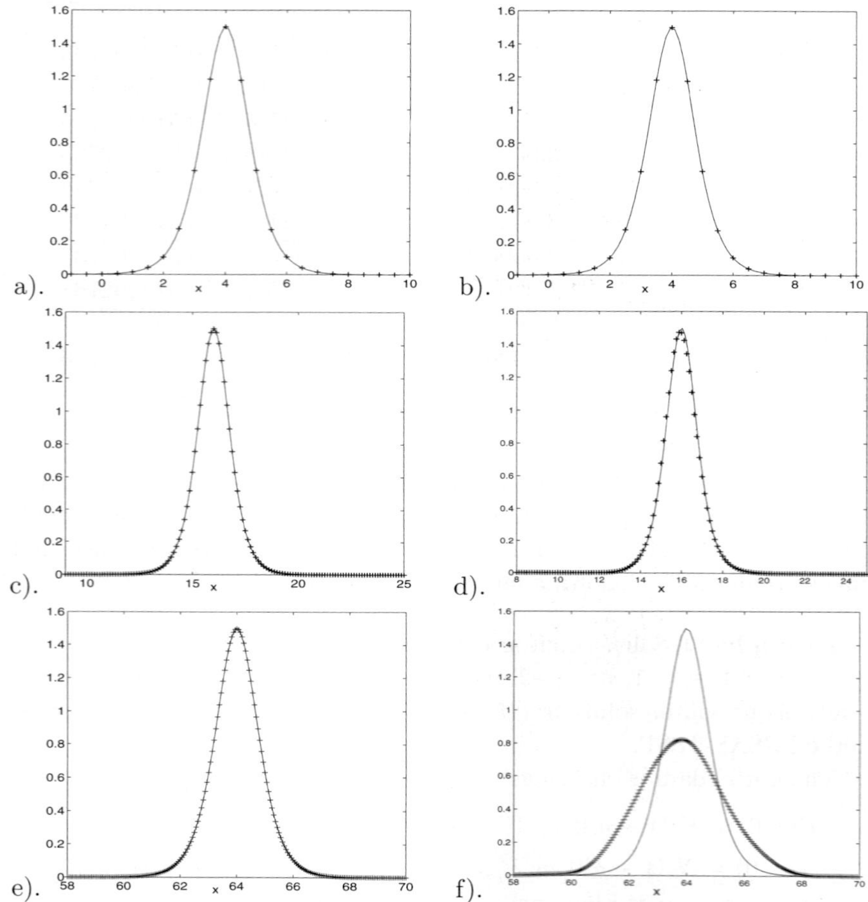

Fig. 9: Numerical solutions of the electric field $|E(x,t)|^2$ at $t = 1$ for Example 10 in the 'subsonic limit' regime by PSAS-TSSP. '—': exact solution, '+ + +': numerical solution. Left column corresponds to $h = O(\varepsilon)$ and $k = O(\varepsilon)$: a). $\mathcal{T}_0 = (\varepsilon_0, h_0, k_0) = (0.125, 0.5, 0.04)$; c). $\mathcal{T}_0/4$; e). $\mathcal{T}_0/16$. Right column corresponds to $h = O(\varepsilon)$ and $k = 0.04$-independent ε: b). $\mathcal{T}_0 = (\varepsilon_0, h_0) = (0.125, 0.5)$; d). $\mathcal{T}_0/4$; f). $\mathcal{T}_0/16$.

10 shows the evolution of the dispersive wave field $|E|^2$ and the acoustic (nondispersive) field N.

Case I corresponds to a soliton-soliton collision when the ratio ν/λ is small, i.e., the GZS (16.1), (16.2) is close to the NLSE. As is seen, the

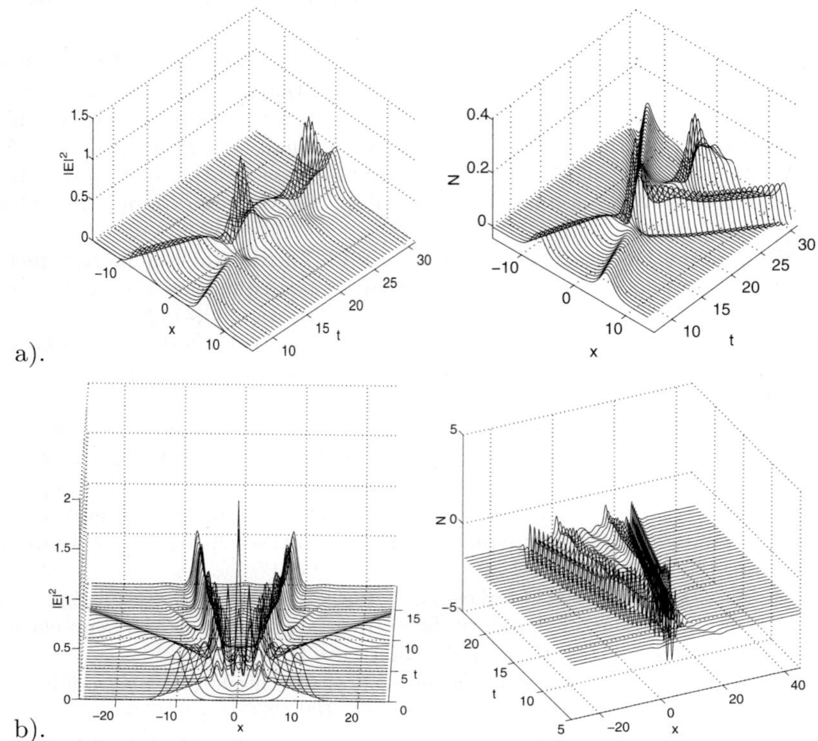

Fig. 10: Evolution of the wave field $|E|^2$ (left column) and acoustic field N (right column) in Example 11. a). For case I; b). For case II.

collision seems quite elastic (cf. Fig. 10a). This also validates the formal reduction from GZS to NLSE in section 2.5. Case II corresponds to the collision of two transonic solitons. Note that the emission of the sound waves is inconspicuous at this value of V (cf. Fig. 10b).

From Figs. 9&10, we can see that the unconditionally stable numerical method PSAS-TSSP can really be applied to solve solitary-wave collisions of GZS.

References

1. S. Abenda, Solitary waves for Maxwell-Dirac and Coulomb-Dirac models, Ann. Inst. H. Poincare Phys. Theor., 68 (1998), pp. 229–244.
2. A. Aftalion and Q. Du, Vortices in a rotating Bose-Einstein condensate: Critical angular velocities and energy diagrams in the Thomas-Fermi regime, Phys. Rev. A, 64 (2001), pp. 063603.

3. M. H. Anderson, J. R. Ensher, M. R. Matthews, C. E. Wieman and E A. Cornell, Science, 269 (1995), 198.

4. W. Bao, Ground states and dynamics of multi-component Bose-Einstein condensates, Multiscale Modeling and Simulation, 2 (2004), pp. 210-236.

5. W. Bao, Numerical methods for nonlinear Schrödinger equation under nonzero far-field conditions, Methods Appl. Anal., 11 (2004), pp. 367-388.

6. W. Bao and Q. Du, Computing the ground state solution of Bose-Einstein condensates by a normalized gradient flow, SIAM J. Sci. Comp., 25 (2004), pp. 1674-1697.

7. W. Bao and D. Jaksch, An explicit unconditionally stable numerical methods for solving damped nonlinear Schrodinger equations with a focusing nonlinearity, SIAM J. Numer. Anal., 41(2003), pp. 1406-1426.

8. W. Bao, D. Jaksch and P.A. Markowich, Numerical solution of the Gross-Pitaevskii equation for Bose-Einstein condensation, J. Comput. Phys., 187 (2003), pp. 318-342.

9. W. Bao, D. Jaksch and P.A. Markowich, Three dimensional simulation of jet formation in collapsing condensates, J. Phys. B: At. Mol. Opt. Phys., 37 (2004), pp. 329-343.

10. W. Bao, S. Jin and P.A. Markowich, On time-splitting spectral approximations for the Schrödinger equation in the semiclassical regime, J. Comput. Phys., 175 (2002), pp. 487-524.

11. W. Bao, S. Jin and P.A. Markowich, Numerical study of time-splitting spectral discretizations of nonlinear Schrödinger equations in the semi-clasical regimes, SIAM J. Sci. Comp., 25 (2003), pp. 27-64.

12. W. Bao and X.-G. Li, An efficient and stable numerical method for the Maxwell-Dirac system, J. Comput. Phys., 199 (2004), pp. 663-687.

13. W. Bao and J. Shen, A fourth-order time-splitting Laguerre-Hermite pseudospectral method for Bose-Einstein condensates, SIAM J. Sci. Comput., 26 (2005), pp. 2010-2028.

14. W. Bao, F. Sun, Efficient and stable numerical methods for the generalized and vector Zakharov system, SIAM J. Sci. Comput., 26 (2005), pp. 1057-1088.

15. W. Bao, F.F. Sun and G.W. Wei, Numerical methods for the generalized Zakharov system, J. Comput. Phys. 190 (2003), pp. 201 - 228.

16. W. Bao and W. Tang, Ground state solution of trapped interacting Bose-Einstein condensate by directly minimizing the energy functional, J. Comput. Phys., 187(2003), pp. 230-254.

17. W. Bao and Y. Zhang, Dynamics of the ground state and central vortex states in Bose-Einstein condensation, Math. Mod. Meth. Appl. Sci., 66 (2006), pp. 758-786.

18. W. Bao, H.Q. Wang and P.A. Markowich, Ground, symmetric and central vortex states in rotating Bose-Einstein condensates, Commun. Math. Sci., 3 (2005), pp. 57-88.

19. G. Baym and C. J. Pethick, Phys. Rev. Lett., 76 (1996), 6.

20. P. Bechouche, N. J. Mauser and F. Poupaud, (Semi)-nonrelativistic limits of the Dirac equation with external time-dependent electromagnetic field, Commun. Math. Phys., 197 (1998), pp. 405-425.

21. J. Bolte and S. Keppeler, A semiclassical approach to the Dirac equation, Ann. Phys., 274 (1999), pp. 125-162.

22. H. S. Booth, G. Legg and P. D. Jarvis, Algebraic solution for the vector potential in the Dirac equation, J. Phys. A: Math. Gen., 34 (2001), pp. 5667-5677.

23. J. Bourgain, On the Cauchy and invariant measure problem for the periodic Zakharov system, Duke Math. J. 76 (1994), pp. 175-202.

24. J. Bourgain and J. Colliander, On wellposedness of the Zakharov system, Internat. Math. Res. Notices, 11 (1996), pp. 515-546.

25. N. Bournaveas, Local existence for the Maxwell-Dirac equations in three space dimensions, Comm. Partial Differential Equations, 21 (1996), pp. 693-720.

26. C.C. Bradley, C.A. Sackett, R.G. Hulet, Bose-Einstein condensation of lithium: Observation of limited condensate number, Phys. Rev. Lett. 78 (1997), pp. 985-989.

27. B. M. Caradoc-Davis, R. J. Ballagh, K. Burnett, Coherent dynamics of vortex formation in trapped Bose-Einstein condensates, Phys. Rev. Lett., 83 (1999), 895.

28. Q. Chang and H. Jiang, A conservative difference scheme for the Zakharov equations, J. Comput. Phys., 113 (1994), pp. 309-319.

29. J. M. Chadam, Global solutions of the Cauchy problem for the (classical) coupled Maxwell-Dirac equations in one space dimension, J. Funct. Anal., 13 (1973), pp. 173-184.

30. T. F. Chan, D. Lee and L. Shen, Stable explicit schemes for equations of the Schrödinger type, SIAM J. Numer. Anal., 23 (1986), pp. 274-281.

31. T. F. Chan and L. Shen, Stability analysis of difference scheme for variable coefficient Schrödinger type equations, SIAM J. Numer. Anal., 24 (1987), pp. 336-349.

32. Q. Chang, B. Guo and H. Jiang, Finite difference method for generalized Zakharov equations, Math. Comp., 64 (1995), pp. 537-553.

33. J. Colliander, Wellposedness for Zakharov systems with generalized nonlinearity, J. Diff. Eqs., 148 (1998), pp. 351-363.

34. F. Dalfovo S. Giorgini, L.P. Pitaevskii, and S. Stringari, Rev. Mod. Phys., 71 (1999), 463.

35. A. Das, General solutions of Maxwell-Dirac equations in 1+1-dimensional space-time and a spatially confined solution, J. Math. Phys, 34 (1993), pp. 3986-3985.

36. A. Das, An ongoing big bang model in the special relativistic Maxwell-Dirac equations, J. Math. Phys., 37 (1996), pp. 2253-2259.

37. A. Das and D. Kay, A class of exact plane wave solutions of the Maxwell-Dirac equations, J. Math. Phys., 30 (1989), pp. 2280-2284.

38. K.B. Davis, M.O. Mewes, M.R. Andrews, N.J. van Druten, D.S. Durfee, D.M. Kurn, and W. Ketterle, Phys. Rev. Lett, 75 (1995), 3969.

39. A.S. Davydov, Phys. Scr., 20 (1979), pp. 387.

40. L.M. Degtyarev, V.G. Nakhan'kov and L.I. Rudakov, Dynamics of the formation and interaction of Langmuir solitons and strong turbulence, Sov. Phys. JETP, 40 (1974), pp. 264-268.

41. B. Desjardins, C. K. Lin and T. C. Tso, Semiclassical limit of the derivative nonlinear Schrödinger equation, M^3AS, 10 (2000), pp. 261-285.

42. P. A. M. Dirac, Principles of Quantum Mechanics, Oxford University Press, London (1958).

43. M. Edwards, P.A. Ruprecht, K. Burnett, R.J. Dodd, and C.W. Clark, Phys. Rev. Lett., 77 (1996), 1671; D.A.W. Hutchinson, E. Zaremba, and A. Griffin, Phys. Rev. Lett., 78 (1997), 1842.

44. M. Edwards and K. Burnett, Numerical solution of the nonlinear Schrödinger equation for small samples of trapped neutral atoms, Phys. Rev. A, 51 (1995), 1382.

45. M. Esteban, E. Sere, An overview on linear and nonlinear Dirac equations, Discrete Contin. Dyn. Syst., 8 (2002), pp. 381-397.

46. C. Fermanian-Kammerer, Semi-classical analysis of a Dirac equaiton without adiabatic decoupling, Monatsh. Math., to appear.

47. A. L. Fetter and A. A. Svidzinsky, Vortices in a trapped dilute Bose-Einstein condensate, J. Phys.: Condens. Matter, 13 (2001), pp. 135-194.

48. G. Fibich, An adiabatic law for self-focusing of optical beams, Opt. Lett., 21 (1996), pp. 1735-1737.

49. G. Fibich, Small beam nonparaxiality arrests self-focusing of optical beams, Phys. Rev. Lett., 76 (1996), 4356-4359.

50. G. Fibich, V. Malkin and G. Papanicolaou, Beam self-focusing in the presence of small normal time dispersion, Phys. Rev. A, 52 (1995), 4218-4228.

51. M. Flato, J. C. H. Simon and E. Taflin, Asymptotic completeness, global existence and the infrared problem for the Maxwell-Dirac equations, Mem. Amer. Math. Soc., 127 (1997), pp. 311.

52. E. Forest and R.D. Ruth, Fourth order symplectic integration, Phys. D, 43 (1990), pp. 105.

53. B. Fornberg and T. A. Driscoll, A fast spectral algorithm for nonlinear wave equations with linear dispersion, J. Comput. Phys., 155 (1999), pp. 456.

54. J. J. Garcia-Ripoll, V. M. Perez-Garcia and V. Vekslerchik, Construction of exact solutions by spatial translations in inhomogeneous nonlinear Schrödinger equaitons, Phys. Rev. E, 64 (2001), 056602.

55. I. Gasser and P.A. Markowich, Quantum hydrodynamics, Wigner transforms and the classical limit, Asymptotic Analysis, 14 (1997), pp. 97-116.

56. P. Gérard, Microlocal defect measures, Comm. PDE., 16 (1991), pp. 1761-1794.

57. V. Georgiev, Small amplitude solutions of the Maxwell-Dirac equations, Indiana Univ. Math. J., 40 (1991), pp. 845-883.

58. P. Gérard, P. A. Markowich, N. J. Mauser and F. Poupaud, Homogenization limits and Wigner transforms, Comm. Pure Appl. Math., 50 (1997), pp. 321-377.

59. J. Ginibre, Y. Tsutsumi and G. Velo, The Cauchy problem for the Zakharov system, J. Funt. Anal., 151 (1997), pp. 384-436.

60. R. Glassey, On the asymptotic behavior of nonlinear Schrödinger equations, Trans. Amer. Math. Soc., 182 (1973), pp. 187-200.

61. R. Glassey, On the blowing-up of solutions to the Cauchy problem for nonlinear Schrödinger equations, J. Math. Phys., 18 (1977), pp. 1794-1797.

62. R. Glassey, Approximate solutions to the Zakharov equations via finite differences, J. Comput. Phys., 100 (1992), pp. 377-383.

63. R. Glassey, Convergence of an energy-preserving scheme for the Zakharov equations in one space dimension, Math. Comp., 58 (1992), pp. 83-102.

64. O. Goubet and I. Moise, Attractor for dissipative Zakharov system, Nonlinear Anal., Theory, Methods & Applications, 31 (1998), pp. 823-847.

65. D. Gottlieb and S.A. Orszag, Numerical analysis of spectral methods, Soc. for Industr. & Appl. Math., Philadelphia, 1977.

66. E. Grenier, Semiclassical limit of the nonlinear Schrödinger equation in small time, Proc. Amer. Math. Soc., 126 (1998), pp. 523-530.

67. M. Greiner, O. Mandel, T. Esslinger, T.W. Hänsch, and I. Bloch, Quantum phase transition from a superfluid to a mott insulator in a gas of ultracold atoms, Nature, 415 (2002), pp. 39-45.

68. A. Griffin, D. Snoke, S. Stringaro (Eds.), Bose-Einstein condensation, Cambridge University Press, New York, 1995.

69. E.P. Gross, Nuovo. Cimento., 20 (1961), pp. 454.

70. L. Gross, The Cauchy problem for coupled Maxwell and Dirac equaitons, Comm. Pure Appl. Math., 19 (1966), pp. 1-15.

71. D.S. Hall, M.R. Mattthews, J.R. Ensher, C.E. Wieman and E.A. Cornell, Dynamics of component separation in a binary mixture of Bose-Einstein condensates, Phys. Rev. Lett., 81 (1998), pp. 1539-1542.

72. H. Hadouaj, B. A. Malomed and G.A. Maugin, Soliton-soliton collisions in a generalized Zakharov system, Phys. Rev. A, 44 (1991), pp. 3932-3940.

73. H. Hadouaj, B. A. Malomed and G.A. Maugin, Dynamics of a soliton in a generalized Zakharov system with dissipation, Phys. Rev. A, 44 (1991), pp. 3925-3931.

74. R.H. Hardin and F.D. Tappert, Applications of the split-step Fourier method to the numerical solution of nonlinear and variable coefficient wave equations, SIAM Rev. Chronicle, 15 (1973), pp. 423.

75. Z.Y. Huang, S. Jin, P.A. Markowich, C. Sparber and C.X. Zheng, A time-splitting spectral scheme for the Maxwell-Dirac system, preprint.

76. W. Hunziker, On the nonrelativistic limit of the Dirac theory, Commun. Math. Phys., 40 (1975), pp. 215-222.

77. D. Jaksch, C. Bruder, J. I. Cirac, C. W. Gardiner, and P. Zoller, Cold bosonic atoms in optical lattices, Phys. Rev. Lett., 81 (1998), pp. 3108-3111.

78. S. Jin, P.A. Markowich and C.X. Zheng, Numerical simulation of a generalized Zakharov system, J. Comput. Phys., to appear.

79. A. Jüngel, Nonlinear problems in quantum semiconductor modeling, Nonlin. Anal., Proceedings of WCNA2000, to appear.

80. A. Jüngel, Quasi-hydrodynamic semiconductor equations, Progress in Nonlinear Differential Equations and Its Applications, Birkhäuser, Basel, 2001.

81. T. Kato, On nonlinear Schrödinger equations, Ann. Inst. H. Poincaré. Phus. Théor, 46 (1987), pp. 113-129.

82. M. Landman, G. Papanicolaou, C. Sulem and P. Sulem, Rate of blowup for solutions of the nonlinear Schrödinger equation at critical dimension, Phys. Rev. A, 38 (1988), pp. 3837-3843.

83. M. Landman, G. Papanicolaou, C. Sulem, P. Sulem and X. Wang, Stability of isotropic singularities for the nonlinear Schrödinger equation, Physica D, 47 (1991), pp. 393-415.

84. L. Laudau and E. Lifschitz, Quantum Mechanics: non-relativistic theory, Pergamon Press, New York, 1977.

85. P. Leboeuf and N. Pavloff, Phys. Rev. A **64**, 033602 (2001); V. Dunjko, V. Lorent, and M. Olshanii, Phys. Rev. Lett., 86 (2001), pp. 5413.

86. A. J. Leggett, Bose-Einstein condensation in the alkali gases: some fundamental concepts, Rev. Mod. Phys., 73 (2001), pp. 307-356.

87. E. H. Lieb, R. Seiringer, J. Yugvason, Bosons in a trap: a rigorous derivation of the Gross-Pitaevskii energy functional, Phys. Rev. A, 61 (2000), pp. 3602.

88. F. Lin and T. C. Lin, Vortices in two-dimensional Bose-Einstein condensates, Geometry and nonlinear partial differential equations (Hangzhou, 2001), pp. 87–114, AMS/IP Stud. Adv. Math., 29, Amer. Math. Soc., Providence, RI, 2002.

89. M.J. Landman, G.C. Papanicolaou, C. Sulem, P.L. Sulem, X.P. Wang, Stability of isotropic singularities for the nonlinear Schrödinger equation, Phys. D, 47 (1991), pp. 393–415.

90. A. G. Lisi, A solitary wave solution of the Maxwell-Dirac equations, J. Phys. A: Math. Gen., 28 (1995), pp. 5385-5392.

91. P.A. Markowich, N.J. Mauser and F. Poupaud, A Wigner function approach to semiclassical limits: electrons in a periodic potential, J. Math. Phys., 35 (1994), pp. 1066-1094.

92. P.A. Markowich, P. Pietra and C. Pohl, Numerical approximation of quadratic observables of Schrödinger-type equations in the semi-classical limit, Numer. Math., 81 (1999), pp. 595-630.

93. N. Masmoudi and N. J. Mauser, The selfconsistent Pauli equaiton, Monatsh. Math., 132 (2001), pp. 19-24.

94. R. McLachlan, Symplectic integration of Hamiltonian wave equations, Numer. Math., 66 (1994), pp. 465.

95. V. Masselin, A result of the blow-up rate for the Zakharov system in dimension 3, SIAM J. Math. Anal., 33 (2001), pp. 440-447.

96. B. Najman, The nonrelativstic limit of the nonlinear Dirac equaiton, Ann. Inst. Henri Poincare Non Lineaire, 9 (1992), pp. 3-12.

97. P.K. Newton, Wave interactions in the singular Zakharov system, J. Math. Phys., 32 (1991), pp. 431-440.

98. G.C. Papanicolaou, C. Sulem, P.L. Sulem, X.P. Wang, Singular solutions of the Zakharov equations for Langmuir turbulence, Phys. Fluids B, 3 (1991), pp. 969-980.

99. A.S. Parkins and D.F. Walls, Physics Reports, 303 (1998), pp. 1.

100. G.L. Payne, D.R. Nicholson, and R.M. Downie, Numerical solution of the Zakharov equations, J. Comput. Phys., 50 (1983), pp. 482-498.

101. N.R. Pereira, Collisions between Langmuir solitons, The Physics of Fluids, 20 (1977), pp. 750-755.

102. C.J. Pethick and H. Smith, Bose-Einstein Condensation in Dilute Gases, Cambridge University Press, 2002.

103. L.P. Pitaevskii, Zh. Eksp. Teor. Fiz., 40 (1961), pp. 646. (Sov. Phys. JETP, 13 (1961), pp. 451).

104. L. Pitaevskii and S. Stringari, Bose-Einstein condensation, Oxford University Press, Oxford, 2002.

105. D. S. Rokhsar, Phys. Rev. Lett., 79 (1997), pp. 2164; R. Dum, J.I. Cirac, M. Lewenstein, and P. Zoller, Phys. Rev. Lett., 80 (1998), pp. 2972; P. O. Fedichev, and G. V. Shlyapnikov, Phys. Rev. A, 60 (1999), pp. 1779.

106. L. Simon, Asymptotics for a class of nonlinear evolution equations, with applications to geometric problems, Annals of Math., 118 (1983), pp. 525-571.

107. E.I. Schulman, Dokl. Akad. Nauk. SSSR, 259 (1981), pp.579 [Sov. Phys. Dokl., 26 (1981), pp. 691].

108. C. Sparber and P. Markowich, Semiclassical asymptotics for the Maxwell-Dirac system, J. Math. Phys., 44 (2003), pp. 4555–4572.

109. H. Spohn, Semiclassical limit of the Dirac equaiton and spin precession, Ann. Phys., 282 (2000), pp. 420-431.

110. L. Stenflo, Phys. Scr., 33 (1986), pp. 156.

111. G. Strang, On the construction and comparison of differential schemes, SIAM J. Numer. Anal., 5 (1968), pp. 506.

112. C. Sulem and P.L. Sulem, Regularity properties for the equations of Langmuir turbulence, C. R. Acad. Sci. Paris Sér. A Math., 289 (1979), pp. 173-176.

113. C. Sulem and P.L. Sulem, The nonlinear Schrödinger equation, Springer, 1999.

114. F.F. Sun, Numerical studies on the Zakharov system, Master thesis, National University of Singapore, 2003.

115. T.R. Taha and M.J. Ablowitz, Analytical and numerical aspects of certain nonlinear evolution equations, II. Numerical, nonlinear Schrödinger equation, J. Comput. Phys., 55 (1984), pp. 203.

116. B. Thaller, The Dirac equation, New York, Springer, 1992.

117. M. Weinstein, Nonlinear Schrödinger equations and sharp interpolation estimates, Comm. Math. Phys., 87 (1983), pp. 567-576.

118. M. Weinstein, Modulational stability of ground states of nonlinear Schrödinger equations, SIAM J. Math. Anal., 16 (1985), pp. 472-490.

119. M. Weinstein, The nonlinear Schrödinger equations-singularity formation, stability and dispersion, Contemporary mathematics, 99 (1989), pp. 213-232.

120. H. Yoshida, Construction of higher order symplectic integrators, Phys. Lett. A., 150 (1990), pp. 262-268.

121. V.E. Zakharov, Zh. Eksp. Teor. Fiz., 62 (1972), pp. 1745 [Sov. Phys. JETP, 35 (1972), pp. 908].

Introduction to Constitutive Modeling of Macromolecular Fluids

Qi Wang

Department of Mathematics
Florida State University
Tallahassee, FL 32306-4510, USA
E-mail: wang@mail.math.fsu.edu

Contents

1. Introduction

Macromolecular or polymeric liquids consist of polymer solutions, polymer melts, particle suspension fluids, many manmade and biological fluids. Polymer solutions are made of polymers dissolved in solutions or solvent; polymer melts are molten polymers; particle suspension fluids consist of solid particles suspended in a matrix fluid which may be a viscous or viscoelastic fluid; blood flows consist of deformable suspensions (red and white blood cells) in viscous fluids; and mucus consisting of live macromolecules is characteristically viscoelastic. Given the large molecular weight and size in polymers, polymeric liquids are capable of forming a variety of meso-phases, in which partial positional as well as orientational order can be present. These meso-phases are termed liquid crystalline phases. The polymeric liquids in the meso-phases are called liquid crystalline polymers. When the meso-phases are created above some critical concentration in solutions, the polymeric liquids are called lyotropic liquid crystal polymers. When the phases are attained at certain low temperature in melts, the materials are called thermotropic liquid crystal polymers. Not only miscible polymeric solutions and melts are capable of forming the liquid crystalline phases, immiscible polymer blends, emulsions, polymer-particle nanocomposites, which are liquid mixtures of polymer solutions or melts and solid nano-sized particles, are all candidates for forming liquid crystalline phases.

Polymeric liquids exhibit a host of distinctive features from the isotropic liquids consisting of small molecules like water, cooking oils, etc. in flows due to their large molecular weight and conformation. There have been several hallmark phenomena in the polymeric fluids well-documented such as rod-climbing, extruded swell, and tubeless siphon [3]; when a gas bubble is trapped within a polymeric liquid, its geometry is distinct from that of a gas bubble trapped within a Newtonian fluid. These fascinating phenomena, distinctive of the polymeric liquids, have spurted a significant amount of research activities over the past few decades. Theories and models developed for polymers are now applied to biofluids and biomaterials, nanocomposite materials, making it a fast growing interdisciplinary research area.

In the lecture notes, we will give a crash course on the basics needed to model the complex fluids in fundamental thermodynamics, statistic mechanics, polymer physics and continuum mechanics, survey the existing models for various polymeric liquids and explore a systematic approach for flexible polymers and liquid crystal polymers within the framework of the kinetic theory. We hope the notes give you a brief introduction to the vast and vibrant subject. More detailed exhibition of the subject can be found in the references cited at the end.

2. Introduction to equilibrium thermodynamics

We introduce the basic thermodynamic variables that will be used through out the lecture, especially, the free energy potentials and the basic thermodynamic laws which will guide our development of dynamic equations for complex fluid systems. A thermodynamical system is any macroscopic matter/substance system. Any measurable macroscopic quantities associated with the system can be called thermodynamical variables. For instance, the temperature T, pressure P, and volume V of the matter system etc. are thermodynamic variables. The space of the thermodynamic variables is called the phase space, i.e., the space of $(T, P, V, \cdots)$ in the previous example. The equation of state, which characterizes the material properties of the thermodynamical system, is a functional relationship among the thermodynamical variables in the phase space. It is normally given as a level surface in the phase space. For example, in a system where the thermodynamical variables are T, P, V, the equation of state is expressed as an algebraic equation

$$f(P, V, T) = 0. \tag{2.1}$$

The equation of state dictates that the thermodynamic process defined therein can only take place in a confined subspace or manifold in the phase space.

A scalar function or functional can be associated to the thermodynamical system as the internal energy. Heat Q is a form of energy, which usually varies with the temperature. Hence, the rate of change of the heat with respect to the temperature is an important thermodynamic quantity, which is termed the heat capacity or specific heat

$$C = \frac{\Delta Q}{\Delta T}, \tag{2.2}$$

where $\Delta(\bullet)$ represents an infinitesimal amount of change in the quantity $(\bullet)$. In another word, the heat capacity is the amount of heat the system absorbs (or releases) when temperature is raised (or lowered) by one degree. Some thermodynamical variables are called extensive if they are proportional to the amount of the substance in the system, e.g., energy and volume, etc.; others, independent of the amount of the substance, are called intensive like temperature, pressure, etc.. The extensive variables are additive while the intensive ones are not.

In equilibrium thermodynamics, the thermodynamic transformation describes a change of state in the phase space. If the initial state in the phase space is an equilibrium state, the transformation can be brought about only by changes in external conditions of the system. The transformation is said to be quasi-static if the external condition changes so slowly that at any moment, the system is approximately in equilibrium. In the phase space, a thermodynamical transformation from point (or state) A to point B is said to be reversible if when the external condition is reversed, the state is transformed back from B to A.

The equilibrium thermodynamics is built upon a set of basic thermodynamical laws which are developed based on the accumulative empirical evidence. The first law of thermodynamics reveals the essential relationship among the heat, the internal energy and the work done to the system.
The first law of thermodynamics:

Let U be the internal energy of the substance system (an extensive thermodynamical variable), dQ the heat absorbed by the system, dW the work done by the system (to its surroundings). Then,

$$dU = dQ - dW \qquad (2.3)$$

and the change in the internal energy only depends on the initial and the final state of the system. Thus, $dQ = dU + dW$.

The first law implies that dU is a complete differential, i.e., the integral

$$\int_L dU \qquad (2.4)$$

is independent of the path L in the phase space of the thermodynamic variables. We remark that the thermodynamical variables are not necessarily defined in the entire phase space. Therefore, the physical laws we discuss here apply to the common domain of the thermodynamical variables. This should be understood throughout the notes since we will not make any special efforts to articulate this over and over again. Clearly dW is not a complete differential since the work done does depend on the path. Therefore, dQ is not a complete differential either. The first law also indicates that a part of the heat absorbed by the matter system is used to increase the internal energy while the other part is used to do work to its surroundings, which can be associated with the energy loss due to frictions, etc.

The second law of thermodynamics is much more subtle. It addresses the issue of dissipation in a thermodynamical system. There is not a single version that is easy to use in practice or even to phrase. One version of the second law is phrased as follows [11].

The second law of thermodynamics

There exists a unique thermodynamic transformation whose sole effect is to extract a quantity of heat from a cooler reservoir and deliver it to a hotter reservoir.

Clearly, this is not easy to apply directly to modeling a thermodynamic process. The Clausius' theorem, which is based on the second law, gives an important consequence of it that one often uses. The proof of the theorem can be found for example in [11].

Clausius' Theorem:

In any cyclic transformation, where a cyclic transformation means a thermodynamic transformation in the phase space whose starting and ending point coincide,

$$\int_C \frac{dQ}{T} \leq 0. \tag{2.5}$$

For reversible processes however, the equality holds

$$\int_C \frac{dQ}{T} = 0. \tag{2.6}$$

It follows from the Clausius' theorem that the ratio $\frac{dQ}{T}$ is a complete differential despite that dQ is not. We introduce another thermodynamic variable called entropy S whose differential is defined as

$$dS = \frac{dQ}{T}. \tag{2.7}$$

The second law guarantees that dS is a complete differential so that the entropy is well defined

$$S(A) = \int_O^A \frac{dQ}{T}, \tag{2.8}$$

where O is a fixed point in the phase space and OA is any reversible path in the phase space joining O and A.

Corollary

(i). For an arbitrary thermodynamic transformation from A to B in the phase space,

$$\int_A^B \frac{dQ}{T} \leq S(B) - S(A) \qquad (2.9)$$

and the equality is held if the path is reversible.

(ii). For a thermally isolated system ($dQ = 0$),

$$dS \geq 0 \text{ or } \int_A^B \frac{dQ}{T} = 0 \leq S(B) - S(A). \qquad (2.10)$$

This implies that entropy is always increasing in an isolated thermodynamic system. It reaches a local maximum at the equilibrium state.

Another direct consequence of the second law is the Clausius-Duhem inequality. In a compact material domain $\mathcal{P}$, we denote S the entropy of the material volume, r the rate of heat production per unit volume, $\mathbf{h}$ the heat flux through the boundary $\partial \mathcal{P}$. The Clausius-Duhem inequality states that the time rate of change of the entropy in the volume is greater than or equal to the heat production per unit temperature minus the heat flux flown out of the domain per unit temperature.

$$\frac{d}{dt} S \geq \int_\mathcal{P} \frac{r}{T} d\mathbf{x} - \int_{\partial \mathcal{P}} \frac{\mathbf{h}}{T} \cdot d\mathbf{S}. \qquad (2.11)$$

This inequality will be used exclusively for the second law in continuum mechanics theories.

Now, we have introduced two thermodynamic variables: internal energy and the entropy, that are potential functions. In addition to the internal energy and entropy there are other scalar extensive thermodynamic variables that may be more convenient to use at certain situations. It turns out the Legendre transformation plays an essential role in relating the extensive thermodynamic variables to internal energy and entropy.

Definition:

Let $y = f(x)$ be a function with $f''(x) \neq 0$. Then, $f'(x)$ is a monotonic function. The equation $p = f'(x)$ has a unique solution $x = f'^{(-1)}(p)$ for any given values of p. We define $g(p) = xp - f(x(p))$ as the Legendre transformation of the function $f(x)$.

A good property of the Legendre transformation is that it transforms a convex function into another convex function. In addition, the Legendre transformation of $g(p)$ is $f(x)$.

When the internal energy is given as a function of the entropy and other thermodynamic variables. Its Legendre transformation gives the Helmholtz free energy

$$A(T, \cdots) = U(S, \cdots) - TS. \qquad (2.12)$$

The Gibbs free energy is defined as

$$G = A + pV, \qquad (2.13)$$

which can be thought of a Legendre transformation of the Helmholtz free energy relative to pressure p. The enthalpy is the Legendre transformation of the internat energy with respect to the volume.

$$H = U + pV. \qquad (2.14)$$

For a mechanically isolated system kept at a constant temperature, the second law yields the following two useful results.

Theorem:

The Helmholtz free energy never increases during any thermodynamical processes in a mechanically isolated system kept at constant temperature. I.e., $\Delta A \leq 0$. Therefore, the state of the thermodynamical equilibrium, if it exists, corresponds to a state of the minimum Helmholtz free energy.

Theorem:

For the mechanically isolated thermodynamical system kept at a constant temperature and pressure, the Gibbs free energy never increases during any thermodynamical processes. Thus, the state of equilibrium is at the state of minimum Gibbs free energy.

The above theorems give the criteria for thermal equilibrium. When the pressure is not constrained, the thermal equilibrium is given by the minimum of the Helmholtz free energy. When the pressure is a constant, the equilibrium can be obtained by minimizing the Gibbs free energy. This is why you see one always calculates the system's free energy and minimize it to study the equilibrium and thermodynamical properties of a material system .

With the potentials, the thermodynamic variables can be represented as their derivatives. From the first law, we have

$$dU = dQ - dW = TdS - pdV, \qquad (2.15)$$

where $dW = pdV$ is assumed.

$$T = \frac{\partial U}{\partial S}, p = -\frac{\partial U}{\partial V}, \qquad (2.16)$$

The temperature is the derivative of the internal energy with respect to the entropy while the pressure is the derivative of the internal energy with respect to the matter system's volume.

From the definition of the Helmholtz free energy, it follows that

$$dA = dU - dTS - TdS = dU - SdT - dQ = -SdT - pdV. \quad (2.17)$$

Then,

$$p = -\frac{\partial A}{\partial V}, S = -\frac{\partial A}{\partial T}. \quad (2.18)$$

So, the entropy is the negative of the derivative of the Helmholtz free energy with respect to the temperature while the pressure is also given by a derivative of the free energy with respect to the volume. Similarly, it follows from the definition of the Gibbs free energy that

$$dG = dA + dpV + pdV = dU - TdS - SdT + pdV + Vdp =$$

$$\quad (2.19)$$

$$-dW - SdT + pdV + Vdp = -SdT + Vdp.$$

Then,

$$S = -\frac{\partial G}{\partial T}, V = \frac{\partial G}{\partial p}. \quad (2.20)$$

It follows from the definition of enthalpy,

$$dH = dU + pdV + Vdp = dQ + Vdp = TdS + Vdp. \quad (2.21)$$

So,

$$T = \frac{\partial H}{\partial S}, V = \frac{\partial H}{\partial p}. \quad (2.22)$$

When the thermodynamic variable is given by the density instead of the volume, the partial derivative

$$\frac{\partial(\bullet)}{\partial V} = -\rho^2 \frac{\partial(\bullet)}{\partial \rho}, \quad (2.23)$$

where V is understood as the specific volume, i.e., the volume occupied by a unit mass of the matter. Then, $\rho = \frac{1}{V}$.

The third law of thermodynamics sets the reference value for the entropy.

The third law of thermodynamics

The entropy of a system at absolute zero temperature is a universal constant, which may be taken to be zero.

In the above discuss, the matter system is assumed closed, i.e., the number of the particles or molecules in the system is a constant. When the macroscopic matter system is not closed, the number of the particles in the system is not conserved. The number density becomes a thermodynamic variable, an intensive variable. Corresponding to the variation in number density, a new potential called chemical potential must be introduced. This is the work done to the system by adding one additional particle to the system. The first law is then rewritten as

$$dU = dQ - dW + \mu dN, \tag{2.24}$$

where N is the number of particles and μ is the chemical potential. Using the various potentials, the chemical potential can be calculated by

$$\mu = (\frac{\partial U}{\partial N})_{S,V} = (\frac{\partial A}{\partial N})_{T,V} = (\frac{\partial G}{\partial N})_{p,T}. \tag{2.25}$$

Incorporating the chemical potential, the grand potential, the fourth potential that we have seen so far, is defined by

$$\mathcal{A} = U - TS - \mu N. \tag{2.26}$$

$$d\mathcal{A} = -SdT - pdV - Nd\mu. \tag{2.27}$$

The chemical potential can also be calculated from the derivative of the grand potential,

$$\mu = -(\frac{\partial \mathcal{A}}{\partial N})_{T,V} \tag{2.28}$$

The ideal gas is the most studied substance system in thermodynamics and statistical mechanics. The equation of state for the ideal gas is given by [8]

$$pV = RT, \tag{2.29}$$

where R is the gas constant. For non-ideal gases, the relation is no longer valid. A volume expansion of the pressure can be written as

$$pV/RT = 1 + B(T)/V + C(T)/V^2 + D(T)/V^3 + \cdots, \tag{2.30}$$

where $B, C, D, \cdots$ are referred to as the second, third, fourth,... virial coefficients. The virial coefficients demonstrate the departure of the matter system away from the ideal. In particular, the second virial coefficient $B(T)$ is the most important since it is the leading order perturbation from the ideal gas.

Thermodynamics are empirical. A more fundamental theory which eventually justifies the thermodynamical laws is the Statistical Mechanics, a subject in which statistical methods are used to study the behavior of ensemble particles in the matter system assuming any matter system is comprised of particles.

3. Introduction to statistical mechanics

We treat the matter/substance system as a particle system consisting of N particles with their center of mass positions $\mathbf{r}_i, i = 1, \cdots, N$ and the momenta $\mathbf{p}_i, i = 1, \cdots, N$ in a domain Ω in a canonical coordinate $\mathbf{R}^n$. This is a closed system when the number of the particles is fixed. In contrast, the number of particles varies in an open system. The Hamiltonian of the system is denoted by $H(\{\mathbf{r}_i\}, \{\mathbf{p}_i\})$, which yields the total energy of the system. Here $\{\mathbf{r}_i\}$ and $\{\mathbf{p}_i\}$ are the shorthand notations for the sets $\{\mathbf{r}_i, i = 1, \cdots, N\}$ and $\{\mathbf{p}_i, i = 1, \cdots, N\}$, respectively. The equation of motion for each particle is given by the Hamilton equation of motion,

$$\begin{cases} \dot{\mathbf{r}}_i = \frac{\partial H}{\partial \mathbf{p}_i}, \\[2mm] \dot{\mathbf{p}}_i = -\frac{\partial H}{\partial \mathbf{r}_i}, i = 1, \cdots, N. \end{cases} \tag{3.1}$$

In a Cartesian coordinate system, the Hamiltonian for a conservative system with potential Φ is given by $H(\{\mathbf{r_i}\}, \{\mathbf{p_i}\}) = \sum_{i=1}^{N} \frac{\|\mathbf{p}_i\|^2}{m_i} + \Phi(\mathbf{r_i})$. The Hamilton equation yields the Newton's second law in classical mechanics

$$m_i \frac{d\mathbf{r}_i}{dt^2} = -\nabla_{\mathbf{r}_i} \Phi, i = 1, \cdots, N. \tag{3.2}$$

We define a term called the virial of the force for the ith particle by

$$-\frac{1}{2}\mathbf{r}_i \cdot \mathbf{F}_i, \tag{3.3}$$

where $\mathbf{F}_i$ is the external force acting on the ith particle located at $\mathbf{r}_i$. The total virial of the force for the system is given by

$$\Theta = -\frac{1}{2}\sum_i \mathbf{r}_i \cdot \mathbf{F}_i. \tag{3.4}$$

In classical mechanics,

$$\frac{d}{dt}\mathbf{p}_i = m_i \frac{d^2\mathbf{r}_i}{dt^2} = \mathbf{F}_i. \tag{3.5}$$

Multiplying both sides by $\mathbf{r}_i$, it follows that

$$\mathbf{r}_i \cdot \mathbf{F}_i = m_i \mathbf{r}_i \cdot \frac{d^2 \mathbf{r}_i}{dt^2} = \frac{m_i}{2} [\frac{d}{dt}(\mathbf{r}_i \cdot \frac{d\mathbf{r}_i}{dt}) - (\frac{d\mathbf{r}_i}{dt} \cdot \frac{d\mathbf{r}_i}{dt})]. \qquad (3.6)$$

Taking a time-average of (3.6) in $[0, T]$ and assuming

$$\lim_{T \to \infty} \frac{1}{T} [\mathbf{r}_i \cdot \frac{d\mathbf{r}_i}{dt}|_T - \mathbf{r}_i \cdot \frac{d\mathbf{r}_i}{dt}|_0] = 0, \qquad (3.7)$$

we arrive at

$$-\frac{1}{2}\overline{\mathbf{r}_i \cdot \mathbf{F}_i} = -\frac{1}{2}\frac{1}{T}\int_0^T (\mathbf{r}_i \cdot \mathbf{F}_i)dt = \frac{m_i}{2}\overline{(\frac{d\mathbf{r}_i}{dt} \cdot \frac{d\mathbf{r}_i}{dt})} = \overline{K}. \qquad (3.8)$$

Then, the Virial Theorem states that

$$\overline{\Theta} = \overline{K} = \sum_i \frac{m_i}{2}\overline{\frac{d\mathbf{r}_i}{dt} \cdot \frac{d\mathbf{r}_i}{dt}}. \qquad (3.9)$$

It indicates that the time-averaged kinetic energy equals to the time averaged total virial of the force. In a conservative system,

$$\mathbf{F}_i = -\nabla_{\mathbf{r}_i}\Phi = -\frac{\partial \Phi}{\partial \mathbf{r}_i}, \qquad (3.10)$$

where $\Phi(\mathbf{r}_i)$ is an external potential. If the potential is a homogeneous function of the coordinates of degree n,

$$\overline{\Theta} = \frac{1}{2}\sum_i \overline{\mathbf{r}_i \cdot \frac{\partial \Phi}{\partial \mathbf{r}_i}} = \frac{n}{2}\overline{\Phi} = \overline{K}. \qquad (3.11)$$

This is the Virial theorem represented in another form.

For the N-particle matter system, we define the $\Gamma-$space also known as the phase space as

$$\Gamma = \{(\mathbf{r}_1, \mathbf{p}_1), \cdots, (\mathbf{r}_N, \mathbf{p}_N)\}. \qquad (3.12)$$

A collection of points in the $\Gamma-$space in a volume V is called an ensemble of particles. We denote short-handedly $\mathbf{r} = \{\mathbf{r}_i\}, \mathbf{p} = \{\mathbf{p}_i\}$. We note that if the particles are all distinguishable, there exists a one-to-one correspondence between the particles in physical space $\mathbf{R}^3 \times \mathbf{R}^3$ and a point in the phase space; if, on the other hand, the particles are not all distinguishable, the correspondence is not one-to-one. In the latter case, the particles in the physical space corresponds multiple points in the phase space. The multiple correspondence will be addressed whenever it arises.

Let $\rho(\mathbf{r}, \mathbf{p}, t)$ denote the number density function in the $\Gamma-$space for the phase point $(\mathbf{r}, \mathbf{p})$, i.e., $\int_{\Delta V} \rho d\mathbf{r}d\mathbf{p}$ gives the number of particles in the volume $\Delta V \in V$. The following theorem shows that the number density obeys a conservation law.

Liouville's Theorem

$$\frac{\partial \rho}{\partial t} + \sum_{i=1}^{3N}\left(\frac{\partial \rho}{\partial p_i}\dot{p}_i + \frac{\partial \rho}{\partial r_i}\dot{r}_i\right) = \frac{\partial \rho}{\partial t} - [H,\rho] = 0, \qquad (3.13)$$

where $[H,\rho] = \frac{\partial H}{\partial \mathbf{r}} \cdot \frac{\partial \rho}{\partial \mathbf{p}} - \frac{\partial \rho}{\partial \mathbf{r}} \cdot \frac{\partial H}{\partial \mathbf{p}}$. It follows from the equation of motion in Hamiltonian mechanics. The bracket $[,]$ is known as the Poisson bracket [1].

As a matter of fact, the density function satisfies the conservation law

$$\frac{\partial \rho}{\partial t} + \sum_{i=1}^{N}\left[\frac{\partial}{\partial \mathbf{r}_i}(\dot{\mathbf{r}}_i\rho) + \frac{\partial}{\partial \mathbf{p}_i}(\dot{\mathbf{p}}_i\rho)\right] = 0. \qquad (3.14)$$

The Liouville'e theorem is obtained after applying the Hamilton's equation of motion.

For a function of $g = g(\mathbf{r},\mathbf{p})$ defined in the phase space, we define the ensemble average as

$$\langle g \rangle = \frac{\int_V g\rho d\mathbf{r}d\mathbf{p}}{\int_V \rho d\mathbf{r}d\mathbf{p}}. \qquad (3.15)$$

The normalized number density function

$$p(\mathbf{r},\mathbf{p}) = \frac{\rho(\mathbf{r},\mathbf{p})}{\int_V \rho(\mathbf{r},\mathbf{p})d\mathbf{r}d\mathbf{p}} \qquad (3.16)$$

is a probability density or distribution function.

In the 6N dimensional $\Gamma-$space, we will introduce three ensembles applicable in statistical sampling. These ensembles correspond to a collection of particle realizations in the physical space dictated by their total energy levels.

Microcanonical ensemble

The microcanonical ensemble is one with the uniform probability density in a shell in the $\Gamma-$space given by

$$p(\mathbf{r},\mathbf{p}) = \begin{cases} const, & \text{if } E < H(\mathbf{r},\mathbf{p}) < E + \Delta, \\ 0, & \text{Otherwise,} \end{cases} \qquad (3.17)$$

where E is the energy level and $\Delta << 1$. The distribution of the particles is concentrated in an energy shell in the phase space. As $\Delta \to 0$, the density function approaches the $\delta-$function centered around the level surface defined by the energy level E.

If we introduce

$$\Gamma(E) = \int_{E<H<E+\Delta} d\mathbf{r}d\mathbf{p},$$ (3.18)

defining the volume of the energy shell in the phase space, then,

$$S = k_B ln\Gamma(E)$$ (3.19)

is identifiable with the entropy defined in the equilibrium thermodynamics following the second law. Define the temperature by

$$\frac{1}{T} = \frac{\partial S}{\partial E},$$ (3.20)

and the pressure of the system by

$$p = T\frac{\partial S}{\partial V}.$$ (3.21)

Then,

$$dS = \frac{\partial S}{\partial E}dE + \frac{\partial S}{\partial V}dV = \frac{1}{T}dE + pdV$$ (3.22)

or

$$dE = TdS - pdV.$$ (3.23)

This is the first law of thermodynamics. Therefore, the first and second law of thermodynamics are recovered from the quantities defined in statistical mechanics.

We denote x_i a component of either $\mathbf{r}_i$ or $\mathbf{p}_i$ and evaluate the ensemble average

$$\langle x_i\frac{\partial H}{\partial x_j}\rangle = \frac{1}{\Gamma(E)}\int_{E<H<E+\Delta} d\mathbf{r}d\mathbf{p}x_i\frac{\partial H}{\partial x_j}$$
$$= \frac{\Delta}{\Gamma(E)}\frac{\partial}{\partial E}\int_{H<E} d\mathbf{r}d\mathbf{p}x_i\frac{\partial H}{\partial x_j} + O(\Delta^2).$$ (3.24)

Notice that

$$\int_{H<E} d\mathbf{r}d\mathbf{p}x_i\frac{\partial H}{\partial x_j} = \int_{H<E} d\mathbf{r}d\mathbf{p}\frac{\partial x_i(H - E)}{\partial x_j}$$
$$-\delta_{ij}\int_{H<E} d\mathbf{r}d\mathbf{p}(H - E).$$ (3.25)

Then,

$$\langle x_i\frac{\partial H}{\partial x_j}\rangle = \delta_{ij}\frac{\Delta}{\Gamma(E)}\frac{\partial}{\partial E}\int_{H<E} d\mathbf{r}d\mathbf{p}(E - H) = \frac{\delta_{ij}\Delta}{\Gamma(E)}\int_{H<E} d\mathbf{r}d\mathbf{p}$$ (3.26)

$$= \delta_{ij}\frac{\int_{H<E} d\mathbf{r}d\mathbf{p}}{\frac{\partial}{\partial E}\int_{H<E} d\mathbf{r}d\mathbf{p}}.$$

It can be shown that [11]

$$ln \int_{H<E} d\mathbf{r}d\mathbf{p} - ln \int_{E<H<E+\Delta} d\mathbf{r}d\mathbf{p} \tag{3.27}$$

is a constant. Thus, the denominator in the above equation can be recognized as

$$\frac{\int_{H<E} d\mathbf{r}d\mathbf{p}}{\frac{\partial}{\partial E}\int_{H<E} d\mathbf{r}d\mathbf{p}} = k_B(\frac{\partial S}{\partial E})^{-1} = k_B T. \tag{3.28}$$

Finally, we arrive at

$$\langle \mathrm{x}_i \frac{\partial H}{\partial \mathbf{x}_j} \rangle = \delta_{ij} k_B T. \tag{3.29}$$

This is called the generalized equipartition theorem. The virial theorem follows

$$-\langle \sum_i \mathbf{r}_i \frac{\partial H}{\partial \mathbf{r}_i} \rangle = \langle \sum_i \mathbf{r}_i \dot{\mathbf{p}}_i \rangle = -3N k_B T. \tag{3.30}$$

$3k_B T$ equals the kinetic energy for the ith particle. Many matter systems have Hamiltonians of the form

$$H = \sum_i (\alpha_i \|\mathbf{r}_i\|^2 + \beta_i \|\mathbf{p}_i\|^2), \tag{3.31}$$

where α_i, β_i are the coefficients. For this type of Hamiltonian,

$$\sum_i \mathbf{r}_i \frac{\partial H}{\partial \mathbf{r}_i} + \mathbf{p_i} \frac{\partial H}{\partial \mathbf{p}_i} = 2H. \tag{3.32}$$

This equality along with the generalized equipartition theorem implies

$$\langle H \rangle = \frac{3n}{2} k_B T, \tag{3.33}$$

where n is the number of nonzero coefficients α_i, β_i. This is known as the theorem of equipartition of energy [11].

Canonical ensemble

A canonical ensemble is given by the probability density function defined by

$$p(\mathbf{r}, \mathbf{p}) = \frac{1}{Z_N N! h^{3N}} e^{-\frac{H(\mathbf{r},\mathbf{p})}{k_B T}}, \tag{3.34}$$

where h is the Planck constant, k_B is the Boltzmann constant, N is the number of particle in the system, and Z_N is the partition function given by

$$Z_N = \int \frac{1}{N! h^{3N}} e^{-\frac{H(\mathbf{r},\mathbf{p})}{k_B T}} d\mathbf{r}d\mathbf{p}. \tag{3.35}$$

This is motivated by the Maxwell-Boltzmann most probable distribution theory [11].

The Helmholtz free energy is defined as

$$A = -k_B T ln Z_N. \tag{3.36}$$

Deferentiate (3.36) with respect to $\frac{1}{k_B T}$, yielding

$$A - \int Hp d\mathbf{r} d\mathbf{p} - T\frac{\partial A}{\partial T} = 0. \tag{3.37}$$

Define the entropy by

$$S = -\frac{\partial A}{\partial T} \tag{3.38}$$

and identify the internal energy by

$$U = \langle H \rangle = \int pH d\mathbf{r} d\mathbf{p}. \tag{3.39}$$

This gives the first law of thermodynamics

$$A = E - TdS. \tag{3.40}$$

The other thermodynamic variables are calculated from the Maxwell relations [11], for instance,

$$p = -\frac{\partial A}{\partial V}. \tag{3.41}$$

Grand canonical ensemble

When the system is an open system (such as a system where chemical reactions take place), the grand canonical ensemble is more appropriate, in which the probability density function for the subsystem with N particles is given by

$$p(\mathbf{r}, \mathbf{p}, N) = \frac{1}{N! h^{3N}} e^{-\frac{H(\mathbf{r},\mathbf{p})}{k_B T}} e^{\frac{N\mu}{k_B T}} e^{-\frac{pV}{k_B T}}, \tag{3.42}$$

where

$$\mu = \frac{\partial A}{\partial N}, p = -\frac{\partial A}{\partial V} \tag{3.43}$$

are the chemical potential and pressure, respectively. We define Z as the grand partition function,

$$Z = \sum_{N=0}^{\infty} e^{\frac{\mu N}{k_B T}} Z_N, \tag{3.44}$$

It follows from (3.42) and $\sum_{N=0}^{\infty} p(\mathbf{r}, \mathbf{p}, N) = 1$,

$$\frac{pV}{k_B T} = lnZ \text{ or } Z = e^{\frac{pV}{k_B T}}. \tag{3.45}$$

The probability density function for the subsystem of N particles in grand canonical ensemble can be rewritten into

$$p(\mathbf{r}, \mathbf{p}, N) = \frac{1}{ZN!h^{3N}} e^{-\frac{H(\mathbf{r},\mathbf{p})}{k_B T}} e^{\frac{N\mu}{k_B T}}. \tag{3.46}$$

The internal energy is defined by

$$U = \langle H \rangle = \sum_{N}^{\infty} \int H p(\mathbf{r}, \mathbf{p}, n) d\mathbf{r} d\mathbf{p}. \tag{3.47}$$

All others are obtained from the Maxwell relations, for example,

$$S = \int_0^T \frac{1}{T} \frac{\partial U}{\partial T} dT \tag{3.48}$$

defines the entropy.

These ensembles are frequently used in statistical and stochastic simulations such as Monte-carlo simulations. The statistical mechanics reformulates the equilibrium thermodynamics on a solid foundation. We next discussion another cornerstone for formulating theories for macromolecular or polymer fluids, the continuum mechanics notions of material deformation, symmetry, and motion.

4. Introduction to continuum mechanics

Continuum mechanics study the motion and deformation of a continuum (macroscopic matter system), which can be a solid or a liquid, following a set of well-established principles. It provides a self-consistent way to establish a continuum theory describing the deformation and motion of the continuum for various materials at specified length and time scales. Multiscale continuum mechanics theories are also developed to handle multiple scales and multiple physics such as electromagnetics and thermodynamics in the continuum [6, 7]. In this notes, we present the one scale continuum mechanics theory with emphasis on the basic material symmetry and thermo-mechanical principles at the macroscopic scale for continua.

4.1. *Material, referential, and spatial description of motion, and deformation tensors*

We consider a volume of continua occupying a region $\mathbf{P} \in \mathbf{R}^3$. Let $\mathbf{Y}$ denote the coordinate describing the material configuration in $\mathbf{P}$, called the material coordinate. We assume there exists a referential configuration occupying $\mathbf{P}_R \in R^3$, which can be the material configuration at certain time or completely irrelevant to the material configuration at all as long as there exists a bijection between $\mathbf{P}$ and $\mathbf{P}_R$. Let $\mathbf{X}$ describe the coordinate for the referential configuration of the material called the Lagrangian coordinate, and $\mathbf{x}$ the coordinate describing the spatial configuration of the material in the current spatial region $\mathcal{P} \in \mathbf{R}^3$, called the Eulerian coordinate. The fundamental assumption in continuum mechanics is that there exists a bijective mapping (one-to-one) $\mathcal{K}$ such that

$$X = \mathcal{K}(Y) \tag{4.1}$$

and one between $\mathbf{x}$ and $\mathbf{X}$, $\mathbf{x} = \mathbf{x}(\mathbf{X}, t)$, which denotes the position of the material with referential coordinate $\mathbf{X}$ at time t, for some short time t. We note that vectors are defined as column vectors throughout the notes. So, the inner product is defined as

$$\mathbf{a} \cdot \mathbf{b} = \mathbf{a}^T \cdot \mathbf{b}. \tag{4.2}$$

We assume the mappings are differentiable. The gradient tensor

$$\mathbf{F} = \frac{\partial \mathbf{x}}{\partial \mathbf{X}} = \frac{\partial x_i}{\partial X_j} \tag{4.3}$$

is called the deformation gradient tensor, which is assumed invertible consistent with the bijection assumption. We remark that we interchangeably use matrix notation and the index notation in the notes. Similarly, the gradient tensor

$$\mathcal{H} = \frac{\partial \mathbf{X}}{\partial \mathbf{Y}} \tag{4.4}$$

is assumed invertible as well. The Green's deformation tensor is defined by

$$\mathbf{C} = \mathbf{F}^T \cdot \mathbf{F}. \tag{4.5}$$

We note that differentials in the Eulerian and Lagrangian coordinate are related by

$$d\mathbf{x} = \mathbf{F} \cdot d\mathbf{X}. \tag{4.6}$$

The square of the length of $d\mathbf{x}$ is given by

$$\|d\mathbf{x}\|^2 = d\mathbf{X}^T \cdot \mathbf{F}^T \cdot \mathbf{F} \cdot d\mathbf{X} = d\mathbf{X}^T \cdot \mathbf{C} \cdot d\mathbf{X}. \tag{4.7}$$

So, the Green's tensor measures the change of the measure (metric) from the Lagrangian coordinate to the Eulerian coordinate. It is a metric tensor in the Lagrange coordinate. Since $\mathbf{F}^{-1}$ always exists, the Cauchy deformation tensor is defined by

$$\mathbf{c} = (\mathbf{F}^{-1})^T \cdot \mathbf{F}^{-1}. \tag{4.8}$$

Notice that

$$\|d\mathbf{X}\|^2 = d\mathbf{x}^T \cdot \mathbf{c} \cdot d\mathbf{x}. \tag{4.9}$$

So, the Cauchy tensor measures the change of measure from the Eulerian to Lagrangian coordinate. It is also a metric tensor in Eulerian coordinate. Another very useful tensor that measures deformation is called the Finger tensor. The finger tensor is defined by

$$\mathbf{B} = \mathbf{c}^{-1} = \mathbf{F} \cdot \mathbf{F}^T, \tag{4.10}$$

which is the inverse of the Cauchy tensor. To some extent, the Finger tensor can be viewed as a measure for the change from the Lagrangian coordinate to the Eulerian coordinate as well.

Normally, the strain tensors are used to describe the deformation in short time given that the mapping between the Eulerian coordinate and the Lagrange or the material coordinate are not guaranteed bijective for all time. Assuming the referential coordinate is the spatial coordinate at time t_0, the deformation gradient tensor $\mathbf{F}$ is the isotropic or identity tensor at $t = t_0$. The relative strain tensors are then introduced to measure the relative deformation with respect to the configuration at t_0.

$$\mathbf{E} = \frac{1}{2}(\mathbf{C} - \mathbf{I}) \tag{4.11}$$

is the Lagrangian strain tensor measuring the deformation relative to the Lagrangian coordinate (or referential coordinate at t_0).

$$\mathbf{e} = \frac{1}{2}(\mathbf{I} - \mathbf{c}) \tag{4.12}$$

is the Eulerian strain tensor measuring the deformation relative to the Eulerian coordinate (or the spatial coordinate at t_0). With the relative strain tensors, the relative change of the deformation can be measured by the relative strain tensors. We note that

$$\|d\mathbf{x}\|^2 - \|d\mathbf{X}\|^2 = 2d\mathbf{X} \cdot \mathbf{E} \cdot d\mathbf{X} \tag{4.13}$$

and

$$\|d\mathbf{x}\|^2 - \|d\mathbf{X}\|^2 = 2d\mathbf{x} \cdot \mathbf{e} \cdot d\mathbf{x}. \tag{4.14}$$

We define the displacement vector as

$$\mathbf{u} = \mathbf{x} - \mathbf{X} \tag{4.15}$$

In terms of the displacement vector, the relative strain tensors can be written as

$$\mathbf{E} = \frac{1}{2}\left(\frac{\partial \mathbf{u}}{\partial \mathbf{X}} + \frac{\partial \mathbf{u}}{\partial \mathbf{X}}^T\right) + \frac{1}{2}\frac{\partial \mathbf{u}}{\partial \mathbf{X}} \cdot \frac{\partial \mathbf{u}}{\partial \mathbf{X}}^T, \tag{4.16}$$

$$\mathbf{e} = \frac{1}{2}\left(\frac{\partial \mathbf{u}}{\partial \mathbf{x}} + \frac{\partial \mathbf{u}}{\partial \mathbf{x}}^T\right) - \frac{1}{2}\frac{\partial \mathbf{u}}{\partial \mathbf{x}} \cdot \frac{\partial \mathbf{u}}{\partial \mathbf{x}}^T. \tag{4.17}$$

Dropping the quadratic terms, we arrive at the linear strain measure for the relative deformation

$$\mathbf{E}_0 = \frac{1}{2}\left(\frac{\partial \mathbf{u}}{\partial \mathbf{X}} + \frac{\partial \mathbf{u}}{\partial \mathbf{X}}^T\right), \mathbf{e}_0 = \frac{1}{2}\left(\frac{\partial \mathbf{u}}{\partial \mathbf{x}} + \frac{\partial \mathbf{u}}{\partial \mathbf{x}}^T\right). \tag{4.18}$$

We note that theories built around the linear strain tensors are called linear elastic theories.

In continuum mechanics theories, the constitutive equations are normally formulated in a coordinate system that deforms with the material, called the convected coordinate. We next introduce the useful notion of the convected coordinate in a general curvilinear coordinate. Let $\mathbf{Y} = (\theta^1, \theta^2, \theta^3)$ be the material coordinate. It describes the convected coordinate in the material system, i.e., the coordinate $\theta_i, i = 1, 2, 3$ do not change in deformation. Then, the material point in the reference configuration is denoted by

$$\mathbf{X} = \mathbf{X}(\theta^1, \theta^2, \theta^3), \tag{4.19}$$

and in the spatial configuration by

$$\mathbf{x} = \mathbf{x}(\theta^1, \theta^2, \theta^3, t). \tag{4.20}$$

A curvilinear basis in the Lagrangian or referential coordinate is defined by

$$\mathbf{G}_i = \frac{\partial \mathbf{X}}{\partial \theta^i}, i = 1, 2, 3, \tag{4.21}$$

and in the Eulerian or spatial coordinate by

$$\mathbf{g}_i = \frac{\partial \mathbf{x}}{\partial \theta^i}, i = 1, 2, 3. \tag{4.22}$$

We denote the basis vectors in each coordinate system by a second order tensor, respectively,

$$\mathcal{F} = \frac{\partial \mathbf{x}}{\partial \mathbf{Y}} = \left(\frac{\partial \mathbf{x}}{\partial \theta^1}, \frac{\partial \mathbf{x}}{\partial \theta^2}, \frac{\partial \mathbf{x}}{\partial \theta^3}\right) = (\mathbf{g}_1, \mathbf{g}_2, \mathbf{g}_3),$$

$$\mathcal{H} = \frac{\partial \mathbf{X}}{\partial \mathbf{Y}} = \left(\frac{\partial \mathbf{X}}{\partial \theta^1}, \frac{\partial \mathbf{X}}{\partial \theta^2}, \frac{\partial \mathbf{X}}{\partial \theta^3}\right) = (\mathbf{G}_1, \mathbf{G}_2, \mathbf{G}_3). \tag{4.23}$$

In the Eulerian coordinate and the Lagrangian coordinate, the differentials are denoted by

$$d\mathbf{x} = \mathbf{g}_i d\theta^i, \, d\mathbf{X} = \mathbf{G}_i d\theta^i, \tag{4.24}$$

respectively, where the Einstein notation for index contraction is used. The change in length from the Lagrange to the Eulerian coordinate is given by

$$\|d\mathbf{x}\|^2 - \|d\mathbf{X}\|^2 = d\theta^i (\mathbf{g}_i \cdot \mathbf{g}_j - \mathbf{G}_i \cdot \mathbf{G}_j) d\theta^j. \tag{4.25}$$

The convected coordinate is defined as the material coordinate with base vectors given in the Eulerian coordinate, in which the coordinates of the material particles are held constant, the relative deformation from the reference to the current spatial configuration is effectively measured by the convected strain tensor

$$\gamma_{ij} = (\mathbf{g}_i \cdot \mathbf{g}_j - \mathbf{G}_i \cdot \mathbf{G}_j), \tag{4.26}$$

where

$$(\mathbf{g}_i \cdot \mathbf{g}_j) = \mathcal{F}^T \cdot \mathcal{F}, (\mathbf{G}_i \cdot \mathbf{G}_j) = \mathcal{H}^T \cdot \mathcal{H} \tag{4.27}$$

are called the metric tensors. This is the analogue of the Lagrange strain tensor.

The reciprocal basis in the Lagrange and Eulerian coordinates are given respectively by

$$\mathbf{g}^i = \nabla \theta^i = \frac{\partial \theta^i}{\partial \mathbf{x}}, \mathbf{G}^i = \nabla \theta^i = \frac{\partial \theta^i}{\partial \mathbf{X}}, i = 1, 2, 3, \tag{4.28}$$

where

$$\mathbf{g}_i \cdot \mathbf{g}^j = \delta_{ij}, \mathbf{G}_i \cdot \mathbf{G}^j = \delta_{ij}. \tag{4.29}$$

In fact, we have

$$(\mathbf{g}^1, \mathbf{g}^2, \mathbf{g}^3)^T = \mathcal{F}^{-1}, (\mathbf{G}^1, \mathbf{G}^2, \mathbf{G}^3)^T = \mathcal{H}^{-1}. \tag{4.30}$$

For the differentials, we have

$$d\mathbf{x} = (d\mathbf{x} \cdot \mathbf{g}_i)\mathbf{g}^i = (d\mathbf{x} \cdot \mathbf{g}^i)\mathbf{g}_i,$$

$$d\mathbf{X} = (d\mathbf{X} \cdot \mathbf{G}_i)\mathbf{G}^i = (d\mathbf{X} \cdot \mathbf{G}^i)\mathbf{G}_i. \tag{4.31}$$

Another strain measure, analogous to the Eulerian strain tensor, can be defined as

$$\gamma^{ij} = -(\mathbf{g}^i \cdot \mathbf{g}^j - \mathbf{G}^i \cdot \mathbf{G}^j) = -(\mathcal{F}^{-1} \cdot \mathcal{F}^{-T} - \mathcal{H}^{-1} \cdot \mathcal{H}^{-T}). \quad (4.32)$$

The two strain measures (tensors) are the variables used to describe the deformation of the material in convected coordinates. The constitutive equations for fluids must be formulated in the convected coordinate.

The base vectors $\mathbf{g}_i$ are called covariant base vectors while $\mathbf{g}^i$ are called contravariant base vectors. For an arbitrary vector $\mathbf{v}$,

$$\mathbf{v} = v^i \mathbf{g}_i = \mathcal{F} \cdot \begin{pmatrix} v^1 \\ v^2 \\ v^3 \end{pmatrix}, v^i = \mathbf{v} \cdot \mathbf{g}^i, i = 1, 2, 3. \quad (4.33)$$

$v^i, i = 1, 2, 3$ are called the contravariant coordinates (or components) since $v^i = \mathbf{v} \cdot \mathbf{g}^i$ varies with respect to $\mathbf{g}^i$ and $v_i = \mathbf{v} \cdot \mathbf{g}_i$ the covariant coordinates (or components) since they vary with respect to $\mathbf{g}_i$.

The tensors $g_{ij} = \mathbf{g}_i \cdot \mathbf{g}_j = \mathcal{F}^T \cdot \mathcal{F}, g_i^j = \mathbf{g}_i \cdot \mathbf{g}^j = \mathcal{F}^T \cdot (\mathcal{F}^{-1})^T, g_j^i = \mathbf{g}^i \cdot \mathbf{g}_j = \mathcal{F}^{-1} \cdot \mathcal{F}, g^{ij} = \mathbf{g}^i \cdot \mathbf{g}^j = \mathcal{F}^{-1} \cdot \mathcal{F}^{-T}$ are called metric tensors that define measures in the space with curvilinear base vectors. In fact,

$$\mathbf{g}_i = g_{ij} \mathbf{g}^j, \mathbf{g}^i = g^{ij} \mathbf{g}_j. \quad (4.34)$$

For any second order tensor

$$\tau = \tau^{ij} \mathbf{g}_i \mathbf{g}_j = \mathcal{F} \cdot (\tau^{ij}) \cdot \mathcal{F}^T = \tau_{ij} \mathbf{g}^i \mathbf{g}^j = \mathcal{F}^{-T} \cdot (\tau_{ij}) \cdot \mathcal{F}^{-1}, \quad (4.35)$$

where $\mathbf{g}_i \mathbf{g}_j$ denotes the tensor product. In some books, the tensor product is denoted by $\mathbf{g}_i \otimes \mathbf{g}_j$ or $\mathbf{g}_i \mathbf{g}_j^T$, where $\mathbf{g_i}$ is a column vector. Similarly in the Lagrange coordinate,

$$\sigma = \sigma^{ij} \mathbf{G}_i \mathbf{G}_j = \sigma_{ij} \mathbf{G}^i \mathbf{G}^j. \quad (4.36)$$

We next define various derivatives. Let $\Phi(\mathbf{x}, t)$ be a function of $\mathbf{x}$, we define the derivative of the function with respect to vector $\mathbf{x}$ as

$$\frac{\partial \Phi}{\partial \mathbf{x}} \cdot \mathbf{u} = \lim_{\alpha \to 0} \frac{\Phi(\mathbf{x} + \alpha \mathbf{u}, t) - \Phi(\mathbf{x}, t)}{\alpha}, \quad (4.37)$$

where $\mathbf{u}$ is an arbitrary nonzero vector.

Let $\mathbf{v}$ be a vector-valued function of $\mathbf{x}$. We can define the derivative of $\mathbf{v}$ with respect to $\mathbf{x}$ as follows:

$$\frac{\partial \mathbf{v}}{\partial \mathbf{x}} \cdot \alpha = \lim_{t \to 0} \frac{\mathbf{v}(\mathbf{x} + t\alpha) - \mathbf{v}(\mathbf{x})}{t}, \quad (4.38)$$

for an arbitrary vector α. In the convected coordinate, let

$$\mathbf{v} = v^i \mathbf{g}_i = v_i \mathbf{g}^i \tag{4.39}$$

be a vector function of $\mathbf{x}$. Its derivative with respect to $\mathbf{x}$ is given by

$$\frac{\partial \mathbf{v}}{\partial \mathbf{x}} = [\frac{\partial v^i}{\partial \theta^j} + \Gamma^i_{mj} v^m] \mathbf{g}_i \mathbf{g}^j = [\frac{\partial v_i}{\partial \theta^j} - \Gamma^m_{ij} v_m] \mathbf{g}^i \mathbf{g}^j, \tag{4.40}$$

where $d\mathbf{x} = d\theta^j \mathbf{g}_j$ and

$$\frac{\partial \mathbf{g}_i}{\partial \theta^j} = \Gamma^k_{ij} \mathbf{g}_k, \frac{\partial \mathbf{g}^i}{\partial \theta^j} = -\Gamma^i_{kj} \mathbf{g}^k, \tag{4.41}$$

Γ^k_{ij} are called the Christoffel symbols of the second kind [3]. Analogously, we can define the derivative with respect to higher order tensors.

4.2. *Transformation under the motion* $\mathbf{x}(\mathbf{X}, t)$

We discuss how the differential elements transform under the motion $\mathbf{x}(\mathbf{X}, t)$ in the line integral, surface integral, and volume integral, respectively. These transformation formulae will be useful in the derivation of conservation equations for various physical quantities.

4.2.1. *Line element*

Consider a curve in $P_R \in \mathbf{R^3}$ parametrized by its own arclength L and the corresponding one in the Eulerian coordinate, mapped by $\mathbf{x}(\mathbf{X}, t)$, is parametrized by l. The line element is denoted as $d\mathbf{x} = \mathbf{t} dl$ in the Eulerian coordinate, where $\mathbf{t} = \frac{d\mathbf{x}}{dl}$ is the tangent vector, and the one in the Lagrangian coordinate is given by $d\mathbf{X} = \mathbf{T} dL$, where $\mathbf{T}$ is the tangent vector in the Lagrange coordinate of the curve. By definition

$$\mathbf{F} \cdot d\mathbf{X} = d\mathbf{x}. \tag{4.42}$$

We arrive at

$$\mathbf{F} \cdot \mathbf{T} = \mathbf{t} \frac{dl}{dL}. \tag{4.43}$$

$\mathbf{F}$ maps the tangent vector in the Lagrangian coordinate to another tangent vector (not necessarily a unit vector though) in the Eulerian coordinate.

4.2.2. *Surface element*

Consider a surface in $P_R \in R^3$. Let $d\mathbf{X}_i, i = 1, 2$ be two surface vectors defining the surface element $\mathbf{N}dS = d\mathbf{X}_1 \times d\mathbf{X}_2$, where dS is the surface area element. The differentials $d\mathbf{X}_i, i = 1, 2$ are transformed by

$$\mathbf{F} \cdot d\mathbf{X}_i = d\mathbf{x}_i, i = 1, 2. \tag{4.44}$$

The mapped surface element in the Eulerian coordinate is

$$\mathbf{F} \cdot d\mathbf{X}_1 \times \mathbf{F} \cdot d\mathbf{X}_2 = d\mathbf{x}_1 \times d\mathbf{x}_2 = \mathbf{n}ds, \tag{4.45}$$

where ds is the surface area element in the Eulerian coordinate. We find

$$\mathbf{F}^T \cdot \mathbf{n}ds = det(\mathbf{F})\mathbf{N}dS. \tag{4.46}$$

We denote $J = det(\mathbf{F})$; then

$$\mathbf{F}^T \cdot \mathbf{n} = J\mathbf{N}\frac{dS}{ds}, (\mathbf{F}^T)^{-1} \cdot \mathbf{N} = \frac{1}{J}\mathbf{n}\frac{ds}{dS}. \tag{4.47}$$

It follows that

$$ds = \frac{J}{\|\mathbf{F}^T \cdot \mathbf{n}\|}dS \tag{4.48}$$

We note that the following identity is used in the above derivation:

$$F_{i\alpha}F_{j\beta}F_{k\gamma}\epsilon_{ijk} = \epsilon_{\alpha\beta\gamma}det(\mathbf{F}). \tag{4.49}$$

4.2.3. *Volume element*

Let $d\mathbf{X}_i, i = 1, 2, 3$ be the three differential vectors in the Lagrange coordinate defining the volume element

$$d\mathbf{V} = (d\mathbf{X}_1 \times d\mathbf{X}_2) \cdot d\mathbf{X}_3. \tag{4.50}$$

The volume element in the Eulerian coordinate in $\mathbf{R}^3$ is given by

$$d\mathbf{v} = (d\mathbf{x}_1 \times d\mathbf{x}_2) \cdot d\mathbf{x}_3. \tag{4.51}$$

Then,

$$d\mathbf{v} = (d\mathbf{x}_1 \times d\mathbf{x}_2) \cdot d\mathbf{x}_3 = (\mathbf{F} \cdot d\mathbf{X}_1 \times \mathbf{F} \cdot d\mathbf{X}_2) \cdot \mathbf{F} \cdot d\mathbf{X}_3$$

$$\tag{4.52}$$

$$= J(d\mathbf{X}_1 \times d\mathbf{X}_2) \cdot d\mathbf{X}_3 = JdV.$$

It follows that $J = 1$ if the transformation $\mathbf{x}(\mathbf{X}, t)$ is volume preserving.

4.2.4. *Material derivative*

The material time derivative of a function f is defined as the partial derivative with respect to time while the material coordinate $\mathbf{Y}$ or the reference coordinate $\mathbf{X}$ is held fixed:

$$\dot{f} = \frac{d}{dt}f = (\frac{\partial}{\partial t} + \mathbf{v} \cdot \frac{\partial}{\partial \mathbf{x}})f = \frac{\partial}{\partial t}f|_{\mathbf{Y}} = \frac{\partial}{\partial t}f|_{\mathbf{x}}, \qquad (4.53)$$

where $\mathbf{v} = \frac{\partial}{\partial t}\mathbf{x}(\mathbf{X}, \mathbf{t})$ is the velocity vector. We denote the velocity gradient tensor by $\mathbf{L} = \nabla\mathbf{v}$. The time derivative of the deformation tensor and the Jacobian are given by

$$\dot{\mathbf{F}} = \mathbf{L} \cdot \mathbf{F},$$

$$\dot{J} = J div\mathbf{v}. \qquad (4.54)$$

Thus, the volume preserving transformation requires

$$div\mathbf{v} = \nabla \cdot \mathbf{v} = 0. \qquad (4.55)$$

We derive the equations that the material derivative of the deformation tensor and the associated strain tensors satisfy. We recall the material derivative of the deformation tensor

$$(\frac{\partial}{\partial t} + \mathbf{v} \cdot \nabla)\mathbf{F} = \mathbf{L} \cdot \mathbf{F}. \qquad (4.56)$$

The material derivative of its inverse is

$$\dot{\mathbf{F}}^{-1} = -\mathbf{F}^{-1} \cdot \mathbf{L}. \qquad (4.57)$$

From this, we can easily calculate the time derivative of the Finger tensor and the Cauchy tensor:

$$(\tfrac{\partial}{\partial t} + \mathbf{v} \cdot \nabla)\mathbf{B} - [\mathbf{L} \cdot \mathbf{B} + \mathbf{B} \cdot \mathbf{L}^T] = 0,$$

$$(\tfrac{\partial}{\partial t} + \mathbf{v} \cdot \nabla)\mathbf{c} + [\mathbf{L}^T \cdot \mathbf{c} + \mathbf{c} \cdot \mathbf{L}] = 0. \qquad (4.58)$$

We decompose the strain rate tensor $\mathbf{L}$ into the symmetric $(\mathbf{D})$ and anti-symmetric part $(\mathbf{W})$:

$$\mathbf{L} = \mathbf{D} + \mathbf{W}, \mathbf{D}^T = \mathbf{D}, \mathbf{W}^T = -\mathbf{W}. \qquad (4.59)$$

(4.58) can be rewritten into

$$(\tfrac{\partial}{\partial t} + \mathbf{v} \cdot \nabla)\mathbf{B} - \mathbf{W} \cdot \mathbf{B} + \mathbf{B} \cdot \mathbf{W} - [\mathbf{D} \cdot \mathbf{B} + \mathbf{B} \cdot \mathbf{D}] = 0,$$

$$(\tfrac{\partial}{\partial t} + \mathbf{v} \cdot \nabla)\mathbf{c} - \mathbf{W} \cdot \mathbf{c} + \mathbf{c} \cdot \mathbf{W} + [\mathbf{D} \cdot \mathbf{c} + \mathbf{c} \cdot \mathbf{D}] = 0. \qquad (4.60)$$

The left hand side expressions are called the upper-convected and lower convected derivative, respectively. A linear interpolation of the upper and lower convected derivative yields a new derivative

$$(\frac{\partial}{\partial t} + \mathbf{v} \cdot \nabla)\mathbf{E} - \mathbf{W} \cdot \mathbf{E} + \mathbf{E} \cdot \mathbf{W} - a[\mathbf{D} \cdot \mathbf{E} + \mathbf{E} \cdot \mathbf{D}], \qquad (4.61)$$

where $\mathbf{E}$ is a second order tensor. For $-1 \le a \le 1$, this is known as the Gordon-Showalter derivative or Jaumann derivative [10].

Using the analogy between $\mathbf{F}$ and $\mathcal{F}$, we have

$$\dot{\mathcal{F}}^T = \mathbf{L} \cdot \mathcal{F}^T, \dot{\mathcal{F}} = \mathcal{F} \cdot \mathbf{L}^T. \qquad (4.62)$$

The time derivative of τ is

$$\check{\tau} = \overline{\mathcal{F} \cdot (\tau^{ij}) \cdot \mathcal{F}^T} = \dot{\mathcal{F}} \cdot (\tau^{ij}) \cdot \mathcal{F}^T + \mathcal{F} \cdot (\dot{\tau}^{ij}) \cdot \mathcal{F}^T + \mathcal{F} \cdot (\tau^{ij}) \cdot \dot{\mathcal{F}}^T \qquad (4.63)$$
$$= \mathcal{F} \cdot [\dot{\tau} + \mathbf{L}^T \cdot \tau + \tau \cdot \mathbf{L}]^{ij} \cdot \mathcal{F}^T.$$

Similarly,

$$\dot{\mathcal{F}}^{-T} = -\mathcal{F}^{-T} \cdot \mathbf{L}, \dot{\mathcal{F}}^{-1} = -\mathbf{L}^T \cdot \mathcal{F}^{-1} \qquad (4.64)$$

and

$$\hat{\tau} = \overline{\mathcal{F}^{-1^T} \tau_{ij} \mathcal{F}^{-1}} = \mathcal{F}^{-1^T}[\dot{\tau}_{ij} - [\mathbf{L} \cdot \tau_{ij} + \tau_{ij} \cdot \mathbf{L}^T]] \cdot \mathcal{F}^{-1}. \qquad (4.65)$$

$\hat{\tau}$ is also known as the contravariant convected derivative and $\check{\tau}$ is the covariant convected derivative, which are essentially the upper-convected and lower convected derivatives.

4.2.5. *Transport theorems*

Let $\mathcal{G}$ be a finite material volume in the Eulerian coordinate and $f(\mathbf{x}, t)$ be a differentiable function. Then,

$$\frac{d}{dt} \int_{\mathcal{G}} f(\mathbf{x}, t)dx = \int_{\mathcal{G}} (\frac{d}{dt}f + divvf)dx = \int_{\mathcal{G}} (\frac{\partial}{\partial t}f + \nabla \cdot (\mathbf{v}f))dx. \qquad (4.66)$$

The integral on the right hand side can be rearranged into

$$\frac{d}{dt} \int_{\mathcal{G}} f dx = \int_{\mathcal{G}} \frac{\partial}{\partial t} f dx + \int_{\partial \mathcal{G}} f \mathbf{v} \cdot \mathbf{n} ds, \qquad (4.67)$$

where $\mathbf{n}$ is the unit external normal of $\partial \mathcal{G}$. This can easily be generalized to tensors. For example for a vector $\mathbf{u}$,

$$\frac{d}{dt} \int_{\mathcal{G}} \mathbf{u}(\mathbf{x}, t)dx = \int_{\mathcal{G}} (\frac{d}{dt}\mathbf{u} + divv\mathbf{u})dx = \int_{\mathcal{G}} (\frac{\partial}{\partial t}\mathbf{u} + \nabla \cdot (\mathbf{v}\mathbf{u}))dx. \qquad (4.68)$$

4.3. *Conservation laws*

In mechanics, several conservation laws are postulated, including the conservation of mass, linear momentum, angular momentum, and energy. We present them in both the Eulerian coordinate and the Lagrange coordinate, respectively.

4.3.1. *Eulerian description*

Let $\mathcal{P}$ be a material volume in $\mathbf{R}^3$. We denote the velocity by $\mathbf{v}$, the Cauchy stress tensor by τ and the internal energy density per unit mass by ϵ.

Conservation of mass

$$\frac{d}{dt}\int_{\mathcal{P}} \rho d\mathbf{x} = 0, \tag{4.69}$$

where ρ is the density in the Eulerian coordinate.

Conservation of linear momentum

$$\frac{d}{dt}\int_{\mathcal{P}} \rho \mathbf{v} d\mathbf{x} = \int_{\partial \mathcal{P}} \tau \cdot \mathbf{n} ds + \int_{\mathcal{P}} \mathbf{b} d\mathbf{x}, \tag{4.70}$$

where $\mathbf{b}$ is the body force density per unit volume.

Conservation of angular momentum

$$\frac{d}{dt}\int_{\mathcal{P}} \rho \mathbf{x} \times \mathbf{v} d\mathbf{x} = \int_{\partial \mathcal{P}} \mathbf{x} \times (\tau \cdot \mathbf{n}) ds + \int_{\mathcal{P}} \mathbf{x} \times \mathbf{b} d\mathbf{x}, \tag{4.71}$$

where $\mathbf{x}$ is the position vector.

Conservation of energy

$$\frac{d}{dt}\int_{\mathcal{P}} \rho[\frac{1}{2}\mathbf{v} \cdot \mathbf{v} + \epsilon] d\mathbf{x} = \int_{\mathcal{P}} (\mathbf{b} \cdot \mathbf{v} + r) dv + \int_{\partial P} (\mathbf{t} \cdot \mathbf{v} - h) ds. \tag{4.72}$$

where h is the heat flux per unit volume and r is the heat source density. The traction and the heat flux are given, respectively, by

$$\mathbf{t} = \tau \cdot \mathbf{n}, h = \mathbf{q} \cdot \mathbf{n}, \tag{4.73}$$

where $\mathbf{q}$ is the flux vector.

Using the transport theorem, we can write the conservation laws in differential forms.

$$\frac{d\rho}{dt} + \rho \nabla \cdot \mathbf{v} = 0,$$

$$\rho \frac{d\mathbf{v}}{dt} = \nabla \cdot \tau + \mathbf{b},$$

$$\tau^T = \tau,$$

$$\rho \frac{d\epsilon}{dt} = r - \nabla \cdot \mathbf{q} + \tau : \nabla \mathbf{v}.$$

(4.74)

4.3.2. *Lagrangian description*

We denote the density in the Lagrange coordinate by ρ_R and same notation for velocity and internal energy density.

Conservation of mass

$$\rho_R(t) = \rho_R.$$

(4.75)

Conservation of linear momentum

$$\frac{d}{dt} \int_{\mathcal{P}_R} \rho_R \mathbf{v} d\mathbf{X} = \int_{\mathcal{P}_R} \mathbf{b}_R d\mathbf{X} + \int_{\partial \mathcal{P}_R} \mathbf{t}_R dS,$$

(4.76)

where $\mathbf{t}_R$ is the traction force and $\mathbf{b}_R$ is the body force density per unit volume in the Lagrange coordinate.

Conservation of angular momentum

$$\frac{d}{dt} \int_{\mathcal{P}_R} \rho_R \mathbf{x}(\mathbf{X}, t) \times \mathbf{v} d\mathbf{X} = \int_{\partial \mathcal{P}_R} \mathbf{x}(\mathbf{X}, t) \times \mathbf{t}_R dS + \int_{\mathcal{P}_R} \mathbf{x}(\mathbf{X}, t) \times \mathbf{b}_R d\mathbf{X},$$

(4.77)

where $\mathbf{X}$ is the position vector in the Lagrange coordinate.

Conservation of energy

$$\frac{d}{dt} \int_{\mathcal{P}_R} \rho_R [\frac{1}{2} \mathbf{v} \cdot \mathbf{v} + \epsilon] d\mathbf{X} = \int_{\mathcal{P}_R} (\mathbf{b}_R \cdot \mathbf{v} + r) d\mathbf{X} + \int_{\partial \mathcal{P}_R} (\mathbf{t}_R \cdot \mathbf{v} - h_R) dS,$$

(4.78)

where h_R is the heat flux.

$$\mathbf{t}_R = \mathbf{P} \cdot \mathbf{N}, h_R = \mathbf{q} \cdot \mathbf{N},$$

(4.79)

$\mathbf{P}$ is the Piola-Kirchhoff stress tensor. There is a relationship between the Cauchy stress tensor τ and the Piola-Kirchhoff tensor $\mathbf{P}$:

$$\mathbf{P} = J\tau \cdot \mathbf{F}^{-T}.$$

(4.80)

As you can see, the Piola-Kirchhoff tensor is normally not symmetric even though τ is.

Again, the localized conservation laws are given by

$$\frac{\partial \rho_R}{\partial t} = 0,$$

$$\rho_R \frac{\partial \mathbf{v}}{\partial t} = \nabla \cdot \mathbf{P} + \mathbf{b}_R,$$

$$\mathbf{F} \cdot \mathbf{P}^T = \mathbf{P} \cdot \mathbf{F}^T,$$

$$\rho_R \frac{\partial \epsilon}{\partial t} = r - \nabla \cdot \mathbf{q} + \mathbf{P} : \frac{\partial \mathbf{v}}{\partial \mathbf{X}}. \tag{4.81}$$

4.4. *Superimposed rigid body motion (SRBM) and invariant principles*

Two motions $(\mathbf{x} = \mathbf{x}(\mathbf{X}, t), t)$ and $(\mathbf{x}^+ = \mathbf{x}^+(\mathbf{X}, t^+), t^+)$ are said to be related by a superimposed rigid body motion (SRBM) if

$$\mathbf{x}^+ = \mathbf{Q}(t) \cdot \mathbf{x} + \mathbf{c}(t), t^+ = t + c. \tag{4.82}$$

The gradient operator between the two Eulerian coordinates is transformed according to

$$\nabla^+ = \mathbf{Q} \cdot \nabla. \tag{4.83}$$

Under the SRBM, we have

$$\mathbf{v}^+ = \Omega \cdot \mathbf{Q} \cdot \mathbf{x} + \mathbf{Q} \cdot \mathbf{v} + \dot{\mathbf{c}}(t), \mathbf{F}^+ = \mathbf{Q} \cdot \mathbf{F}, \mathbf{C}^+ = \mathbf{C}, \mathbf{B}^+ = \mathbf{F} \cdot \mathbf{B} \cdot \mathbf{F}^T,$$

$$\mathbf{L}^+ = \mathbf{Q} \cdot \mathbf{L} \cdot \mathbf{Q} + \Omega, \mathbf{D}^+ = \mathbf{Q} \cdot \mathbf{D} \cdot \mathbf{Q}^T, \mathbf{W}^+ = \mathbf{Q} \cdot \mathbf{W} \cdot \mathbf{Q}^T + \Omega, \tag{4.84}$$

where

$$\Omega = \dot{\mathbf{Q}} \cdot \mathbf{Q}^T, \tag{4.85}$$

is an antisymmetric tensor. Let $\mathbf{n}$ be a vector and $\mathbf{n}^+$ the vector in the " $+$ " coordinate with the same coordinates. Then,

$$\mathbf{n}^+ = \mathbf{Q} \cdot \mathbf{n}. \tag{4.86}$$

The time derivative of the vector $\mathbf{n}^+$ is given by

$$\dot{\mathbf{n}}^+ = \Omega \cdot \mathbf{Q} \cdot \mathbf{n} + \mathbf{Q} \cdot \dot{\mathbf{n}}. \tag{4.87}$$

Let $\mathbf{G}$ be a second order tensor in the coordinate system $\mathbf{x}$ and $\mathbf{G}^+$ the "same" tensor with the same components in the coordinate system $\mathbf{x}^+$, i.e.,

$$\mathbf{G} = g^{ij} \mathbf{e}_i \mathbf{e}_j, \mathbf{G}^+ = g^{ij} \mathbf{e}_i^+ \mathbf{e}_j^+, \tag{4.88}$$

where

$$\mathbf{e}_i^+ = \mathbf{Q} \cdot \mathbf{e}_i. \tag{4.89}$$

Then

$$\mathbf{G}^+ = \mathbf{Q} \cdot \mathbf{G} \cdot \mathbf{Q}^T, G_{kl}^+ = Q_{ki}Q_{lj}G_{ij}. \tag{4.90}$$

Let $\mathcal{L}_{i_1 \cdots i_n}$ be an n-th order tensor and $\mathcal{L}_{i_1 \cdots i_n}^+$ the same order in the transformed space with the same coordinates.

$$\mathcal{L}_{k_1 \cdots k_n}^+ = Q_{k_1 i_1} \cdots Q_{k_n i_n} \mathcal{L}_{i_1 \cdots i_n}. \tag{4.91}$$

If $\mathcal{L}^+$ and $\mathcal{L}$ satisfy the above equation, they are said to transform like an nth order tensor under the SRBM. Any constitutive law must be invariant under the SRBM. In addition, we impose the following invariant assumptions.

For the physical variables, the scalar and the angle between any two vectors should be invariant. In addition, the difference between the inertia force and the body force transforms like a vector. The following invariance is assumed in continuum mechanics:

$$\mathbf{t} \cdot (\mathbf{x} - \mathbf{y}) = \mathbf{t}^+ \cdot (\mathbf{x}^+ - \mathbf{y}^+), \mathbf{t}^+ = \mathbf{Q} \cdot \mathbf{t},$$

$$\tag{4.92}$$

$$h^+ = h, \rho^+ = \rho, \epsilon^+ = \epsilon, r^+ = r, \mathbf{b}^+ - \rho\ddot{\mathbf{x}}^+ = \mathbf{Q} \cdot (\mathbf{b} - \rho\ddot{\mathbf{x}}), \eta^+ = \eta, \theta^+ = \theta.$$

Here $\mathbf{t}$ is the traction vector, h is the heat flux, η is the entropy and θ the temperature.

4.5. *Invariant time derivatives*

Let $\mathbf{S}$ be a second order tensor. The corotational derivative defined by

$$\hat{\mathbf{S}} = \dot{\mathbf{S}} - \mathbf{W} \cdot \mathbf{S} + \mathbf{S} \cdot \mathbf{W}, \tag{4.93}$$

is an invariant time derivative in that it transforms like a second order tensor under a SRBM, i.e.

$$\hat{\mathbf{S}}^+ = \mathbf{Q} \cdot \hat{\mathbf{S}} \cdot \mathbf{Q}^T. \tag{4.94}$$

Let $\mathcal{L}$ be the third order tensor. Then,

$$\dot{\mathcal{L}}_{ijk}^+ - \mathbf{W}_{i\alpha}^+ \mathcal{L}_{\alpha jk}^+ - \mathbf{W}_{j\alpha}^+ \mathcal{L}_{i\alpha k}^+ - \mathbf{W}_{k\alpha}^+ \mathcal{L}_{ij\alpha}^+ = Q_{in}Q_{jm}Q_{kp}[\dot{\mathcal{L}}_{nmp} -$$

$$\tag{4.95}$$

$$\mathbf{W}_{n\alpha}\mathcal{L}_{\alpha mp} - \mathbf{W}_{m\alpha}\mathcal{L}_{n\alpha p} - \mathbf{W}_{p\alpha}\mathcal{L}_{nm\alpha}].$$

The invariant time derivative is defined as

$$\hat{\mathcal{L}}_{ijk} = \dot{\mathcal{L}}_{ijk} - \mathbf{W}_{i\alpha}\mathcal{L}_{\alpha jk} - \mathbf{W}_{j\alpha}\mathcal{L}_{i\alpha k} - \mathbf{W}_{k\alpha}\mathcal{L}_{ij\alpha}. \tag{4.96}$$

For n-th order tensor, the invariant time derivative is defined as

$$\hat{\mathcal{L}}_{i_1\cdots i_n} = \dot{\mathcal{L}}_{i_1\cdots i_n} - \sum_{k=1}^{n} W_{i_k\alpha}\mathcal{L}_{i_1\cdots\alpha\cdots i_n}. \tag{4.97}$$

Namely, it transforms like an n-th order tensor.

The invariant time derivative for a vector is

$$\mathbf{N} = \dot{\mathbf{n}} - \mathbf{W}\cdot\mathbf{n}, \tag{4.98}$$

namely, $\mathbf{N}$ transforms like a vector:

$$\mathbf{N}^+ = \mathbf{Q}\cdot\mathbf{N}. \tag{4.99}$$

In constitutive equations, only the invariant derivatives can be used.

4.6. *Material symmetry*

Consider two referential descriptions of the same material volume denoted by the coordinate $\mathbf{X}$, $\mathcal{X}$, respectively, that yields the same constitutive function or functional f. We denote

$$\mathbf{H} = \frac{\partial \mathcal{X}}{\partial \mathbf{X}}. \tag{4.100}$$

$\mathbf{H}$ is called a symmetry transformation. All symmetry transformations form a group called the symmetry group, denoted by $\mathcal{G}$. The density in the two Lagrangian coordinates are related by

$$\rho_R |det(H)| = \rho_{\mathcal{R}}. \tag{4.101}$$

By requiring $\rho_R = \rho_{\mathcal{R}}$, we arrive at

$$det(H) = \pm 1. \tag{4.102}$$

So the symmetry group consists of all uni-modular matrices. Let $f(\mathbf{F}, ...)$ be a constitutive function depending on the deformation tensor $\mathbf{F}$ with respect to the two referential descriptions. Then,

$$f(\mathbf{F}, ...) = f(\mathbf{F}\cdot\mathbf{H}, ...). \tag{4.103}$$

We define

$$\mu = \{\mathbf{H} | det(\mathbf{H}) = \pm 1\}, O(3) = \{\mathbf{H} | \mathbf{H}^T = \mathbf{H}^{-1}\}. \tag{4.104}$$

If $\mathcal{G} = \mu$, the material is called a fluid; if $\mathcal{G} = O(3)$, it is called an isotropic solid; if $\mathcal{G} \subset O(3)$, it is called an anisotropic solid. If $O(3) \not\subset \mathcal{G}$, the material is neither a fluid nor a solid.

The invariant assumptions along with the material symmetry can be used together to derive constitutive relations for the material in question.

4.7. *Clausius-Duhem inequality*

The conservation laws, material symmetry and the invariant principles are not sufficient to determine the necessary relations among physical variables. Thermodynamical considerations must be accounted for next. Let η be the entropy density per unit mass. The second law of thermodynamics in the form of the Clausius-Duhem inequality states that

$$\frac{d}{dt}\int_{\mathcal{P}}\rho\eta d\mathbf{x} \geq \int_{\mathcal{P}}\frac{r}{\theta}d\mathbf{x} - \int_{\partial\mathcal{P}}\frac{h}{\theta}ds, \qquad (4.105)$$

where $\mathcal{P}$ is an arbitrary material volume in the spatial configuration. This states that the total entropy production rate exceeds the rate of heat generated internally per unit temperature minus the heat loss per unit temperature through the boundary in material volume $\mathcal{P}$. The equivalent Lagrangian form is

$$\frac{d}{dt}\int_{\mathcal{P}_R}\rho_R\eta d\mathbf{x} \geq \int_{\mathcal{P}_R}\frac{r_R}{\theta}d\mathbf{x} - \int_{\partial\mathcal{P}_R}\frac{h_R}{\theta}ds, \qquad (4.106)$$

where $\mathcal{P}_R$ is the material volume in the referential configuration. For a thermodynamical process, both the internal energy and the entropy are fundamental physical variables. However, the absolute temperature in many occasions is more convenient to apply than the entropy in formulating a theory. In order to use the absolute temperature, we need to adopt the Helmholtz free energy:

$$\psi = \epsilon - \theta\eta. \qquad (4.107)$$

Namely, the relationship between η and θ is through $\theta = \frac{\partial\epsilon}{\partial\eta}$. Solving the implicit equation for η and substitute it into the definition, we obtain the functional dependence of ψ on θ. With the Helmholtz free energy, the Clausius-Duhem inequality in its point-wise or localized form becomes

$$-\rho\dot{\psi} - \rho\eta\dot{\theta} + \mathbf{T}\cdot\mathbf{D} - \frac{1}{\theta}\mathbf{q}\cdot\nabla\theta \geq 0. \qquad (4.108)$$

The invariant principles, the material symmetry, the second law of thermodynamics, and a constitutive assumption allow us to derive the constitutive relation for the stress and other thermodynamical variables. As examples, we illustrate how to derive the stress constitutive law for viscous fluids under isothermal and nonisorthermal conditions, respectively.

Example 1: isothermal viscous fluids

We consider a purely mechanical theory for isothermal viscous fluids. We assume the stress tensor depends on the history of motion only through

$\rho, \mathbf{v}, \mathbf{L}$:

$$\mathbf{T} = \mathbf{T}(\rho, \mathbf{v}, \mathbf{L}) = \mathbf{T}(\rho, \mathbf{v}, \mathbf{D}, \mathbf{W}). \tag{4.109}$$

We apply SRBM on the stress tensor to obtain

$$\mathbf{T}^{+} = \mathbf{Q} \cdot \mathbf{T} \mathbf{Q}^{T} = \mathbf{T}(\rho, \Omega \cdot \mathbf{Q}\mathbf{x} + \mathbf{Q} \cdot \mathbf{v} + \dot{\mathbf{c}}, \mathbf{Q} \cdot \mathbf{D} \cdot \mathbf{Q}^{T}, \mathbf{Q} \cdot \mathbf{W} \cdot \mathbf{Q}^{T} + \Omega) \tag{4.110}$$

for an arbitrary orthogonal function of $\mathbf{Q}(t)$, where $\Omega = \dot{\mathbf{Q}} \cdot \mathbf{Q}$.

- We first choose thermomechanical processes such that $\mathbf{Q}(t) = I$, $\dot{\mathbf{Q}}(0) = 0$ and $\dot{\mathbf{c}}$ arbitrary. Then, $\mathbf{T}(\rho, \mathbf{v} + \dot{\mathbf{c}}, \mathbf{D}, \mathbf{W}) = \mathbf{T}(\rho, \mathbf{v}, \mathbf{D}, \mathbf{W})$ for an arbitrary vector $\dot{\mathbf{c}}$. This implies that $\mathbf{T}$ does not depend on $\mathbf{v}$.
- Next, we consider $\mathbf{Q} = e^{\Omega t}$. Then, $\mathbf{T}(\rho, \mathbf{D}, \mathbf{W}) = \mathbf{T}(\rho, \mathbf{D}, \mathbf{W} + \Omega)$ for an arbitrary antisymmetric tensor Ω at $t = 0$. Hence, $\mathbf{T}$ does not depend on $\mathbf{W}$ either, i.e., $\mathbf{T} = \mathbf{T}(\rho, \mathbf{D})$. Moreover,

$$\mathbf{T}(\rho, \mathbf{Q} \cdot \mathbf{D} \cdot \mathbf{Q}^{T}) = \mathbf{Q} \cdot \mathbf{T}(\rho, \mathbf{D}) \cdot \mathbf{Q}^{T}, \tag{4.111}$$

according to SRBM. Namely, $\mathbf{T}$ must be an isotropic tensor function of $\mathbf{D}$.

Reiner-Rivlin fluids: if we assume $\mathbf{T}$ is given by a power series of $\mathbf{D}$, we arrive at the Reiner-Rivlin fluid. From Cailey-Hamilton theorem, we have

$$\mathbf{T} = \alpha_0 \mathbf{I} + \alpha_1 \mathbf{D} + \alpha_2 \mathbf{D}^2, \tag{4.112}$$

where $\alpha_i, i = 0, 1, 2$ are functions of ρ and invariants of $\mathbf{D}$ $(tr(\mathbf{D}), \mathbf{D} : \mathbf{D}, det(\mathbf{D}))$.

Linear viscous fluids: if we assume

$$\mathbf{T} = \mathcal{C}(\rho) : \mathbf{D} - \mathbf{A}(\rho), \tag{4.113}$$

where $\mathcal{C}$ is a fourth order tensor; then

$$\mathcal{C} : \mathbf{Q} \cdot \mathbf{D} \cdot \mathbf{Q}^{T} - \mathbf{A} = \mathbf{Q} \cdot [\mathcal{C} : \mathbf{D} - \mathbf{A}] \cdot \mathbf{Q}^{T} \tag{4.114}$$

for all orthogonal tensors $\mathbf{Q}$. A sufficient condition for this is

$$\mathcal{C} : \mathbf{Q} \cdot \mathbf{D} \cdot \mathbf{Q}^{T} = \mathbf{Q} \cdot \mathcal{C} : \mathbf{D} \cdot \mathbf{Q}^{T}, \mathbf{A} = \mathbf{Q} \cdot \mathbf{A} \cdot \mathbf{Q}^{T}. \tag{4.115}$$

This implies

$$\mathbf{A} = -p\mathbf{I}, \mathcal{C}_{ijkl} = a\delta_{ij}\delta_{kl} + b\delta_{il}\delta_{jk} + c\delta_{il}\delta_{jk}, \tag{4.116}$$

where p, a, b, c are functions of ρ. Finally, the stress tensor can be written into

$$\mathbf{T} = -p + \lambda(tr\mathbf{D})\mathbf{I} + 2\mu\mathbf{D}, \tag{4.117}$$

with p, λ, μ functions of ρ. The theory as is is not complete, an equation of state, a relation between p and ρ, is needed to complete the theory. This can only be done by accounting for the thermodynamics.

Example 2: nonisothermal viscous fluids

We assume the Cauchy stress tensor $\mathbf{T}$, the heat flux $\mathbf{q}$, the free energy ψ, and the entropy are functions of $(\rho, \mathbf{v}, \mathbf{L}, \theta, \mathbf{g})$, where $\mathbf{g} = \nabla\theta$ is the temperature gradient. Using the same SRBM argument, we deduce that these thermodynamical quantities only depend on $(\rho, \mathbf{D}, \theta, \mathbf{g})$. Substitution of the thermodynamical quantities into the Clausius-Duhem inequality yields

$$-\rho\left(\frac{\partial\psi}{\partial\theta} + \dot{\eta}\right)\dot{\theta} + \left(\mathbf{T} + \rho^2\frac{\partial\psi}{\partial\rho}\mathbf{I}\right)\cdot\mathbf{D} - \rho\frac{\partial\psi}{\partial\mathbf{D}}\cdot\dot{\mathbf{D}} - \rho\frac{\partial\psi}{\partial\mathbf{g}}\cdot\dot{\mathbf{g}} - \frac{1}{\theta}\mathbf{q}\cdot\mathbf{g} \geq 0, \quad (4.118)$$

where we have used the conservation of mass: $\dot{\rho} = -\rho tr\mathbf{D} = -\rho\mathbf{I} : \mathbf{D}$. This inequality should be valid for any thermomechanical processes. We can choose the special thermomechanical process such as

$$\dot{\theta} = \text{arbitrary constant}, \mathbf{D} = \dot{\mathbf{D}} = \mathbf{0}, \dot{\mathbf{g}} = \mathbf{g} = \mathbf{0}. \quad (4.119)$$

to deduce

$$\eta = -\frac{\partial\psi}{\partial\theta}. \quad (4.120)$$

Analogously, we find

$$\frac{\partial\psi}{\partial\mathbf{D}} = \mathbf{0}, \frac{\partial\psi}{\partial\mathbf{g}} = \mathbf{0}. \quad (4.121)$$

Thus,

$$\psi = \psi(\rho, \theta) \quad (4.122)$$

and the Clausius-Duhem inequality reduces to

$$\left(\mathbf{T} + \rho^2\frac{\partial\psi}{\partial\rho}\mathbf{I}\right)\cdot\mathbf{D} - \frac{1}{\theta}\mathbf{q}\cdot\mathbf{g} \geq 0. \quad (4.123)$$

We note that the free energy is invariant under the SRBM: $\psi^+ = \psi$. For linear viscous fluids, we assume

$$\mathbf{T} + \rho^2\frac{\partial\psi}{\partial\rho}\mathbf{I} = \mathcal{A}_1 + \mathcal{A}_2\cdot\mathbf{g} + \mathcal{A}_3 : \mathbf{D},$$

$$\quad (4.124)$$

$$\mathbf{q} = \mathcal{K}_1 + \mathcal{K}_2\cdot\mathbf{g} + \mathcal{K}_3 : \mathbf{D},$$

where the coefficients are functions of (ρ, θ), $\mathcal{A}_1$ is a second order tensor, $\mathcal{A}_2$ is a third order tensor, $\mathcal{A}_3$ a fourth order tensor, and $\mathcal{K}_1$ is a vector, $\mathcal{K}_2$

a second order tensor, $\mathcal{K}_3$ a third order tensor, respectively. We consider the impact of a SRBM:

$$\mathbf{T}^+ = \mathbf{Q} \cdot \mathbf{T} \cdot \mathbf{Q}^T, \mathbf{q}^+ = \mathbf{Q} \cdot \mathbf{q}, \mathbf{g}^+ = \mathbf{Q} \cdot \mathbf{g}, \mathbf{D}^+ = \mathbf{Q} \cdot \mathbf{D} \cdot \mathbf{Q}^T \quad (4.125)$$

for all orthogonal tensors $\mathbf{Q}$. This demands

$$\mathcal{A}_{1,ij} = -p\delta_{ij}, \mathcal{A}_{2,ijk} = c\epsilon_{ijk},$$

$$\mathcal{A}_{3,ijkl} = \lambda\delta_{ij}\delta_{kl} + \mu\delta_{ik}\delta_{jl} + \gamma\delta_{il}\delta_{jk}, \quad (4.126)$$

$$\mathcal{K}_1 = \mathbf{0}, \mathcal{K}_{2,ij} = -K\delta_{ij}, \mathcal{K}_{3,i,j,k} = E\epsilon_{ijk}.$$

This gives us

$$\mathbf{T} + \rho^2 \frac{\partial \psi}{\partial \rho} = -p\mathbf{I} + \lambda(tr\mathbf{D})\mathbf{I} + 2\mu\mathbf{D}, \mathbf{q} = -K\mathbf{g}, \quad (4.127)$$

where we have replaced $\mu + \lambda$ by 2μ in the final expression of the stress tensor and assumed $c = 0$.

From (4.123), we have

$$\mathbf{q} \cdot \mathbf{g} \leq 0, [\mathbf{T} + \rho^2 \frac{\partial \psi}{\partial \rho}] : \mathbf{D} \geq 0, \quad (4.128)$$

which implies

$$K \geq 0. \quad (4.129)$$

We rescale the second inequality by $\alpha\mathbf{D}$ so that

$$[-p\mathbf{I} + \rho^2 \frac{\partial \psi}{\partial \rho}\mathbf{I} + \alpha\lambda tr(\mathbf{D})\mathbf{I} + 2\alpha\mu\mathbf{D}] : \mathbf{D} \geq 0 \quad (4.130)$$

for all α. Then, we have

$$(-p + \rho^2 \frac{\partial \psi}{\partial \rho})\mathbf{I} : \mathbf{D} \geq 0,$$

$$(4.131)$$

$$(\lambda tr(\mathbf{D})\mathbf{I} + 2\mu\mathbf{D}) : \mathbf{D} \geq 0.$$

Notice that

$$\mathbf{D} = \frac{1}{3}tr(\mathbf{D})\mathbf{I} + \mathbf{S}, \quad (4.132)$$

where $tr(\mathbf{S}) = 0$.

$$[\lambda tr(\mathbf{D})\mathbf{I} + 2\mu\mathbf{D}] : \mathbf{D} = (\lambda + \frac{2\mu}{3})(tr(\mathbf{D})^2 + 2\mu\mathbf{S} : \mathbf{S}) \geq 0. \quad (4.133)$$

Thus,

$$p = -\rho^2 \frac{\partial \psi}{\partial \rho}, \lambda + \frac{2\mu}{3} \geq 0, \mu \geq 0. \tag{4.134}$$

This completes the theory for compressible (linear) viscous fluids. Without the second law of thermodynamics, we would never be able to derive the equation of state for the viscous fluid [2].

5. Some constitutive models for flexible polymers

In the modeling of complex fluids, we assume the materials incompressible. The governing system of equations for the material system consists of the continuity equation, the balance of linear momentum equation, and the constitutive equation for the elastic stress tensor.

Continuity equation

$$\nabla \cdot \mathbf{v} = 0, \tag{5.1}$$

where $\mathbf{v}$ is the velocity vector.

Balance of linear momentum equation

$$\rho \frac{\partial}{\partial t} \mathbf{v} = \nabla(-p\mathbf{I} + \tau) + \mathbf{b}, \tag{5.2}$$

where ρ is the density for the polymeric fluid, p is the pressure, τ is the extra stress tensor, and $\mathbf{b}$ is the external force per unit volume.

Constitutive equation of stress

$$f(\mathbf{D}, \tau) = 0, \tag{5.3}$$

where f is a function or functional of the strain rate tensor $\mathbf{D}$ and extra stress tensor τ.

The constitutive modeling is the process of establishing the functional relationship between $\mathbf{D}$ and τ. There are several approaches to modeling of complex fluids. These include the models of the generalized Newtonian Fluids, linear viscoelastic models, retarded-motion expansion, memory-integral expansion, continuum mechanics models, kinetic (molecular) theories, and GENERIC/Poisson bracket formulation. We list the first a few in a chronological order below.

Generalized Newtonian Fluids:

The shear viscosity defined by

$$\eta = \frac{\tau_{xy}}{\dot\gamma},$$
(5.4)

where $\dot\gamma$ is the shear strain rate, is not a constant in polymeric fluids. If it is a monotone decreasing function of $\dot\gamma$, the fluid is said to exhibit shear-thinning behavior. In order to model the strain-rate dependent viscosity exhibited in the polymeric fluids, the viscosity in the extra stress is expressed as a nonlinear function of the strain rate tensor $\mathbf{D}$ and the extra stress is proportional to the strain rate nonlinearly:

$$\tau = \eta(\mathbf{D})\mathbf{D}.$$
(5.5)

Example: (i). Carreau-Yasuda model:

$$\eta(\mathbf{D}) = \eta_\infty + (\eta_0 - \eta_\infty)[1 + (\lambda\sqrt{\mathbf{D}:\mathbf{D}/2})^a]^{(n-1)/a},$$
(5.6)

where η_0 is the zero-shear-rate viscosity, η_∞ is the infinite-shear-rate viscosity, λ is a time constant, n is the power-law-exponent, a is a dimensionless parameter that characterizes the transitional region between the zero-shear-rate region and the power-law region [3].

(ii). Power law model

$$\eta = m(\mathbf{D}:\mathbf{D})^{(n-1)/2},$$
(5.7)

where n is the power-law-exponent and m is dimensional parameter.

Both models can be fine-tuned to capture the shear thinning behavior in shear. However, they are after all phenomenological models, thus often fail to model the complex relaxation phenomenon in the complex fluids such as in shear reversal, step-strain, step-stress shear experiments, which is tied to elastic relaxation behavior.

Linear viscoelastic models

Generalizing the linear viscoelastic theories developed for solids to complex fluids, one can derive the linear viscoelastic models, in which the extra stress tensor is a linear functional of the strain rate history or strain history.

$$\tau = \int_{-\infty}^{t} G(t-t')\mathbf{D}(t')dt' = -\int_{-\infty}^{t} M(t-t')\gamma(t,t')dt', \qquad (5.8)$$

where $G(t)$ is the elastic relaxation modulus, $M(t)$ is the memory function, and $\gamma(t,t') = \int_{t}^{t'} \mathbf{D}(t'')dt''$ is the strain tensor [3].

Example: (i). The linear Maxwell model is given by $G(t) = \frac{\eta}{\lambda}e^{-t/\lambda}$, where η is called the polymer viscosity and λ is the relaxation time.

(ii). The Jeffreys model is given by $\frac{\eta}{\lambda_1}(1 - \frac{\lambda_2}{\lambda_1})e^{-t/\lambda_1} - \frac{2\eta\lambda_2}{\lambda_1}\delta(t)$, where the additional time parameter λ_2 is called the retardation time.

Quasilinear viscoelastic models

The quasilinear viscoelastic models are obtained by replacing the time derivative in the linear viscoelastic models by the convected time derivatives. [3].

Retarded-motion expansion

In the retarded-motion expansion, the extra stress tensor is expanded in the convected derivatives of the rate of strain tensor:

$$\tau = b_1\mathbf{D} + b_2\mathbf{D}_{(2)} + b_{11}\mathbf{D} \cdot \mathbf{D} + b_3\mathbf{D}_{(3)} + b_{12}(\mathbf{D} \cdot \mathbf{D}_{(2)} + \mathbf{D}_{(2)} \cdot \mathbf{D})$$
$$+ b_{111}\mathbf{D} : \mathbf{DD} + \cdots, \qquad (5.9)$$

where $\mathbf{D}_{(i)}$ is the ith convected derivative of the strain tensor with $\mathbf{D} = \mathbf{D}_{(1)}$ and b_j are the retarded-motion constants. $\mathbf{D}_{(i+1)} = \frac{d\mathbf{D}_{(i)}}{dt} - [\nabla\mathbf{v} \cdot \mathbf{D}_{(i)} + \mathbf{D}_{(i)} \cdot \nabla\mathbf{v}^T]$. For example, the model for the second order fluid is given by the first three terms in the retarded-motion expansion [3].

Memory-integral expansion

The stress is assumed a functional of the strain history, in which the stress is represented as the Frechet series of the convected derivatives of the strain tensor. The retarded-motion expansion can be derived from this more general method as a matter of fact.

$$\tau = -\int_{-\infty}^{t}\int_{-\infty}^{t} M_{II}(t-t',t-t'')(\gamma'\cdot\gamma''+\gamma''\cdot\gamma')dt''dt'$$
$$+\int_{-\infty}^{t}\int_{-\infty}^{t}\int_{-\infty}^{t} M_{III}(t-t',t-t'',t-t''')\gamma'\gamma'':\gamma'''dt'''dt''dt'$$
$$-\int_{-\infty}^{t} M(t-t')\gamma(t')dt' + \cdots, \tag{5.10}$$

where $\gamma' = \gamma(t,t')$ is the strain tensor and M's are the memory functions [3].

Macroscopic models:

A large class of constitutive equations for the stress tensor can be written into the form

$$\tau = 2\eta\mathbf{D} + \tau_p,$$

$$\tau_p + \lambda\hat{\tau}_p + f(\mathbf{D},\tau_p) = 2\eta_p\mathbf{D}, \tag{5.11}$$

where

$$\hat{\tau}_p = \frac{d\tau_p}{dt} - \mathbf{w}\cdot\tau_p + \tau_p\cdot\mathbf{W} - a[\mathbf{D}\cdot\tau_p + \tau_p\cdot\mathbf{D}] \tag{5.12}$$

is the Gordon-Schowalter derivative; a=-1: lower convected, a=1: upper convected, a=0: corotational; $\mathbf{W}$ is the vorticity tensor; η_p is the polymeric viscosity; the choice of f specifies the model.

Example:

(i). f=0 yields the Johnson-Segalman model.

(ii). $f = c\tau_p\cdot\tau_p$ gives the Giesekus model, where c is a constant.

(iii). $f = c\mathbf{D}:\tau_p(\tau_p+G\mathbf{I})$, where c and G are constants, gives the Larson model.

(iv). $f = 0$, $\lambda = \lambda(\sqrt{\mathbf{D}:\mathbf{D}})$ and $\eta_p = \eta_p(\sqrt{\mathbf{D}:\mathbf{D}})$ yield the white metzner model.

(v). Phan-Thien/Tanner model corresponds to $f = Y(tr(\tau))\tau$, where Y is a scalar function.

The details on the Poisson bracket formulation and the Generic formulation can be found in the book by Beris and Edwards [1].

6. Introduction to polymer physics

Polymers are macromolecules comprised of a number of subunits called monomers. The conformation of a polymer can be a linear chain, a side chain, a branched chain, a network of chains, etc. Therefore the stating point for studying polymer physics is the construction of polymer models using simple geometric objects like beads, spheres, rods, springs, etc.. For a linear polymer chain, we can coarse-grain it into a bead rod or bead spring chains. The dynamics of the polymer is then mimicked by the dynamics of the bead rod or bead spring chains.

6.1. *Equilibrium distribution of the end-to-end vector in simple polymer models*

We first consider the bead rod chain with $N+1$ beads located at $\mathbf{r}_i, i = 0, \cdots, N$. Each rod is of the same length $b = \|\mathbf{r}_{i+1} - \mathbf{r}_i\|$. The simplest model is the freely joint model where the distributions of the rods are independent. Let $\phi(\mathbf{q})$ be the pdf for rod or bond $\mathbf{q}$

$$\phi(\mathbf{q}) = \frac{\delta(\|\mathbf{q}\| - b)}{4\pi b^2} \tag{6.1}$$

the joint pdf is

$$f(\mathbf{q}_1, \cdots, \mathbf{q}_N) = \Pi_1^N \phi(\mathbf{q}_i), \tag{6.2}$$

where we denote $\mathbf{q}_i = \mathbf{r}_{i+1} - \mathbf{r}_i$ [5]. It can be easily shown that

$$\langle \|\mathbf{q}_i\|^2 \rangle = b^2, \tag{6.3}$$

where $\langle \rangle$ denotes the ensemble average with respect to the pdf f. We denote the end-to-end vector as

$$\mathbf{R} = \mathbf{r}_n - \mathbf{r}_0 = \sum_{i=1}^{N} \mathbf{q}_i. \tag{6.4}$$

Then,

$$\langle \mathbf{R} \rangle = \mathbf{0}, \langle \|\mathbf{R}\|^2 \rangle = \sum_i^N \langle \|\mathbf{q}_i\|^2 \rangle = Nb^2. \tag{6.5}$$

Next we examine the freely rotating model where the (i+1)-th bond $\mathbf{q}_{i+1}$ is connected to the i-th bond $\mathbf{q}_i$ with a rotational degree of freedom, i.e., it can rotate freely with respect to $\mathbf{q}_i$ subject to

$$\mathbf{q}_{i+1} \cdot \mathbf{q}_i = \cos\theta. \tag{6.6}$$

While $\mathbf{q}_{i-1}, \cdots, \mathbf{q}_1$ are held fixed, the conditional probability distribution is uniform with respect to the rotational angle θ. So,

$$\langle \mathbf{q}_{i+1} \rangle = \cos \theta \mathbf{q}_i. \tag{6.7}$$

Recursively, we can show that

$$\langle \mathbf{q}_i \cdot \mathbf{q}_j \rangle = \cos^{|i-j|} \theta b^2. \tag{6.8}$$

It then follows that

$$\langle \|\mathbf{R}\|^2 \rangle = \sum_{ij=1}^{N} \langle \mathbf{q}_i \mathbf{q}_j \rangle = \sum_{i=1}^{N} \sum_{j=1-i}^{N-i} \langle \mathbf{q}_i \cdot \mathbf{q}_{i+j} \rangle$$
$$\approx \sum_{i=1}^{N} \sum_{j=-\infty}^{\infty} \langle \mathbf{q}_i \mathbf{q}_j \rangle = Nb^2 (1 + 2 \sum_{i=1}^{\infty} \cos^i \theta) = Nb^2 \frac{1+\cos\theta}{1-\cos\theta}. \tag{6.9}$$

For the freely joint chain, we denote $\phi(\mathbf{R}, N)$ the pdf that the end-to-end vector of the chain consisting of N links is $\mathbf{R}$. Then,

$$\phi(\mathbf{R}, N) = \int \delta(\mathbf{R} - \sum_{n=1}^{N} \mathbf{q}_n) \prod_{n=1}^{N} \phi(\mathbf{q}_n) d\mathbf{q}_1 \cdots d\mathbf{q}_N. \tag{6.10}$$

We use the identity

$$\delta(\mathbf{q}) = \frac{1}{(2\pi)^3} \int e^{i\mathbf{k} \cdot \mathbf{q}} d\mathbf{k}. \tag{6.11}$$

Then,

$$\phi(\mathbf{R}, N) = \frac{1}{(2\pi)^3} \int e^{i\mathbf{k} \cdot \mathbf{R}} d\mathbf{k} [\int e^{-\mathbf{k} \cdot \mathbf{r}} d\mathbf{r}]^N$$

$$= \frac{1}{(2\pi)^3} \int e^{i\mathbf{k} \cdot \mathbf{R}} d\mathbf{k} [\frac{1}{4\pi b^2} \int_0^\infty r^2 dr \int_0^{2\pi} d\phi \int_0^\pi \sin\theta e^{-ikr\cos\theta} \delta(r - b) d\theta]^N$$

$$= \frac{1}{(2\pi)^3} \int e^{i\mathbf{k} \cdot \mathbf{R}} d\mathbf{k} [\frac{\sin kb}{kb}]^N \tag{6.12}$$

$$\approx \frac{1}{(2\pi)^3} \int e^{i\mathbf{k} \cdot \mathbf{R}} d\mathbf{k} e^{-\frac{Nk^2 b^2}{6}}$$

$$= (\frac{3}{2\pi Nb^2})^{3/2} e^{-\frac{3\|\mathbf{R}\|^2}{2Nb^2}}.$$

This is a Gaussian distribution. The pdf for the end-to-end vector in the freely rotational chain model is also Gaussian. In fact, the end-to-end distribution is Gaussian provided

$$\psi(\mathbf{q}_n) = \Pi_n \psi(\mathbf{q}_n, \mathbf{q}_{n+1}, \cdots, \mathbf{q}_{n+k}), \tag{6.13}$$

where k is a fixed integer [5].

6.2. *Flory-Huggins Theory*

Polymer solutions or polymer melt mixtures are comprised of two types of materials. To study their thermodynamic properties, we have to calculate their bulk free energy density. The Flory-Huggins theory provides a method to obtain the mixing free energy density using the volume fraction of each component in the mixture. We use a lattice model to derive the Flory-huggins theory for polymer and solvent mixture first. We assume all polymers are consisted of the same number of monomers N and each monomer occupies a lattice site in 3-D space. Let M be the total number of the lattice sites and n_p the total number of polymers in the system. the remaining lattice sites are occupied by the solvent molecules and the number is $M - Nn_p = n_s$. We assume the volume of the monomer is the same as that of the solvent molecule and equal to v_0. The volume fraction of the polymer in the system is given by

$$\phi = \frac{n_p N}{M}. \tag{6.14}$$

The volume fraction of the solvent is given by $1 - \phi$ assuming the system is incompressible.

The partition function of the system is given by, up to a constant,

$$Z = \sum_\alpha e^{-\beta E_\alpha}, \tag{6.15}$$

where α denotes a configuration for the arrangement of n_p polymers on lattice sites and E_i is the energy of that configuration, and the integral is replaced by a sum. To simply the derivation, we approximate E_i by a mean value $\bar{E}$. We next estimate $\bar{E}$. We denote c the lattice coordinate number, i.e., the number of sites that can be connected to the current site by a bond. On average, each lattice site is surrounded by $c\phi$ polymer segments and $(1-\phi)c$ solvent molecules and the number of pairs of polymer segments is $N_{pp} = n_p N c\phi/2 = cM\phi^2/2$. Similarly, the number of neighboring pairs of solvent molecules is $N_{ss} = n_s c(1 - \phi)/2 = cM(1 - \phi)^2/2$. Finally, there are $N_{ps} = Mc/2 - N_{pp} - N_{ss} = cM\phi(1-\phi)$ pairs of neighboring polymer segments and solvent molecules. Denoting the interaction energy associated with pairs as e_{pp}, e_{ps}, e_{ss} for the polymer-polymer, polymer-solvent molecule, solvent-solvent molecule pair, respectively. We can write

$$\begin{aligned}
\bar{E} &= -[N_{pp}e_{pp} + N_{ps}e_{ps} + N_{ss}e_{ss}] \\
&= -\frac{cM}{2}[e_{pp}\phi^2 + 2e_{ps}\phi(1 - \phi) + e_{ss}(1 - \phi)^2].
\end{aligned} \tag{6.16}$$

$$Z = W e^{\beta \bar{E}}, \tag{6.17}$$

where W is the total number of allowed configurations for n_p polymers.

To calculate W, we consider placing polymer chains on the lattice one after another. When we place the first polymer, the first segment can be placed on M sites, the second can be put on c sites neighboring to the first one; and the third and the following ones can be put on $c - 1$ sites. Thus, the number of ways to place the first polymer is

$$w_1 = Mc(c-1)^{N-2} \approx M(c-1)^{N-1}. \tag{6.18}$$

Next, we consider the number of ways w_{j+1} of placing the (j+1)th polymer when the jth polymer have already been placed. Now, Nj lattice sites have been taken. There are $M - Nj$ ways to place the first segment. Due to the occupation of Nj lattice sites, some of the neighboring sites may have been taken. The probability that a neighboring site is available is $(1 - Nj/M)$. So, the ways to place the second segment is $c(1 - Nj/M)$ and those to place the third and the rest is $(c - 1)(1 - Nj/M)$.

$$w_{j+1} \approx (M - Nj)((c-1)(1 - Nj/M))^{N-1} \approx w_1(1 - Nj/M)^N. \tag{6.19}$$

Noticing that the monomers or polymer segments are indistinguishable, the total number of ways of placing n_p polymers on the lattice is given by

$$W = \frac{1}{n_p!} \prod_{j=1}^{n_p} w_j. \tag{6.20}$$

Then,

$$lnW = \sum_{j=1}^{n_p} ln(w_j/j)$$

$$\approx \int_0^{np} dj[ln(M(c-1)^{N-1}(1 - Nj/M)^N) - lnj] \tag{6.21}$$

$$= M[-\frac{\phi}{N}ln\phi - (1 - \phi)ln(1 - \phi) + \frac{\phi}{N}(1 - ln(c-1) + lnN) + \phi ln(\frac{c-1}{e})].$$

The free energy

$$A(M, \phi) = -k_B T lnZ = -k_B T lnW + \bar{E}. \tag{6.22}$$

We need to find the mixing free energy which is defined as the free energy of the mixed state minus the sum of the free energies of the pure components:

$$A_m = A(M, \phi) - A(M\phi, 1) - A(M(1 - \phi), 0) = Mk_B T f(\phi), \tag{6.23}$$

where

$$f(\phi) = \frac{\phi}{N}ln\phi + (1 - \phi)ln(1 - \phi) + \chi\phi(1 - \phi) \tag{6.24}$$

is the mixing free energy per lattice site or per monomer and

$$\chi = \frac{c}{2k_B T}(e_{pp} + e_{ss} - 2e_{ps})$$ (6.25)

is the mixing parameter. The free energy for the solution can be expressed as

$$A(M,\phi) = Mk_B T f(\phi) + A(M\phi, 1) + A(M(1-\phi), 0)$$
$$= -k_B T M\phi[\frac{1}{N}(1 - ln(c-1) + lnN) + ln\frac{c-1}{e}]$$
$$- \frac{e_{pp}}{2}M\phi c - \frac{e_{ss}}{2}M(1-\phi)c + Mk_B T f(\phi).$$ (6.26)

We next calculate the chemical potential for the polymer and solvent, respectively, and the osmotic pressure in equilibrium. The free energy obtained above is for a closed system with a fixed volume $V = v_0 M$. Let P be the pressure of the system. The Gibbs free energy of the system is given by

$$G(n_p, n_s, P, T) = A + PV = A + P(Nn_p + n_s)v_0.$$ (6.27)

The chemical potential μ_s for the solvent is defined as

$$\mu_s = G(n_p, n_s + 1, P, T) - G(n_p, n_s, P, T)$$
$$= \frac{\partial A}{\partial M}\frac{\partial M}{\partial n_s} + \frac{\partial A}{\partial \phi}\frac{\partial \phi}{\partial n_s} + Pv_0.$$ (6.28)

Using the fact that

$$\frac{\partial M}{\partial n_s} = 1, \qquad \frac{\partial \phi}{\partial n_s} = -\frac{\phi}{M},$$ (6.29)

$$A(M,\phi) = A_m + A(\phi M, 1) + A(M(1-\phi), 0),$$ (6.30)

we obtain

$$\mu_s = k_B T(f - \phi\frac{\partial f}{\partial \phi}) + Pv_0.$$ (6.31)

Similarly, the chemical potential for the monomer or polymer segment is given by

$$\mu_p = \frac{1}{N}[G(n_p + 1, n_s, P, T) - G(n_p, n_s, P, T)]$$
$$= k_B T(f + (1-\phi)\frac{\partial f}{\partial \phi}) + Pv_0.$$ (6.32)

$$\mu_p - \mu_s = k_B T\frac{\partial f}{\partial \phi},$$ (6.33)

where

$$\frac{\partial f}{\partial \phi} = \frac{1}{N} ln\phi - ln(1-\phi) + \frac{1}{N} - 1 - \chi(1-2\phi). \tag{6.34}$$

We remark that if we treat ϕ and $1-\phi$ as two separate variables in the free energy, the chemical potential for the corresponding species can be obtained by differentiating the free energy with respect to each volume fraction while holding the other fixed; the difference in the two chemical potential differs from the one obtained above only by a function of temperature. In a binary system, the force or torque is related to the difference of the two chemical potential in the mixture. In multi-component system, the chemical potential calculation becomes subtle.

The osmotic pressure is defined as the extra pressure needed across a semi-permeable membrane to maintain the equilibrium of solvent molecules. Hence,

$$\mu_s(\phi, P + \Pi, T) = \mu_s(0, P, T). \tag{6.35}$$

This yields

$$\Pi = \frac{k_B T}{v_0}(\phi \frac{\partial f}{\partial \phi} - f) = \frac{k_B T}{v_0}\phi^2 \frac{\partial}{\partial \phi}(\frac{f}{\phi}). \tag{6.36}$$

The chemical potential for each species can be rewritten using the osmotic pressure

$$\mu_s = (P - \Pi)v_0, \mu_p = (P - \Pi)v_0 + k_B T\frac{\partial f}{\partial \phi}. \tag{6.37}$$

Hence, the osmotic pressure is the extra pressure due to the presence of polymer chains in the mixture in order to maintain equilibrium.

Let m_0 be the mass per unit monomer. Then,

$$\frac{m_0}{v_0}\phi = \frac{Nn_p m_0}{Mv_0} = \frac{Nn_p m_0}{V} = \rho_p \tag{6.38}$$

is the mass density of the polymer. The osmotic pressure can be rewritten into

$$\Pi = \frac{k_B T}{v_0}\rho^2 \frac{\partial}{\partial \rho}(\frac{f}{\rho}). \tag{6.39}$$

$\frac{k_B T f}{v_0 \rho}$ can be identified as the free energy per unit monomer volume. Substituting in the form of the Flory-Huggins free energy, the osmotic pressure can be written into

$$\Pi = \frac{k_B T}{v_0}(\frac{\phi}{N} - ln(1-\phi) - \phi - \chi\phi^2). \tag{6.40}$$

The free energy of the system can be written as the sum of the mixing free energy and the free energy for each component at its respect volume fraction

$$A(M, \phi) = A_m(M, \phi) + A(M\phi, 1) + A(M(1 - \phi), 0)$$
$$= A_m(M, \phi) + A_l, \tag{6.41}$$

where

$$A_l(M, \phi) = A(M\phi, 1) + A(M(1 - \phi), 0) = \tag{6.42}$$

$$M\phi[\tfrac{c}{2}(e_{ss} - e_{pp}) - k_B T(\tfrac{1}{N}(1 + lnN - ln(c - 1)) + ln\tfrac{(c-1)}{e})] - \tfrac{Mce_{ss}}{2}$$

is the sum of the free energy for each component at its respective volume fraction, a linear function of ϕ. We consider the phase separate kinetics next.

Suppose the solution separates into two phases P and Q with volume fraction ϕ_P and ϕ_Q, respectively. The volumes M_P and M_Q of each phase are determined from the following equations

$$M_P\phi_P + M_Q\phi_Q = M\phi, \, M_P + M_Q = M. \tag{6.43}$$

This gives

$$M_P = \frac{\phi_P - \phi}{\phi_Q - \phi_P}M, \, M_Q = \frac{\phi - \phi_P}{\phi_Q - \phi_P}M. \tag{6.44}$$

The free energy of the phase separated system is

$$A^{sep} = k_B T[M_P A_m(\phi_P) + M_Q A_m(\phi_Q)] + A_l(M_p, \phi_P) + A_l(M_Q, \phi_Q) \tag{6.45}$$

$$= M k_B T[\tfrac{\phi_Q - \phi}{\phi_Q - \phi_P} A_m(\phi_P) + \tfrac{\phi - \phi_P}{\phi_Q - \phi_P} A_m(\phi_P)] + A_l(M, \phi).$$

Notice that A^{sep} is a linear interpolating function and A^{sep} and A_m share the same linear component A_l. Thus, we can simply compare the mixing free energy when discussing phase separation kinetics. If $A_m(M, \phi)$ is concave up in $\phi \in (0, 1)$, the mixing free energy of the separated system is always bigger than that of the mixed system. So, there will be no spontaneous phase separation. On the other hand, if A_m has two local minima in $[0, 1]$, phase separated system can have a lower mixing free energy for certain values of ϕ_P, ϕ_Q. The separated system reaches the minimum mixing free energy when the function A^{sep} is tangent to A_m at the neighborhood of the local minima, i.e., the two volume fractions ϕ_A and ϕ_B satisfy

$$\frac{\partial A_m}{\partial \phi}\Big|_{\phi_A} = \frac{\partial A_m}{\partial \phi}\Big|_{\phi_B} = \frac{A_m(\phi_A) - A_m(\phi_B)}{\phi_A - \phi_B}. \tag{6.46}$$

This implies

$$\mu_{p,s}(\phi_A) = \mu_{p,s}^0 + Pv_0 + A_m(\phi_A) - \phi_A \frac{\partial A_m(\phi_A)}{\partial \phi}$$

$$= \mu_{p,s}^0 + Pv_0 + A_m(\phi_B) - \phi_B \frac{\partial A_m(\phi_B)}{\partial \phi} = \mu_{p,s}(\phi_B). \tag{6.47}$$

Namely, the chemical potential for both solvent and polymers are equal at the two phases. When A_m has two local minima, a solution of volume fraction $\phi_A < \phi < \phi_B$ will phase separate into two phases with volume fractions ϕ_A and ϕ_B. The coexistence region in the phase space (ϕ, T) is called a phase diagram. The spinodal line in the phase diagram is defined by the points in (ϕ, T) space where ϕ is the inflection point of the mixing free energy A_m:

$$\frac{\partial^2 A_m}{\partial \phi^2} = 0. \tag{6.48}$$

When two inflexion points coincide, a critical point in the phase diagram emerges

$$\frac{\partial^3 A_m}{\partial \phi^3} = \frac{\partial^2 A_m}{\partial \phi^2} = 0. \tag{6.49}$$

The critical point for the polymer-solvent system is given by

$$\phi_c = \frac{1}{1 + \sqrt{N}}, \chi_c = \frac{1}{2}(1 + 1/\sqrt{N})^2. \tag{6.50}$$

When the solvent in the system is another type of polymers, the Flory-Huggins theory can also be derived. We will omit the detail of the derivation. We consider a mixture of two polymers of polymerization indices N_A and N_B, respectively. Let ϕ_A and ϕ_B denote the local concentration or volume fraction, respectively. For an incompressible system,

$$\phi_A(\mathbf{r}) + \phi_B(\mathbf{r}) = 1. \tag{6.51}$$

The mixing free energy for two different polymers A and B is given by:

$$A_m = \frac{\phi_A}{N_A} ln\phi_A + \frac{\phi_B}{N_B} ln\phi_B + \chi\phi_A\phi_B. \tag{6.52}$$

This function has two local minima as N_A and N_B are comparable. Therefore, the system is prone to phase separation.

7. Kinetic theory and the Rouse model for flexible polymers

In this section, we give a crash course on the development of kinetic theories for polymeric liquids. We derive the kinetic theory using a phenomenological approach, which can be justified from the Liouville theorem. We begin with the conservation of polymer number density, an analog of the Liouville theorem. This is the most fundamental conservation law in the development of kinetic theories. Let ψ be the number density of some polymer and $\mathbf{F}$ the flux of the polymer flow in the generalized coordinate or phase space Γ, where $x \in \Gamma$. The conservation law for the number density of polymers in any "material volume" in the phase space yields

$$\frac{\partial \psi}{\partial t} + \frac{\partial \mathbf{F}}{\partial x} = 0. \tag{7.1}$$

Using the instantaneous velocity $\mathbf{v}$ at x, we can rewrite the flux as

$$\mathbf{F} = \mathbf{v}\psi. \tag{7.2}$$

Assume the motion of polymers is due to a force generated by an external field U and the Brownian force. The inertialess force balance equation reads

$$L^{-1} \cdot \mathbf{v} + \frac{\partial}{\partial x}\mu = 0, \mu = kTln\psi + U, \tag{7.3}$$

where μ is the chemical potential and L^{-1} the friction coefficient matrix, which is assumed invertible. Then,

$$\mathbf{v} = -L \cdot \frac{\partial \mu}{\partial x}. \tag{7.4}$$

(7.1) becomes

$$\frac{\partial \psi}{\partial t} - \frac{\partial}{\partial x} \cdot (L \cdot \frac{\partial \mu}{\partial x}\psi) = 0. \tag{7.5}$$

This equation is called the Smoluchowski equation or the kinetic equation.

For a system in which the molecule is described by $x = \{\mathbf{x}_i\}_1^N$, where $\mathbf{x}_i \in \mathcal{R}^3$, the Smoluchowski equation is usually written as

$$\frac{\partial \psi}{\partial t} + \sum_{n=1}^{N} \frac{\partial}{\partial \mathbf{x_n}} \cdot (\mathbf{v}_n\psi) = 0, \mathbf{v}_n = -\sum_{m} L_{nm}\frac{\partial \mu}{\partial \mathbf{x_m}}, \tag{7.6}$$

where $L = (L_{nm})$ is the mobility matrix and $L_{nm} = L_{mn}, (L_{mn}) > 0$.

When the particle system x is immersed in a viscous solvent, each particle is going to be subject to a drag exerted by the solvent and additional forces on each particle caused by the perturbation of the flow field due to the motion of the particles. This is called the hydrodynamic effect. When hydrodynamic effect is included, the total instantaneous velocity consists of two parts:

$$\mathbf{v}_n = \mathbf{v}_n^e + \mathbf{v}_n^v, \quad \mathbf{v}_n^e = -\sum_m L_{nm} \cdot \frac{\partial \mu}{\partial \mathbf{x}_m}, \tag{7.7}$$

where the mobility matrix depends on the location of the particles and the second part $\mathbf{v}_n^v$ is due to the existence of the macroscopic flow field and often it is approximated by

$$\mathbf{v}_n^v = \nabla \mathbf{v} \cdot \mathbf{x}_n, \tag{7.8}$$

where $\mathbf{v}$ is the mass-averaged velocity field. We assume $\nabla \mathbf{v}$ is a slowly varying function in space in the length scale of the system x. Thus, in the process of finding the velocity due to external forces, it is assumed a space-independent function.

In the following, we assume the particle is approximated by a sphere. Due to the presence and motion of the spheres, the flow field around each particle is perturbed, which in turn affect the motion of the other spheres. For very dilute solution where the distance between spheres are sufficiently far so that the hydrodynamic interaction can be neglected, the velocity of the sphere is determined by the external force acting on it alone and the mobility matrix is given by

$$L_{mn} = \frac{\delta_{mn}}{\zeta} \mathbf{I}, \tag{7.9}$$

where $\zeta = 6\pi\eta_s a$ for spheres of radius a. In the general case, we have to solve the fluid velocity $\mathbf{v}(\mathbf{x})$ with the given external force $\mathbf{F}_n$ exerted on the sphere at $\mathbf{x}_n$, which can be expressed as

$$\mathbf{g}(\mathbf{x}) = \sum_n \mathbf{F}_n \delta(\mathbf{x} - \mathbf{x}_n). \tag{7.10}$$

We simplify the problem by treating the sphere as a point mass.

We assume the solvent is incompressible $\nabla \cdot \mathbf{v} = 0$ and governed by the Stokes equation

$$\nabla \cdot \tau + \mathbf{g} = 0, \tag{7.11}$$

where the stress tensor is given by

$$\tau = -p + 2\eta_s \mathbf{D}. \tag{7.12}$$

A solution of the equation is given by

$$\mathbf{v} = K \cdot \mathbf{x} + \sum_m \mathbf{H}(\mathbf{x} - \mathbf{x}_m) \cdot \mathbf{F}_m, \tag{7.13}$$

where $K = \nabla \mathbf{v}$ is treated as a spatially homogeneous tensor and

$$\mathbf{H}(\mathbf{x}) = \frac{1}{8\eta_s \pi x}\left(\mathbf{I} + \frac{\mathbf{x}\mathbf{x}}{x^2}\right) \tag{7.14}$$

is called the Oseen tensor, where $x = \|\mathbf{x}\|$. This tensor is singular at $\mathbf{x} = 0$, which is due to the assumption that the particle is a point. For a finite size sphere, this singularity would be removed. However, the exact solution of the stokes equation is not feasible. A compromise is to approximate the Oseen tensor for point mass by

$$\tilde{\mathbf{H}}(\mathbf{x}) = \begin{cases} \frac{\mathbf{I}}{\zeta}, & \mathbf{x} = 0, \\ \mathbf{H}(\mathbf{x}), & \mathbf{x} \neq 0. \end{cases} \tag{7.15}$$

The mobility matrix is then calculated by

$$L_{mn} = \tilde{\mathbf{H}}(\mathbf{x}_n - \mathbf{x}_m). \tag{7.16}$$

The Smoluchowski equation then becomes

$$\frac{\partial \psi}{\partial t} = \sum_{n=1}^{N} \frac{\partial}{\partial \mathbf{x_n}} \cdot \left[\sum_m L_{nm}\frac{\partial \mu}{\partial \mathbf{x_m}} - \nabla \mathbf{v} \cdot \mathbf{x}_n\right]. \tag{7.17}$$

The elastic stress in the polymer system can be derived from a virtual work principle. Let

$$\mathcal{A} = \int_V \int \mu\psi dx dv \tag{7.18}$$

be the free energy over the material volume V. (We remark that this free energy formulation is based on that the potential U is independent of ψ; if U depends on ψ as well, $\mathcal{A}$ has to be modified such that $\frac{\delta \mathcal{A}}{\delta \psi} = \mu$.) We consider an infinitesimal deformation given by

$$\delta\psi = \frac{d\psi}{dt}\delta t = -\sum_n \frac{\partial}{\partial \mathbf{x}_n} \cdot [\nabla \mathbf{v} \cdot \mathbf{x}_n]\delta t. \tag{7.19}$$

Take the variation of the free energy and assume $vol(V)$ finite, we have

$$\delta\mathcal{A} = vol(V)\tau^e : \delta t\nabla \mathbf{v} = -\delta t \nabla \mathbf{v}_{\alpha\beta}\sum_n \langle \mathbf{F}_{n\alpha}\mathbf{x}_{n\beta}\rangle, \tag{7.20}$$

where

$$\mathbf{F}_n = -\frac{\partial \mu}{\partial \mathbf{x}_n} \tag{7.21}$$

is the force exerted on the part of the polymer at $\mathbf{x}_n$. The elastic part of stress is then given by

$$\tau^e = -\frac{1}{vol(V)} \sum_n \langle \mathbf{F}_n \mathbf{x}_n \rangle, \qquad (7.22)$$

where

$$\langle (\bullet) \rangle = \int_V \int (\bullet) d\mathbf{x} dv. \qquad (7.23)$$

7.1. *Langevin equation*

An alternative description of the molecular motion is the Langevin equation. For each particle $\mathbf{x}_n$, there is a stochastic differential equation

$$\dot{\mathbf{x}}_n = \sum_m L_{nm} \left(-\frac{\partial U}{\partial \mathbf{x}_n} + f_m(t) \right) + \frac{1}{2} kT \frac{\partial}{\partial \mathbf{x}_m} L_{nm}, \qquad (7.24)$$

where $f_m(t)$ is a random force subject to the Gaussian distribution and

$$\langle f_m \rangle = 0, \langle f_n f_m \rangle = 2(L^{-1})_{nm} kT \delta(t - t'). \qquad (7.25)$$

This Langevin equation implies the Smoluchowski equation.
Lemma:

The Langevin equation

$$\frac{dx}{dt} = -\frac{1}{\xi} \frac{\partial}{\partial x} U(x) + \sqrt{\frac{kT}{\xi}} g(t) + \frac{1}{2} \frac{d}{dx} \left(\frac{kT}{\xi} \right) \qquad (7.26)$$

implies the distribution density function of x satisfies

$$\frac{\partial}{\partial t} \psi = \frac{\partial}{\partial x} \frac{1}{\xi} \left(kT \frac{\partial}{\partial x} \psi + \frac{\partial U}{\partial x} \psi \right) = \frac{\partial}{\partial x} \frac{1}{\xi} \left(\frac{\partial}{\partial x} \mu \psi \right), \qquad (7.27)$$

where

$$\langle g(t) \rangle = 0, \langle g(t) g(t') \rangle = 2\delta(t - t'). \qquad (7.28)$$

7.2. *System of constraint*

Assume the motion of the particles is subject to a set of constraints

$$C_p(x) = 0, p = 1, \cdots, P. \qquad (7.29)$$

Then, the forces exerted on each particle must include the constraining forces

$$\mathbf{F}_n^{(c)} = \lambda_p \frac{\partial}{\partial \mathbf{x}_n} C_p, \qquad (7.30)$$

where λ_p is the Lagrange multiplier. The velocity of the particle is calculated from

$$\mathbf{v}_m = K \cdot \mathbf{x}_m + H_{mn} \cdot (\mathbf{F}_n + \mathbf{F}_n^{(c)}),\qquad(7.31)$$

where $\mathbf{F}_n = -\frac{\partial}{\partial \mathbf{x}_n}\mu$. Taking the time derivative on the constraints, we have

$$\frac{\partial C_p}{\partial \mathbf{x}_n} \cdot \mathbf{v}_n = 0.\qquad(7.32)$$

Combining the above equations together, we can solve the Lagrange multiplier

$$\lambda_p = (h^{-1})_{pq}\Big[\frac{\partial C_q}{\partial \mathbf{x}_m} \cdot \mathbf{H}_{mn} \cdot \frac{\partial}{\partial \mathbf{x}_n}\mu - \frac{\partial C_q}{\partial \mathbf{x}_n} \cdot K \cdot \mathbf{x}_n\Big].\qquad(7.33)$$

The Smoluchowski equation with constraints is then

$$\frac{\partial}{\partial t}\psi + \frac{\partial}{\partial \mathbf{x}_n} \cdot (\mathbf{v}_n\psi) = 0.\qquad(7.34)$$

Kirkwood showed that the stress formula is quite general and it applies to any forces acting on the particle [5]. Therefore, it can also be used to calculate the viscous stress due to the particle constraints. We assume each particle is subject to a drag or constraint force linear to the velocity gradient and the velocity gradient is slowly varying,

$$F_{mn}^{(c)} = C_{mnkl}K_{kl},\qquad(7.35)$$

where m is the index for the particle and n is the index for the component of the force $\mathbf{F}_m$. The viscous stress is given by

$$\tau_{ij}^{(v)} = -\frac{1}{vol(V)}\sum_{m=1}^{N}\langle C_{mikl}\mathbf{x}_{mj}\rangle K_{kl}.\qquad(7.36)$$

In addition, we need to add the viscous stress from the contribution of the solvent

$$\tau_s^{(v)} = 2\eta_s\mathbf{D}.\qquad(7.37)$$

Next, we give a specific model for polymers modeled as beads connected by linear elastic springs. This is called the Rouse model [5].

7.3. *Rouse model*

We assume the bead-spring system is described by the phase space coordinate $\{\mathbf{x}_i\}, i = 1, \cdots, N$. The system can also be uniquely described by the connecting vector and the center of mass

$$\mathbf{x}_c = \frac{1}{N} \sum_{i=1}^{N} \mathbf{x}_i, \mathbf{q}_i = \mathbf{x}_{i+1} - \mathbf{x}_i, i = 1, \cdots, N - 1. \tag{7.38}$$

The elastic potential of the system is

$$U = \frac{\xi kT}{2} \sum_{i=1}^{N-1} \|\mathbf{q}_i\|^2, \tag{7.39}$$

where ξ is an elastic constant. For the Rouse model, we assume

$$\psi = \nu\phi(\{\mathbf{q}\}_1^{N-1}, t)h(\mathbf{x}_c, t), \qquad h \neq 0 \tag{7.40}$$

where ν is the constant number density of polymers. The Smoluchowski equation reduces to

$$h\frac{\partial \phi}{\partial t} + \phi\frac{\partial h}{\partial t} + \sum_{n=1}^{N-1} \frac{\partial}{\partial \mathbf{q}_n} \cdot (\dot{\mathbf{q}}_n\phi)h + \frac{\partial}{\partial \mathbf{x}_c}(\dot{\mathbf{x}}_c h)\phi = 0. \tag{7.41}$$

Assuming

$$\frac{\partial h}{\partial t} + \frac{\partial h}{\partial \mathbf{x}_c} \cdot (\dot{\mathbf{x}}_c h) = 0, \tag{7.42}$$

we end up with a decoupled Smoluchowski equation for ϕ:

$$\frac{\partial \phi}{\partial t} + \sum_{n=1}^{N-1} \frac{\partial}{\partial \mathbf{q}_n} \cdot (\dot{\mathbf{q}}_n\phi) = 0, \tag{7.43}$$

in which

$$\dot{\mathbf{q}}_n = \nabla\mathbf{v} \cdot \mathbf{q}_n - A_{nm}(kT\frac{\partial}{\partial \mathbf{q}_n}ln\phi + \xi kT\mathbf{q}_n),$$

$$A_{nm} = \frac{1}{\zeta} \begin{cases} 2, & n = m, \\ -1, & n = m \pm 1, \\ 0, & otherwise, \end{cases} \tag{7.44}$$

where ζ is the friction coefficient. Since A is symmetric, there exists an eigenvalue-eigenvector decomposition

$$A = \Omega\Lambda\Omega^T, \tag{7.45}$$

where $\Lambda = Diag(\Lambda_{ii})$ with $\Lambda_{ii} = 4\sin^2(\frac{i\pi}{2N})$, $i = 1, \cdots, N-1$, and Ω is the orthogonal matrix whose column vectors are the orthonormal eigenvectors of A. We introduce a new coordinate $\mathbf{q}'_n, n = 1, \cdots, N-1$ such that

$$\frac{\partial}{\partial \mathbf{q}_n} \cdot \Omega = \frac{\partial}{\partial \mathbf{q}'}. \tag{7.46}$$

Namely,

$$\mathbf{q}'_n = \sum_m \Omega_{nm} \cdot \mathbf{q}_m. \tag{7.47}$$

The Smoluchowski equation is transformed into

$$\frac{\partial \phi}{\partial t} + \sum_{n=1}^{N-1} \frac{\partial}{\partial \mathbf{q}'_n} \cdot [\nabla \mathbf{v} \cdot \mathbf{q}'_n - \frac{1}{\zeta}\Lambda_{nn}\frac{\partial \mu}{\partial \mathbf{q}'_n}\phi] = 0. \tag{7.48}$$

Let $\phi = \Pi_n^{N-1}\phi_n(\mathbf{q}_n, t)$, where ϕ_n is the solution of

$$\frac{\partial \phi_n}{\partial t} = -\frac{\partial}{\partial \mathbf{q}'_n} \cdot [\nabla \mathbf{v} \cdot \mathbf{q}'_n \phi_n - \frac{\Lambda_{nn}}{\zeta}\frac{\partial \mu_n}{\partial \mathbf{q}'_n}\phi_n],$$

$$\mu_n = kTln\phi_n + \frac{\xi kT}{2}\|\mathbf{q}'_n\|^2. \tag{7.49}$$

Take the second moment of ϕ_n with respect to $\mathbf{q}'_n$, we obtain

$$\overline{\langle \mathbf{q}'_n \mathbf{q}'_n \rangle} = \frac{2\Lambda_{nn}}{\zeta}[kT\mathbf{I} - \xi kT\langle \mathbf{q}'_n \mathbf{q}'_n \rangle]. \tag{7.50}$$

We note that the left hand side of the equation is the upper convected derivative of the second moment tensor $\langle \mathbf{q}'_n \mathbf{q}'_n \rangle$. The total elastic stress tensor is

$$\tau^{(e)} = \nu\xi kT \sum_{n=1}^{N-1} \langle \mathbf{q}'_n \mathbf{q}'_n \rangle. \tag{7.51}$$

When $N = 2$, this is the Oldroyd-B model which is derived from the linear viscoelastic theory by using the convected derivative [3, 4].

Normally, a viscous stress due to the solvent is added

$$\tau^{(v)} = 2\eta_s\mathbf{D}.$$

So, the total extra stress is given by

$$\tau = \tau^{(e)} + \tau^{(v)}.$$

References

1. A. N. Beris and B. J. Edwards. *Thermodynamics of flowing system with internal microstructure*, Oxford University Press, New York, 1994.
2. S. E. Bechtel, Advanced continuum mechanics, Lecture Notes, Ohio State University, 1990.
3. R. B. Bird, R. C. Armstrong and O. Hassager. *Dynamics of Polymeric Liquids*, v. 1, John Wiley & Sons, New York, 1987.
4. R. B. Bird, C. F. Curtiss, R. C. Armstrong and O. Hassager. *Dynamics of Polymeric Liquids*, v. 2, John Wiley & Sons, New York, 1987.
5. M. Doi and S.F. Edwards. *Theory of Polymer Dynamics*, Oxford University Press (Clarendon), 1986.
6. A. C. Eringen, *Mechanics of Continua*, Robert E. Krieger Publishing Company, Malabar, Florida, 1980.
7. A. C. Eringen, *Microcontinuum Field Theories I: Foundations and Solids, II: Fluent Media,* Springer-Verlag, New York, 1999.
8. J. O. Hirchfelder, C. F. Curtiss, and R. B. Bird, *The Molecular Theory of Gases and Liquids*, John Wiley and Sons, New York, 1964.
9. R. G. Larson. *The structure and Rheology of Complex Fluids*, Oxford University Press, 1999.
10. R. G. Larson. *Constitutive Equations for Polymer Melts and Solutions*, Butterworths, Boston, 1988.
11. K. Huang, Statistical Mechanics, 2nd Edition, John Wiley & Sons, New York, 1987.